Chemistry of Free Atoms
and Particles

Chemistry of Free Atoms and Particles

Kenneth J. Klabunde

Department of Chemistry
The University of North Dakota
Grand Forks, North Dakota

1980

ACADEMIC PRESS

A Subsidiary of Harcourt Brace Jovanovich, Publishers

New York   London   Toronto   Sydney   San Francisco

ACADEMIC PRESS, INC.
111 Fifth Avenue, New York, New York 10003

United Kingdom Edition published by
ACADEMIC PRESS, INC. (LONDON) LTD.
24/28 Oval Road, London NW1 7DX

Library of Congress Cataloging in Publication Data

Klabunde, Kenneth J
    Chemistry of free atoms and particles.

    Includes index.
    1. Chemical reaction, Conditions and laws of.
2. Atoms. I. Title.
QD501.K7545   547.1'394   80–10823
ISBN 0–12–410750–8

PRINTED IN THE UNITED STATES OF AMERICA

80 81 82 83   9 8 7 6 5 4 3 2 1

Contents

Preface

Studies of the chemical reactions of high-temperature species began in the 1920s and 1930s with the investigations of gas-phase sodium organohalide flames. In the early 1960s a new era opened with the studies of C_1, C_2, and C_3 as *macroscale* (> 50 mg) synthons. The carbon vapor work represented the first examples of how new and interesting molecules could be prepared incorporating the high-temperature species. Also in the 1960s, *microscale* (< 50 mg) matrix isolation spectroscopy techniques for investigating "frozen" high-temperature species were developed. New chemistry, but not new syntheses, has resulted.

In the 1970s the field of "metal atom" or "vapor synthesis" chemistry has grown tremendously. Hundreds of macroscale and microscale studies on the chemistry of reactive high-temperature atoms and particles have appeared. And, although there have been a variety of reviews published, there is a definite need for a complete work with proper organization. This book attempts to fill such a need.

The coverage of this volume is as follows. First, metal atoms and metallic molecules or fragments are considered. The chemistry of H·, O:, S:, organic free radicals and carbenes, and halogen atoms is not covered. A separate volume would be needed for H· and O: chemistry alone; and these do not fit particularly well into the present work whose main theme is the use of high-temperature species to prepare novel molecules, on either a macro- or a microscale. Essentially, the entire remaining portions of the periodic chart are covered. This includes all the available metals, as well as B and C. Then, the literature is covered exhaustively (or nearly so) up through 1978 and the early part of 1979. Vaporization properties of the elements, oxides, sulfides, etc., are covered through 1977. A great deal of original literature is referenced regarding vaporization properties and vapor compositions of various materials.

A variety of needs surfaced as the literature search was made. For example, very little is known about vapor compositions for many elements and compounds at high temperature. This is especially true for "free vaporizations" from a normal surface (as opposed to Knudsen cell studies). Also, there is a great need for microscale studies for gaining mechanistic information about macroscale syntheses using high-temperature species.

Grateful acknowledgment must be made at this time to the author's students who carried out a great deal of the work described herein. In addition to this, they provided help in organization and reading of the manuscripts and continued their work quite independently as much of the author's time was expended on the manuscript. These students are James Y. F. Low, Curt White, Howard F. Efner, John S. Roberts, Bruce B. Anderson, Thomas Murdock, William Kennelly, Thomas Groshens, Richard Kaba, Steve Davis, Dan Ralston, Russell Morris, William Martin, Robert Gastinger, and Robert Zoellner. The author also most gratefully acknowledges the agencies that have supported much of his research, especially the National Science Foundation, but also Research Corporation, the Petroleum Research Fund, and the Department of Energy.

Kenneth J. Klabunde*

* Present address: Department of Chemistry, Kansas State University, Manhattan, Kansas 66506.

To Sara

CHAPTER **1**

Introduction

The book deals with the chemistry of free atoms and coordination-difficient molecules. Included are such species as free metal atoms (e.g., V or Ni atoms), molecular salts (e.g., MgF_2 or $NiCl_2$), and molecular sub-halides, oxides, and sulfides (e.g., BF or CS). Generally, high temperature is needed to generate these particles, so the chemistry investigated is necessarily that of high-temperature species. However, the reaction chemistries of these high-temperature atoms and molecules are usually studied at extremely low temperatures.

How many reactive high-temperature particles exist? Counting the atoms of the elements, clearly more than 100. If we then narrowly define a reactive high-temperature molecule (or particle) as something that possesses no more than three atoms (in a few cases, four and five atoms must be considered) and that is coordination or bond deficient, we must in the first analysis consider diatomics and triatomics of all the elements. The number of possible reactive particles becomes astonishingly high. Species such as Mn_2, V_3, MgO, CaCl, BCl, and CS are typical examples. Even in the second analysis, after we consider experimental feasibility, the number of reactive particles studied or to be studied is very large indeed.

The chemistry of these reactive species is a rather young field in that experimental advances in vacuum technology, high-temperature ceramics, and cryochemical techniques have been absolutely necessary. These advancements have only come in recent times, and have thus allowed a major new chemical research field to develop.

I. Extremes in Temperatures, Energies, and Chemistry

In order to generate free atoms of most of the elements, a great deal of energy is required. Simple vaporization of nickel for example requires at least 100 kcal/mole. This energy is required to break the neighboring Ni—Ni bonds and allow Ni atoms to escape, and this requires temperatures in excess of 1400°C under vacuum. Many other elements require much larger

1

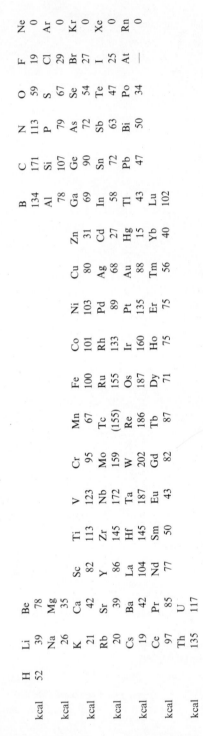

Figure 1-1. Periodic chart showing heats of formation of the elements (kcal/mole).

energies, for example W or Pt. Figure 1-1 shows a periodic chart of the elements and the approximate energies required for their vaporization. Further vaporization data and information can be found in the respective chapters covering the element in question.

A free atom or particle is extremely reactive because it carries a high kinetic energy and orbitals poised for reaction without steric restrictions. Therefore, a high-temperature particle will usually react at very *low* temperature with a substrate of interest. Therefore, temperatures low enough to moderate reaction rates are often desired, and these temperatures are usually in the -50 to $-200°C$ range. Low temperatures also serve to hold down the vapor pressures for incoming reactants, which is a necessity, since almost all reactants of interest must not be allowed to contact the hot source generating the high-temperature species. A variety of experimental techniques have been devised to study the "chemistry of high-temperature species at low temperature"; they are outlined in Table 1-1 with page references to

TABLE 1-1

Techniques Employed for the Study of the Chemistry of High-Temperature Species

Technique	Temperatures involved (°C)	Comments	Page
(1) Diffusion flame	100–500	Gas-phase reactions	9
(2) Life period	100–500	Gas-phase reactions	9
(3) Gas phase flow (macroscale)	200–400	Gas-phase reactions	10
(4) Rotating cryostat (microscale)	500–1500	Cocondensation reactions	10
(5) Stationary codeposition (microscale)	500–2000	Cocodensation reactions	10
(6) Stationary codeposition (macroscale)	300–2000	Cocondensation reactions	35
(7) Rotating codeposition (macroscale)	500–2000	Cocondensation reactions	62
(8) Rotating solution	300–1500	Solution-phase reactions	84
(9) Resistive heating vaporization	25–2000	Used in methods 1–8 above	a
(10) Electron beam vaporization	1000–2500	Used in methods 6 and 7 above	51, 52
(11) Laser vaporization	1000–2000	Used in method 6 above	53, 54
(12) High-temperature disproportionation processes	800–1500	Used for preparation of BF and SiF_2 for method 6 above	172
(13) Discharge processes		Used for preparation of CS for method 6 above	191
(14) Electric arc processes		Used for graphite vaporization	179
(15) High-temperature fast flow reactor	25–1400	Gas-phase reactions	156

[a] Ref. 6.

TABLE 1-2

High-temperature Species of Greatest Interest to Date

Species	Method of generation (from Table 1-1)
Chapter 2	
Li, Na, K, Rb, Cs atoms	1, 2, 3, 4, 5, 6, 7, 8, 9
LiF, NaF, KF, RbF, CsF vapors	5, 9
LiCl, NaCl, KCl, RbCl, CsCl vapors	5, 9
LiBr, NaBr vapors	5, 9
Chapter 3	
Be, Mg, Ca, Sr, Ba atoms	5, 6, 9
BeF_2, $BeCl_2$, $BeBr_2$ vapors	5, 9
MgF_2, $MgCl_2$, $MgBr_2$ vapors	5, 9
CaF_2 vapors	5, 9
SrF_2 vapors	5, 9
BaF_2	5, 9
Chapter 4	
Ti, V, Cr, Mn atoms	5, 6, 7, 9, 10, 11
Zr, Nb, Mo atoms	5, 6, 7, 9, 10, 11
Hf, Ta, W, Re atoms	5, 6, 7, 9, 10, 11
Chapter 5	
Fe, Co, Ni atoms	5, 6, 7, 8, 9, 10, 11
Pd, Pt atoms	5, 6, 7, 8, 9, 10, 11
CoF_2, NiF_2 vapors	6, 9
$NiCl_2$ vapors	6, 9
$NiBr_2$ vapors	6, 9
Chapter 6	
Cu, Zn atoms	5, 6, 7, 8, 9
Ag, Cd atoms	5, 6, 7, 8, 9
Au atoms	5, 6, 7, 8, 9
CuF_2, CuCl vapors	6, 9
Chapter 7	
B, Al, Ga, In atoms	5, 6, 7, 9, 11, 12
BF, AlF	6, 12
BCl, AlCl	6, 12
Chapter 8	
C_1, C_2, C_3 vapors	6, 9, 11, 14
Si, Ge, Sn, Pb atoms	5, 6, 9, 10
Carbenes (not included)	
CS	6, 13
SiF_2, $SiCl_2$	6, 12
Chapter 9	
Se, Te vapors	6, 9
Chapter 10	
Pr, Nd, Sm, Eu, Dy, Ho, Er atoms	6, 9

detailed descriptions throughout this book. Excellent earlier reviews[1-6] are helpful in this area, especially a recent article by Timms.[4]

Studies on the chemistry of the high-temperature species has indicated that a wealth of new reactions and products are available through such pursuits, and that a wide variety of studies await us. Thus, each high-temperature species examined exhibits its own rich and varied chemistry.

Table 1-2 lists the atoms and particles that have been seriously investigated and that are covered in this book. Other species that obviously await study are included throughout the book, to the extent that methods for their generation (usually vaporization properties) are included. The only limitations imposed are that usually only species of one, two, or three atoms are included, and only metal atoms, metal halides or subhalides, metal oxides or suboxides, or metal sulfides or subsulfides (boron and carbon are included). In a practical sense, these limitations on coverage are not detrimental since essentially all of the species that have been studied or that appear experimentally feasible fall into these categories.

II. Organization of the Book

This book is organized on the basis of the Periodic Chart. Each group of elements is separated into a discussion of first the free atoms, followed by a discussion of reactive molecular forms of metal halides, oxides, and sulfides. These sections are further broken down into subsections on "Occurrence, Properties, and Techniques" followed by "Chemistry." This organizational pattern is shown in the Table of Contents. The "Chemistry" sections are further divided into several of the following headings, shown below with explanation and examples. If specific headings are not applicable to a certain group of high temperature species, then that heading is not included in that chapter, e.g., if no Oxidative Addition Processes were found for Alkali Metal Atoms, then that heading is not included, or a brief statement about the absence of work in the area is made.

III. Chemistry

A. Abstraction Processes

A fragment of a molecule is removed by the reactive species in question:

$$\text{e.g., } Au + CH_3Br \longrightarrow AuBr + {}^{\cdot}CH_3$$

B. Electron-Transfer Processes

A nearly complete transfer of an electron, generally from the reactive species to the substrate, takes place:

e.g., $Li + TCNQ \longrightarrow Li^+TCNQ^-$ $TCNQ \equiv \left(\begin{array}{c} NC \\ NC \end{array} C = \text{—} = C \begin{array}{c} CN \\ CN \end{array} \right)$

C. Oxidative Addition Processes

An oxidative insertion of the reactive species into a σ-bond of the substrate takes place:

e.g., $Ni + C_6F_5Br \longrightarrow C_6F_5NiBr$

e.g., $SiF_2 + C_6F_6 \longrightarrow C_6F_5SiF_3$

D. Simple Orbital Mixing Processes

A π- or σ-complex is formed by mixing of π- or nonbonding electrons with the available orbitals of the metal atom or other reactive species. No σ-bonds are made or broken in the substrate:

e.g., $Cr + C_6H_6 \longrightarrow (C_6H_6)_2Cr$

e.g., $Ni + N_2 \longrightarrow Ni(N_2)_4$

e.g., $MgF_2 + CO \longrightarrow F_2MgCO$

E. Substitution Processes

A high-temperature species displaces a fragment of a substrate molecule. Usually two high-temperature species are required for this to occur:

e.g., $8 Li + CCl_4 \longrightarrow CLi_4 + 4 LiCl$

e.g., $2 Ag + (CF_3)_2CFI \longrightarrow AgCF(CF_3)_2 + AgI$

F. Disproportionation and Ligand Transfer Processes

Groups attached to the substrate or an intermediate product are transferred to the reactive species in question:

e.g., $2 Ni + (CH_2{=}CHCH_2)_4Sn \longrightarrow 2 H\text{—}C \begin{array}{c} CH_2 \\ CH_2 \end{array} (\text{—Ni—}) \begin{array}{c} CH_2 \\ CH_2 \end{array} C\text{—}H + (Sn)_n$

e.g., $2 Ni + 2 C_6F_5Br \longrightarrow 2 C_6F_5NiBr \longrightarrow (C_6F_5)_2Ni + NiBr_2$

G. Cluster Formation Processes

The reactive species in question begins re-forming bonds with itself, but only small clusters form, not bulk material, since the small clusters react in their own way:

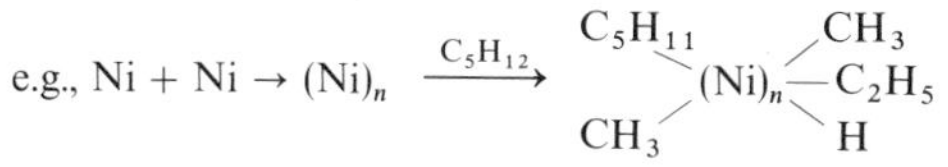

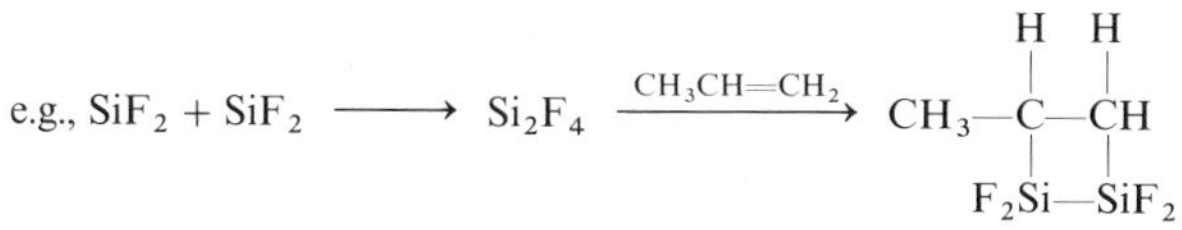

$$\text{e.g., Ni} + \text{Ni} \rightarrow (\text{Ni})_n \xrightarrow{C_5H_{12}} \begin{array}{c} C_5H_{11} \\ \diagdown \\ CH_3 \diagup \end{array} (\text{Ni})_n \begin{array}{c} CH_3 \\ \diagup \\ \diagdown \\ H \end{array} C_2H_5$$

$$\text{e.g., SiF}_2 + \text{SiF}_2 \longrightarrow \text{Si}_2\text{F}_4 \xrightarrow{CH_3CH=CH_2} \begin{array}{c} H \quad H \\ | \quad\; | \\ CH_3-C-CH \\ | \quad\; | \\ F_2Si-SiF_2 \end{array}$$

References

1. P. L. Timms, *Adv. Inorg. Chem. Radiochem.* **14**, 121 (1972).
2. P. S. Skell, J. J. Havel, and M. J. McGlinchey, *Acc. Chem. Res.* **6**, 97 (1973).
3. K. J. Klabunde, *Acc. Chem. Res.* **8**, 393 (1975).
4. P. L. Timms, *in* "Cryochemistry", (M. Moskovits and G. Ozin, ed.), p. 61. Wiley (Interscience), New York, 1976.
5. M. Moskovits and G. Ozin, *in* "Cryochemistry", (M. Moskovits and G. Ozin, eds.), p. 9. Wiley (Interscience), New York, 1976.
6. K. J. Klabunde, *in* "Reactive Intermediates", (R. A. Abramovitch, ed.), Plenum, New York, 1979.

CHAPTER 2

Alkali Metals and Alkali Metal Halides, Oxides, and Sulfides (Group IA)

I. Alkali Metal Atoms Li, Na, K, Rb, and Cs

A. Occurrence, Properties, and Techniques

The natural occurrence of these vapors in the atmosphere, in flames, and in stars is intriguing. There are numerous reports of Li, Na, and K vapor in the twilight sky[1,1a] or twilight airglow (Li 6707 Å absorption, Na 5890 Å, K 7665 Å and 7699 Å). These atoms appear to be responsible for some of the beautiful colors observed in twilight airglow from the upper atmosphere. Donahue[2] has spectroscopically measured the Li atom/Na atom ratio in the upper atmosphere; he has discussed their atmospheric origin in detail and favors a marine origin, based on the ratios of Li/Na in seawater as compared with the Li/Na ratio in the upper atmosphere. Similarly, Lytle and Hunten[2a] have measured the Na/K ratio in the upper atmosphere at 30/1, which compares with 47/1 in seawater, and has concluded that the source of Na and K in the upper atmosphere is marine. However, both of these studies, especially the Li/Na study, were disputed and later work seems strongly to indicate an extraterrestial source for Li and Na in the upper atmosphere. Thus, Kvifte[1a] found a Na/K ratio of 50/1 at 85–130 km and believes meteorites are the source. Gadsden and Salmon[1] made absolute measurements of Li and found 3×10^8 atoms/cm^3 at one level in the upper atmosphere. Ghosh[3] found 900 atoms Na/cm^3 at 93 km and calculated that about 550 Na atoms/cm^3/sec are deposited extraterrestially in the upper atmosphere.

So what is the source of Li, Na, and K in the upper atmosphere, marine or extraterrestial? There are a number of authors who now believe neither of these sources can satisfactorily explain recent data. At 80–90 km over a Pacific island, Delannoy[4] measured Li/Na ratios of 7×10^{-4} to 8×10^{-3} and attributed the origin to meteorites *and* thermonuclear explosions in the upper atmosphere over the Pacific. And indeed, Gadsden[5] found Li/Na ratios of 1.5 and 1.1 shortly after two such explosions. Apparently, a large amount of Li was generated in the upper atmosphere by these blasts.

TABLE 2-1

Vaporization Data for Alkali Metals

Element	mp (°C)	bp (°C)	Vap temp[a] under vac (°C)	Vap method[b]	Vapor composition	References
Li	180	1347	535	Resistive heating of crucibles or Knudsen cell	Li	8c,d
Na	98	883	289	Resistive heating of crucibles or Knudsen cell	Na	8c,d
K	64	774	208	Resistive heating of crucibles or Knudsen cell	K, K_2 (small)	8a–d
Rb	39	688	173	Resistive heating of crucibles or Knudsen cell	Rb	8c,d
Cs	28	678	145	Resistive heating of crucibles or Knudsen cell	Cs	8c,d

[a] Vapor pressure of the metal is approximately 10 μm at this temperature.

[b] Stainless steel crucibles, Al_2O_3 crucibles, and Knudsen cell have all been used satisfactorily.

Alkali metal atoms have also been observed in abundance in stars,[6,6a] star dwarfs,[6a] and flames.[7] Actually, the spectral lines for Li and Na in flames can be used to determine flame temperature.[7]

In the laboratory, alkali metal atoms are readily prepared by thermal vaporization of elements, yielding mainly monoatomic species.[8] However, significant portions of M_2 have been detected, for example 5% K_2 in K vapor at 935°C.[8a,b] Mixed dimers can also be detected spectroscopically when alkali metal alloys are vaporized.[8a] Table 2-1[8a–8d] summarizes some of the vaporization data for the alkali metals.

The vaporizations can be readily carried out in stainless steel crucibles or Al_2O_3–W crucibles by resistive heating methods. Therefore, the production and study of these atoms is experimentally trivial, and they can be conveniently produced even at pressures of several torr. Accordingly, studies have been carried out in flow systems as well as in high vacuum. These techniques are now described.

1. DIFFUSION FLAME METHOD AND LIFE-PERIOD METHOD[9–10] (MACROSCALE)

Hot liquid Na (about 300°C) can be slowly vaporized by passing an inert carrier gas over it and transporting the Na vapor to a spray nozzle. The Na vapor effusing from the nozzle is then allowed to mix with the vapor of a halide, usually organic. At the high temperatures employed (200°–400°C) a violent oxidation–reduction reaction takes place (Na + RX → NaX + R') which, when operating continuously, can be called a sodium flame. The sodium flame area is illuminated by use of an external sodium resonance lamp, making the Na atoms visible. The sodium flame size and vapor cloud are then known, as is the zone of penetration of the halide reactant into the Na cloud; and therefore reaction rates can be determined if the partial pressures of Na and halide are known.[9] This is called the diffusion flame method, and although it has been applied primarily in Na vapor studies, any of the alkali metals could be studied in this way.

The life-period method[9] is very similar except that an excess of halide is employed and the number of Na atoms introduced per unit time is determined. The number of Na atoms in the flame region is calculated (from flame size), and so the average life of a typical Na atom can be determined. Since the concentration of halide is also known, the rate constant for reaction can be determined.

2. GAS-PHASE FLOW SYSTEM (MACROSCALE)

By spraying liquid Na–K alloy through a small orifice, a fine mist of Na–K can be produced. Spraying into a reaction chamber at ∼ 300°C in the presence of organohalides causes rapid abstraction processes to occur, and Wurtz-like coupling products are formed, sometimes on a useful synthetic scale.[10–12] Although this process mimics metal atom studies, the reactive metal species are probably fine droplets of metal (clusters).

3. ROTATING CRYOSTAT (MICROSCALE)

Cocondensation of alkali metal vapors with organohalides or other reactants on a cold rotating drum has been described by Mile (rotating crystat).[13] These were matrix-isolation spectroscopy studies carried out in the range of 4°–77°K. Figure 2-1 illustrates the methodology. By depositing alkali metal vapor on one side of the rotating drum and reactant vapor on the other, an "onion skin" deposition was achieved with excellent matrix mixing of metal and reactant, without prior gas-phase mixing.

4. STATIONARY CODEPOSITION (MICROSCALE)

Simple codeposition of metal vapor and reactant in the same region on a cold wall under good vacuum has been used extensively for matrix isolation spectroscopy studies. The general features are to (1) direct the metal vapor and reactant vapor to the same region of the cold wall (usually 4°–20°K) but allow minimal contact of the reactant vapor with the hot metal vapor source, (2) employ the simultaneous deposition of a huge excess of inert gas

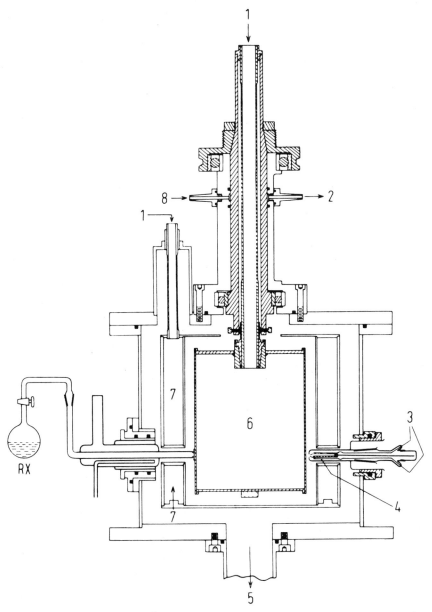

Figure 2-1. Rotating cryostat method for matrix isolation spectroscopy studies (after Mile).[13] (1) Filling tube for liquid nitrogen. (2) Outlet for coolant. (3) Terminals for heater windings. (4) Sodium wire. (5) To vacuum pumps. (6) Spinning drum containing liquid nitrogen. (7) Liquid nitrogen. (8) Inlet for coolant.

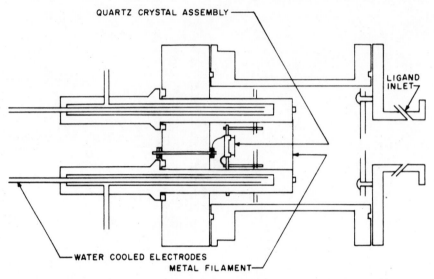

Figure 2-2. Stationary matrix isolation spectroscopy apparatus (after Ozin)[14a].

as well. (A typical deposition ratio of metal:reactant:inert gas is 1:10:1000. The inert gas is generally argon or xenon.), and (3) design the apparatus so that the window to be examined spectroscopically can be rotated to face first the metal–reactant vapor source, and later the spectrometer source (90° rotation). Figure 2-2 illustrates a typical design employed by Ozin and co-workers.[14a]

B. Chemistry

1. ABSTRACTION PROCESSES

Beginning with the work of Ladenburg and Minkowski[15] and Hartel and Polanyi,[16] who devised and perfected the Diffusion Flame and Life-Period methods for studying Na vapor reactions, the first era of metal atom chemistry was initiated. The purpose of these studies, mainly dealing with organohalide–Na reactions, was to elucidate mechanism and energies for abstraction processes. Synthesis of new materials was not a concern at that time.

$$Na + RX \longrightarrow NaX + R^{\cdot}$$

Hundreds of Na–organic substrate, especially Na–organohalide, reactions have been examined, and these have been thoroughly reviewed by Steacie.[9] Benchmark papers by Polanyi and co-workers[16,17] rapidly established reactivity trends, such as $RI > RBr > RCl \gg RF$ as well as $RX_{prim} > RX_{sec} >$

RX_{tert}. Polyhaloalkanes were also studied and found to have greatly increased reactivity over monohaloalkanes.[18-20]

These sodium flame studies were usually carried out in the range of 240°–280°C, although with particularly unreactive substrates such as CH_3F, higher temperatures (500°C) were employed.[16,21] Activation energies for the reactions ranged from <1–25 kcal/mole, while the overall reaction exothermicities ranged from about 10–40 kcal/mole. One of the most interesting features of the work was that reaction exothermicity did not necessarily reflect the Ea (Energy of Activation) or the reaction efficiencies (reactions: collisions); this is illustrated in Table 2-2.[9,16-18,20-23]

Note for example, that the CH_3Br reaction is more exothermic than the CH_3I reaction, and yet the CH_3I reaction is significantly more efficient. Similarly, the CH_3Cl system reacted much less efficiently even though only a small change in predicted exothermicity was shown. Also, CH_3F was very unreactive. Thus, steric factors must play a role in these reactions. This was clearly demonstrated in the C_2H_5Cl, $(CH_3)_2CHCl$, $(CH_3)_3CCl$ work where the primary halide, C_2H_5Cl, reacted most efficiently. Note however, that the secondary halide, $(CH_3)_2CHCl$, reacted least efficiently. Thus, in the approach of the Na atom to the halide, a combination of steric factors is important, but the stability of the radical formed must also be of importance: $(CH_3)_3C^{\cdot} > (CH_3)_2CH^{\cdot} > {}^{\cdot}C_2H_5$, which of course is reflected in Ea. So a combination of steric factors and Ea are required to explain some of these data.

TABLE 2-2

Sodium Atom Reactions with Alkyl and Aryl Halides (Na + RX → NaX + R⋅)[a]

RX	Temp (°C)	Activation energy (Ea, kcal/mole)	Exothermicity (kcal/mole)	Reactions: collision	References
CH_3F	500	25	10	1:100,000	16, 21
CH_3Cl	260	8.8	24.5	1:5000	16, 21
CH_3Br	240	3.2	28.8	1:25	16, 21
CH_3I	240	0.30	26.8	1:1.6	16
C_2H_5Cl	270	7.3	—	1:900	16
$(CH_3)_2CHCl$	275	8.9	—	1:3300	17
$(CH_3)_3CCl$	275	8.0	—	1:1500	17
CH_2Cl_2	250	6.8	—	1:760	18, 20, 22
CCl_4	240	1.7	—	1:5.5	18
C_6H_5F		very slow	—	—	9
C_6H_5Cl	281	8.3	12.0	1:1980	22
C_6H_5Br	255	3.8	28.1	1:36	23
C_6H_5I	240	0.83	36.2	1:2.3	16

[a] See Steacie[9] for a review.

As expected, incorporation of more Cl groups causes the reactions to become more efficient, since Ea and steric factors should become more favorable. Also, the radical stability should increase (Ea lowered) and it would be easier for the Na atom to "find" chlorine (steric factors become more favorable). Thus, CCl_4 reacts more efficiently than CH_2Cl_2 which is in turn more efficient than CH_3Cl.

In the aryl halide series no surprises are found. Note however, that C_6H_5Cl is similar in its reaction efficiency to $(CH_3)_3CCl$. Since the phenyl radical is considered to be higher in energy than the *tert*-butyl radical, it is apparent that the flat aryl ring is less of a problem sterically to the incoming Na atom than is the *tert*-butyl group. This steric explanation seems more plausible than the possible intermediacy of a Na–π-arene complex (which could serve to trap Na and hold it in the reaction sphere for a longer time) prior to the abstraction reaction.

Many other halides have also been examined including aryl, vinyl, and polyhalo systems.[9,25–30] In addition, groups other than halogen have been abstracted by alkali metal atoms. Some examples include CN from $(CN)_2$ and CNCl,[16] oxygen from nitrites[31] and O_2,[32,32a] and sulfur from CS_2.[30]

$$Na + (CN)_2 \longrightarrow NaCN + \cdot CN$$

$$Na + C_5H_{11}ONO \longrightarrow C_5H_{11}ON + NaO$$

$$Ea = 2.8 \text{ kcal/mole}$$
$$\text{Reaction: collision} = 1:14$$

$$Na + O_2 \longrightarrow NaO_2 \xrightarrow{Na} 2\,NaO$$

$$Na + CS_2 \longrightarrow NaS + CS$$

In the case of NaO_2, Haber and Sachsse believe NaO_2 is formed prior to NaO formation.[32] These authors do not believe that direct oxygen abstraction occurred as $Na + O_2 \rightarrow NaO + O$: (studied in 250°–400°C range). However, more recent kinetic investigations on elevated temperatures indicate that $NaO + O$: may indeed be the primary products.[32a] (High-temperature fast-flow reactor techniques were employed and will be described in more detail in Chapter 10.)

What is the exact mechanism of these abstraction processes and in particular of the halide abstractions? Nothing definite can be concluded about mechanism from the diffusion flame or life-period methods of studying gas-phase reactions. However, with the advent of crossed molecular beam technology, some Na–halogen reactions were investigated in mechanistic detail. In these studies, the heavier alkali metal atoms (K, Rb, Cs) were generally employed and were allowed to react with Br_2, I_2, ICl, and IBr. Wilson and co-workers[33] found that the reaction to form alkali metal halide occurs with a very large reaction cross section (~ 100 Å^2). Almost all of the excess vibrational energy for this "stripping reaction" is deposited in the MX molecule as vibrational energy, and MX recoils forward a slight amount.

$$M + X\text{---}X \longrightarrow MX^* + X$$

Likewise, Datz and Minturn,[34] in the Cs–Br$_2$ crossed molecular beam reaction, found that the "stripping" or abstraction occurred by Cs interaction with only one Br, just prior to final product formation. Only 1% of the excess energy went into recoil processes whereas nearly all of the energy went into CsBr internal vibrational excitation. Further work by Minturn, Datz, and Becker[35] indicated that the "stripping" reaction may proceed by prior electron transfer. However, the electron transfer is apparently directionally inhibited, since consideration of adiabatic electron affinity of the halogen molecule would predict even larger reaction cross sections. In still later work, Kwei and Herschbach[36] did find larger reaction cross sections (> 150 Å^2), and they believe that the "electron jump model" (electron transfer) is satisfactory. Some supporting MO calculations predicted that the valence electron of the alkali metal atom would be transferred to the $\bar{\sigma}$-orbital of ICl, which is made up primarily of the 5 pz orbital of the I atom. The charge would shift to the Cl atom as the intermediate ICl$^-$ molecule dissociated in the field of the incoming M$^+$ ion, thereby yielding MCl + I$^{\cdot}$ as final products.

In the 1960s a new era in alkali metal atom abstraction processes began. This work was not concerned with the actual metal atom reaction (rate, mechanism, etc.) but with the fragment molecule generated. That is, the abstraction reaction usually generated a reactive free radical and it was the chemical properties of this radical that were of primary interest. Three methods of study were employed—the Sodium–Potassium Spray Apparatus for macroscale studies, and both the Rotating Cryostat and the Stationary Cocondensation Apparatus for microscale matrix isolation studies.

$$M + RX \longrightarrow MX + R^{\cdot}$$

In the Na–K spray work, Petersen and Skell,[11] and Doerr and Skell[37,38] examined possible modes of generation of novel diradical species by abstraction of halogen from dihalides. Some examples are shown below. Although no direct spectroscopic evidence could be obtained employing the Na–K spray method, careful product analysis allowed certain conclusions regarding the intermediacy of the diradicals to be made. Since the double abstraction should produce excess vibrational energy in the diradical, there should be plenty of energy available for production of both singlet and triplet (usually lower energy) states and for cleavage and/or intra- or intermolecular coupling processes to occur (if the excited species have a finite lifetime under the experimental conditions). It was found that trimethylene methane dimerized

$$X—C \text{\raisebox{0pt}{$\sim\!\sim\!\sim\!\sim$}} C—X \xrightarrow[\text{(gas phase)}]{\text{Na–K spray}} \;\;^\bullet C \text{\raisebox{0pt}{$\sim\!\sim\!\sim\!\sim$}} C^\bullet$$

Examples

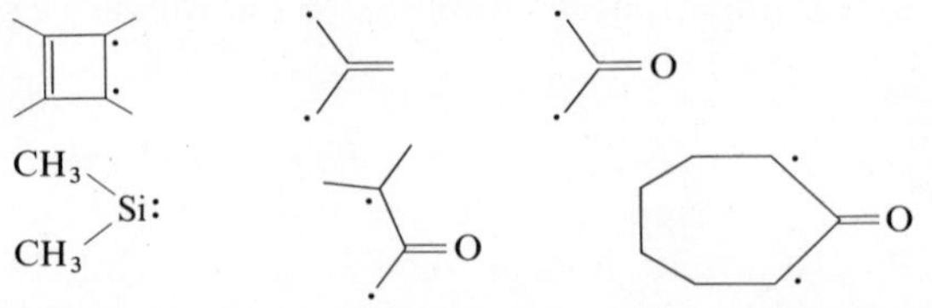

efficiently, probably through a triplet state, and that the presence of excess Na–K vapor apparently allowed facile intersystem crossing between singlet and triplet. In contrast, ketone diradicals generally cleaved to alkene plus CO, most likely through the intermediacy of vibrationally excited cyclopropanones. These results were taken as supportive of prior theoretical work predicting a ground-state triplet for trimethylene methane, but a ground-state singlet for the diradical ketone.[39] Thus, the presense of excess Na–K vapor allowed facile intersystem crossing and rapid collisional deactivation of the diradical species formed.

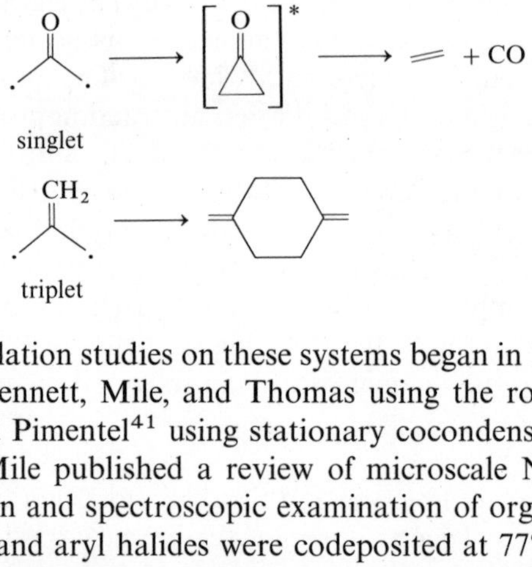

singlet

triplet

Matrix-isolation studies on these systems began in 1966 with independent reports by Bennett, Mile, and Thomas using the rotating cryostat,[40] and Andrews and Pimentel[41] using stationary cocondensation procedures.

In 1968, Mile published a review of microscale Na atom reactions for the generation and spectroscopic examination of organic free radicals.[13,42] Alkyl, vinyl, and aryl halides were codeposited at 77°K, with Na vapor on the rotating cryostat drum. Electron spin resonance (ESR) was used heavily for investigation of the radicals generated, and comparisons were made with the same radicals prepared by matrix radiolysis techniques. Generally the ESR spectra matched those obtained by radiolysis methods, and therefore the radicals so produced were "free" and not strongly complexed to nearby NaX molecules.

$$Na + CH_3CH_2CH_2CH_2CH_2CH_2CH_2X \longrightarrow CH_3CH_2CH_2CH_2CH_2CH_2CH_2{}^\bullet + NaX$$

$$Na + C_6H_5I \longrightarrow C_6H_5{}^\bullet + NaI$$

Different trapping matrices were often employed, and it was found that the phenyl radical could be trapped for long periods at $77°$K in water, benzene, or perfluorocyclohexane.

Many other Na atom reactions were reported by Mile and will be covered under the Electron Transfer Processes section of this chapter.

In 1966, Andrews and Pimentel deposited Li atoms with CH_3I and observed an IR absorption at 730 cm^{-1}, presumably due to $\cdot CH_3$. However, the Milligan and Jacox vacuum UV studies of matrix-isolated CH_4 ($CH_4 \xrightarrow[UV]{vac}$ $\cdot CH_3$ + H$\cdot$) yielded a 611 cm^{-1} band.[43] Further work by Tan and Pimentel[44] showed that the 730 cm^{-1} band was probably due to LiI complexed $\cdot CH_3$ rather than free $\cdot CH_3$. At the very low temperature employed ($<15°$K), where very poor mobility of $\cdot CH_3$ and/or LiI would be expected, it is not surprising that two reactive species such as $\cdot CH_3$ and free LiI would complex. Still, the CH_3I system may be unique since later work by Andrews and co-workers indicated that truly "free" radicals were apparently produced (non-MX complexed) when polyhalo systems were employed. Thus, with Li atoms free $\cdot CCl_3$,[45] $\cdot CBr_3$,[46] $\cdot CHCl_2$,[47] $\cdot CHBr_2$,[47] $\cdot CHF_2$,[48] and $\cdot CH_2I$[49] could be prepared in the matrix. In order to confirm the "free" nature of these species, direct comparisons were made with experiments in which radiolysis or other methods were employed to produce the same radicals.

If it is indeed true that $\cdot CH_3$ complexes LiX whereas the halogenated radicals do not, the variance may be rationalized by noting the greater stability of the halo radicals vs $\cdot CH_3$, and their expected lessened propensity for complexation. Theoretical models for such interactions would be of importance using the gross geometries shown below.

$$LiX\text{------}\cdot R$$

$$XLi\text{------}\cdot R$$

$$\begin{matrix} X\text{---}Li \\ \diagdown \diagup \\ R\cdot \end{matrix}$$

Further studies have shown that alkali metal atoms react with CCl_4 in argon at $15°$K to form $M\text{--}CCl_3$ and $M\text{--}CCl_2$ species according to the spectra observed.[49a] Note the carbeneoid species postulated, which represents a very strong M—C interaction.

Inorganic, highly reactive free radicals have also been generated by Li, Na, and K microscale reactions. For example, OF_2 and OCl_2 have been studied by Andrews[50] and co-workers. However, the production of $\cdot OF$ or $\cdot OCl$ radicals was accompanied by other processes, including electron transfer (cf. next section), to yield biproducts such as Li^+OF^-, O_2, F_2, Li^+OCl^-, and ClO—ClO.

$$Li + OF_2 \longrightarrow LiF + \cdot OF$$

Oxygen abstraction by Li atoms in a matrix was observed when N_2O and N_2 were deposited with Li.[50,51] Curiously, argon matrices did not encourage this abstraction reaction but nitrogen matrices did.

$$3\,Li + N_2O \xrightarrow{\text{N}_2} LiO + Li_2O + N_2$$

One final example was recently published by Margrave and co-workers,[34a] where Li atoms and SiF_4 were allowed to react in a matrix. This was used as a method for generation of diamagnetic, untelomerized $:SiF_2$ (see Chapter 8).

$$2\,Li + SiF_4 \longrightarrow 2\,LiF + :SiF_2$$

2. ELECTRON-TRANSFER PROCESSES

In a low-temperature matrix, electron transfer from alkali metal atoms is more facile than are abstraction processes, where bond making and breaking must take place. The movement of an electron completely to the substrate (or nearly so) is very easy, and probably most abstraction processes by Li, Na, K, Rb, and Cs proceed through a prior electron-transfer step.

$$M + RX \longrightarrow [M^+RX^-] \longrightarrow MX + R^{\boldsymbol{\cdot}}$$

With simple diatomic molecules detection of the charge-transfer species is not difficult. A host of different studies of Li, Na, K, Rb, and Cs atoms cocondensed on microscale with N_2, O_2, NO, O_3, F_2, Cl_2, NO_2, CS_2, $R_2C{=}O$, ROH, H_2O, H_2S, B_2H_6, and $Cr(CO)_6$ have been reported.[10,42,50–54] Table 2-3 summarizes the findings.[42,50,52–66] Note the facile formation of $O_2{}^-$, $N_2{}^-$, $F_2{}^-$, and a variety of other similar molecules, each bonding to M^+ in a characteristic fashion.

Apparently $N_2{}^-$ and $O_2{}^-$ coordinate with different geometries to Li^+, N_2 being end-on and O_2 being side-on. Also, side-on bonded $M^+(O_2)_2{}^-$ systems were believed to have been formed (D_{2d}), since bands near 990–1000 cm^{-1} were found, as well as bands near 1110 cm^{-1} for the $M^+O_2{}^-$ species. Comparisons of the ν_{O-O} stretching frequencies throughout the series of Li—Cs shows correlation for Na, K, Rb, and Cs, but not for Li. The numbers (Table 2-3) might be interpreted to mean that the strongest M^+–$O_2{}^-$ bonding is in the order Cs > Rb > K > Li > Na, if the strength of binding is reflected in a raising of the ν_{O-O} stretch.[60,60a] Similar arguments have been applied to the M^+NO^- system. A good model for the bonding is side-on for Li, but end-on bonding to oxygen (or slightly bent) for Na, K, Rb, and Cs. Again, Li is somewhat anomalous, and again the larger alkali metal cations were more capable of dispersing the antibonding charge density on NO^-, causing a slight but steady increase in ν_{N-O} on progression from Li, Na, K, Rb, to Cs (free NO^- $\nu_{N-O} = 1352$).[63]

TABLE 2-3

Alkali Metal Microscale Electron-Transfer Reactions

Metal and reactant	Products and comments	References
$Li + N_2$	$Li^+N{\equiv}N^-(C_{\infty v})$, $\nu_{N-N} = 1800\ cm^{-1}$	50, 55
$Li + O_2$	$Li^+ \begin{matrix} O^- \\ \Vert \\ O \end{matrix}$ (C_{2v}), $\nu_{O-O} = 1097.4\ cm^{-1}$	50, 56
	$Li_2{}^+O_2{}^-$	56
$Na + O_2$	$Na^+ \begin{matrix} O^- \\ \Vert \\ O \end{matrix}$ (C_{2v}), $\nu_{O-O} = 1094\ cm^{-1}$	50, 57, 58
$K + O_2$	$K^+ \begin{matrix} O^- \\ \Vert \\ O \end{matrix}$ (C_{2v}), $\nu_{O-O} = 1108\ cm^{-1}$	50, 58, 59
	$K^+\left(\begin{matrix} O \\ \Vert \\ O \end{matrix}\right)_2^-$ (D_{2d}), $\nu_{O-O} = 993.4\ cm^{-1}$	50, 58, 59
$Rb + O_2$	$Rb^+ \begin{matrix} O^- \\ \Vert \\ O \end{matrix}$ (C_{2v}), $\nu_{O-O} = 111.3\ cm^{-1}$	50, 58, 59
	$Rb^+\left(\begin{matrix} O \\ \Vert \\ O \end{matrix}\right)_2^-$ (D_{2d}), $\nu_{O-O} = 991.7\ cm^{-1}$	50, 58, 59
$Cs + O_2$	$Cs^+ \begin{matrix} O^- \\ \Vert \\ O \end{matrix}$ (C_{2v}), $\nu_{O-O} = 1115.6\ cm^{-1}$	50, 58
	$Cs^+\left(\begin{matrix} O \\ \Vert \\ O \end{matrix}\right)_2^-$ (D_{2d}), $\nu_{O-O} = 1002.5\ cm^{-1}$	50, 58
$Li + NO$	$Li^+ \begin{matrix} N^- \\ \Vert \\ O \end{matrix}$ (triangular), $\nu_{Li-O} = 651\ cm^{-1}$ $\nu_{N-O} = 1352\ cm^{-1}$	50, 60
$Na + NO$	$Na^+O{=}N^-$ (linear or bent), $\nu_{Na-O} = 361\ cm^{-1}$ $\nu_{N-O} = 1358\ cm^{-1}$	50, 60
$K + NO$	$K^+O{=}N^-$ (linear or bent), $\nu_{K-O} = 280\ cm^{-1}$ $\nu_{N-O} = 1372\ cm^{-1}$	50, 60
$Rb + NO$	$Rb^+O{=}N^-$ (linear or bent), $\nu_{R\,l-O} = 235\ cm^{-1}$ $\nu_{N-O} = 1373\ cm^{-1}$	50, 60
$Cs + NO$	$Cs^+O{=}N^-$ (linear or bent), $\nu_{Cs-O} = 219\ cm^{-1}$ $\nu_{N-O} = 1374\ cm^{-1}$	50, 60
$Na + O_3$	$Na^+ \begin{matrix} O \\ - \\ O \end{matrix}\!\!>\!O(C_{2v})$	50, 61

(*continued*)

TABLE 2-3 (*continued*)

Metal and reactant	Products and comments	References
$Cs + O_3$	$Cs^+ \begin{smallmatrix} O \\ - \\ O \end{smallmatrix} O(C_{2v})$	50, 61
$Li + F_2$	$LiF, Li^+\begin{smallmatrix} F^- \\ \vert \\ F \end{smallmatrix}$ $(C_{2v}), v_{F-F} = 452$ cm^{-1}	52
$Na + F_2$	$NaF, Na^+\begin{smallmatrix} F^- \\ \vert \\ F \end{smallmatrix}$ $(C_{2v}), v_{F-F} = 474.9$ cm^{-1}	52
$K + F_2$	$KF, K^+\begin{smallmatrix} F^- \\ \vert \\ F \end{smallmatrix}$ $(C_{2v}), v_{F-F} = 464.1$ cm^{-1}	52
$Rb + F_2$	$RbF, Rb^+\begin{smallmatrix} F^- \\ \vert \\ F \end{smallmatrix}$ $(C_{2v}), v_{F-F} = 462.4$ cm^{-1}	52
$Cs + F_2$	$CsF, Cs^+\begin{smallmatrix} F^- \\ \vert \\ F \end{smallmatrix}$ $(C_{2v}), v_{F-F} = 458.8$ cm^{-1}	52
$Li + Cl_2$	$Li^+\begin{smallmatrix} Cl^- \\ \vert \\ Cl \end{smallmatrix}$ $(?), v_{Cl-Cl} = 246$ cm^{-1}	50, 52, 62
$Na + Cl_2$	$Na^+\begin{smallmatrix} Cl^- \\ \vert \\ Cl \end{smallmatrix}$ $(?), v_{Cl-Cl} = 225$ cm^{-1}	50, 52, 62
$K + Cl_2$	$K^+\begin{smallmatrix} Cl^- \\ \vert \\ Cl \end{smallmatrix}$ $(?), v_{Cl-Cl} = 264$ cm^{-1}	50, 52, 62
$Rb + Cl_2$	$Rb^+\begin{smallmatrix} Cl^- \\ \vert \\ Cl \end{smallmatrix}$ $(?), v_{Cl-Cl} = 259$ cm^{-1}	50, 52, 62
$Cs + Cl_2$	$Cs^+\begin{smallmatrix} Cl^- \\ \vert \\ Cl \end{smallmatrix}$ $(?), v_{Cl-Cl} = 260$ cm^{-1}	50, 52, 62
$Na + Br_2$	$Na^+\begin{smallmatrix} Br \\ \vert \\ Br \end{smallmatrix}$ (other alkali metals also)	62a,b
$Na + ClF$	Na^+ClF^- (other alkali metals also)	62c
$Na + I_2$	$Na^+I_2^-$ (other alkali metals also)	62b
$Li + NO_2$	$Li^+NO_2^-(?), v_{as-N-O} = 1244$ cm^{-1} $Li_2^+NO_2^-$	50, 63, 64
$Na + NO_2$	$Na^+NO_2^-(?)$ $Na_2^+NO_2^-$	50, 63, 64

TABLE 2-3 (*continued*)

Metal and reactant	Products and comments	References
$K + NO_2$	$K^+NO_2^-$(?)	50, 63, 64
	$K_2^+NO_2^-$	
$Cs + NO_2$	$Cs^+NO_2^-$	50, 63, 64
	$Cs_2^+NO_2^-$	
$K + CO_2/N_2O$	$K^+CO_3^-$ (C_{2v})	60a
$K + CO/O_2$	$K^+CO_3^-$ (C_{2v})	60a
$Na + CO_2$	$Na^+CO_2^-$ (ESR coupling to Na)	42
$Na + CS_2$	$Na^+CS_2^-$ (ESR coupling to Na)	42
$Na + (CH_3)_2C{=}O$	$Na^+[(CH_3)_2C{=}O]^-$	42
$K + (CH_3)_2C{=}O$	$K^+[(CH_3)_2C{=}O]^-$ (same ESR as Na species)	42
$Na + $ 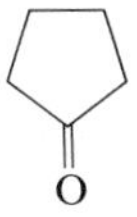 (cyclopentanone)	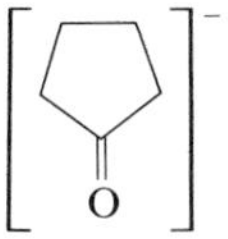 Na^+ [cyclopentanone]$^-$	42
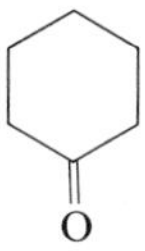 $Na + $ (cyclohexanone)	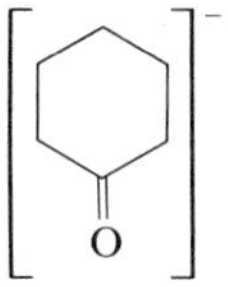 Na^+ [cyclohexanone]$^-$	42
$Na + ROH$	$Na^+(ROH)_n^{\bar{\cdot}}$ (solvated electron)	42
$Na + H_2O$	$Na^+(H_2O)_n^{\bar{\cdot}}$ (solvated electron)	42
$Na + H_2S$	$Na^+H_2S^-$ (low-lying d-orbitals accommodate electron)	42
$Na + B_2H_6$	$Na^+B_2H_6^-$ (low-energy light required)	53
$Na + Cr(CO)_6$	$Na^+[Cr(CO_5)]^-$ (low-energy light required)	54
$Cs + TCNQ^a$	$Cs^+TCNQ^{\bar{\cdot}}$	65
$Li + TCNQ$	$Li^+TCNQ^{\bar{\cdot}}$	66
$Na + TCNQ$	$Na^+TCNQ^{\bar{\cdot}}$	66
$K + TCNQ$	$K^+TCNQ^{\bar{\cdot}}$	66
$Rb + TCNQ$	$Rb^+TCNQ^{\bar{\cdot}}$	66
$Cs + TCNQ$	$Cs^+TCNQ^{\bar{\cdot}}$	65
	$Cs_2^{2+}(TCNQ)_3^{2-}$	66

^{a} TCNQ $\equiv$ (structure: 7,7,8,8-tetracyanoquinodimethane, $(CN)_2C{=}C_6H_4{=}C(CN)_2$)

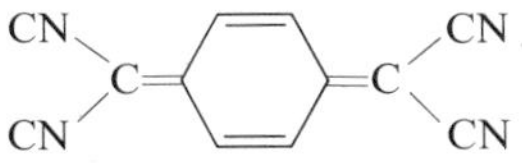

In the studies of pure halogens, F_2 and Cl_2, it is not surprising that both MX and $M^+X_2^-$ are formed in the low-temperature matrix. Side-on bonding of F_2^- to M^+ is most reasonable (C_{2v}). The ν_{F-F} values support the view that the $M^+-F_2^-$ interaction is in the order of Na > K > Rb > Cs > Li, again showing Li to be the "strange one." In the analogous $M^+-Cl_2^-$ system, the ordering of the interaction appears to be K > Cs > Rb > Li > Na, which is very striking in that Na is completely reversed from F_2^- vs Cl_2^-. It is interesting and perhaps meaningful that Na^+ interacts most strongly with F_2^- while K^+, one down the periodic family, interacts most strongly with Cl_2^-, which is one down the family from F_2^-.

All of these variations in binding of M^+ to X_2^- emphasize an important point stressed earlier by Andrews.[50] A charge separated pair $M^+X_2^-$ in a matrix cannot, because of coulombic forces, separate even under very high dilutions and so X_2^- cannot be completely free. Its approximation to being free will depend on its orbital interactions with M^+. This point concerning the strength of interaction is emphasized in the NO_2^- work. Milligan and co-workers[63,64] were able to generate what appeared to be free NO_2^- in a matrix by either vacuum photolysis, electron bombardment, or metal atom reactions. A 1244 cm^{-1}-band was assigned to matrix-isolated NO_2^-. This anion is very stable, and apparently $M^+NO_2^-$ is a good approximation to free NO_2^-. Andrews believes additional bands observed were due to structural isomers of $M^+NO_2^-$ and/or $M_2^+NO_2^-$.

Jacox and Milligan[60a] have deposited K atoms and CO_2–N_2O, and CO–O_2 mixtures, and have generated the interesting CO_3^- molecule at 14°K in argon. This species has C_{2v} geometry, and upon Hg arc irradiation of the matrix, CO_3^{2-} was produced.

Mile's rotating cryostat work has demonstrated the facile generation of organic radical anions by Na or K atom codepositions.[42] Deposition of Na with CO_2 or CS_2 yielded CO_2^- and CS_2^- that were not completely free, as evidenced by ESR coupling to Na^+. Comparisons with the same, but free, radical anions formed by γ-radiolysis methods showed marked differences, and reconfirmed the notion that Na^+ was interacting with CO_2^- in a significant way. However, in the case of ketones, no ESR evidence was found to indicate Na^+ or K^+ interactions. Similarly, depositions of Na with H_2O or ROH yielded solvated electrons $[Na^+(H_2O)_n^{\cdot}]$ not interacting with Na^+, as only a single narrow ESR line was observed. Interestingly, warming from 77°K then yielded a seven-line spectrum, apparently due to splitting by six protons of three water molecules (octahedral arrangement of six H around e^-). And finally, $Na^+H_2S^{\cdot}$ was prepared. In this case a single H_2S accommodated the electron, apparently able to do so, where H_2O could not, because of low-lying d-orbitals.[42]

Some rather peculiar findings of Kasai and McLeod,[53] and Turner[54] and co-workers are of interest. In these studies Na atoms were deposited with B_2H_6 and $Cr(CO)_6$, respectively. Electron transfer took place to yield Na^+ $B_2H_6^-$ and $Na^+[Cr(CO)_5]^- + CO$, but only in the presence of low-energy light. These processes are light-induced and energetically possible because of the coulombic potential energy between the anion–cation pair. This potential energy becomes available because a donor–acceptor complex is formed on close approach (7–8 Å) of the metal atom and the reactant in the matrix.[66a]

There has been a recent resurgence of interest in electron transfer with regard to molecular metals technology. The process of electron transfer through crystals of organic complexes is of critical importance and will be discussed further under transition metal charge transfer studies (Chapter 5). Here we see that alkali metal–TCNQ depositions efficiently yield the salt M^+TCNQ^-.[65,66] With Cs, however, complex $Cs_2(TCNQ)_3$ salts were also formed.[66]

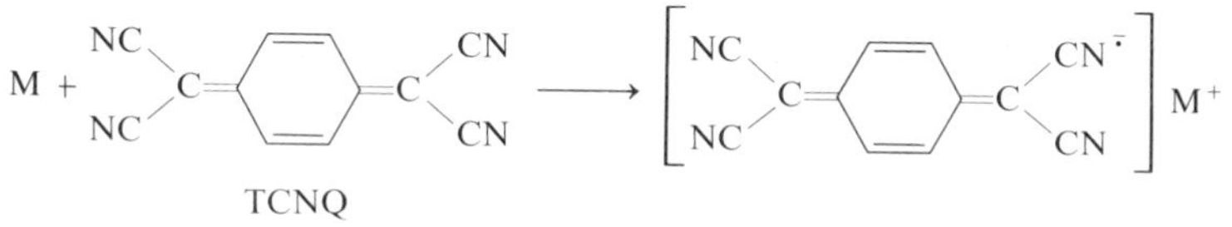

TCNQ

A very significant recent development in alkali metal atom chemistry was communicated by Timms and co-workers.[67] Employing a solution-phase metal atom reaction (see Chapter 5 for a description of equipment used), potassium atoms were allowed to interact with toluene–THF solutions of transition metal salts, for example $MoCl_5$.

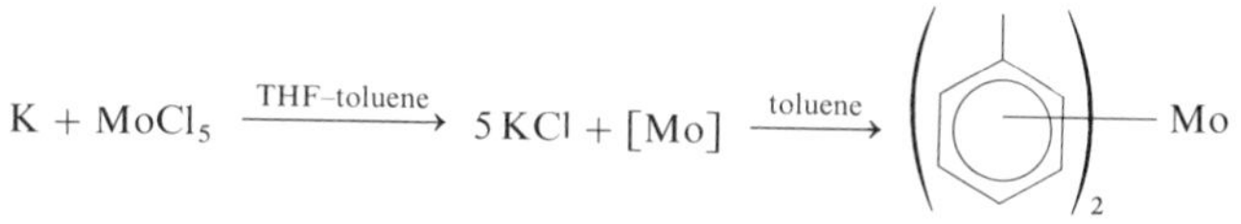

The high reactivity of the K atoms allowed the low-temperature reduction of the $MoCl_5$ to a zero valent form of Mo which complexed toluene to form (bis-π-toluene)molybdenum(0). This serves as an important new synthetic method for production of early transition metal bis-π-arene complexes, and also points out another important use of metal atom chemistry. Other similar π-arene complexes of Ti, V, and Cr were prepared in this way.[67] The fact that K atoms were successful and bulk K metal was not in this reduction process probably indicates that a dilute solution of K or solvated electrons are necessary for the success of this reduction process and for the preservation of product.

3. OXIDATIVE ADDITION PROCESSES

The alkali metals, being monovalent, are not amenable to oxidative addition studies $M + X-Y \rightarrow X-M-Y$.

4. SIMPLE ORBITAL MIXING PROCESSES

Margrave and co-workers have investigated, by matrix isolation spectroscopy, the interactions of Li atoms with Lewis base-type molecules.

Much of this work remains to be published, especially that on Li–CO,[68] Li–C$_2$H$_4$,[68a] and Li–C$_2$H$_2$[68a] systems.

One such study recently published by Meier, Hauge, and Margrave describes the low-temperature (15°K) interaction of Li atoms with H$_2$O and NH$_3$,[69] as studied by ESR spectroscopy. Thus, ESR of the Li(^{2}S) with a nuclear spin of $\frac{3}{2}$ could be used to measure electron density directly, and in this way molecular complexes, 1:1 for Li–NH$_3$, and 1:1 and 1:2 for Li–H$_2$O, were detected. These complexes were stable at 15°K prior to complete electron transfer. Comparison of the spin densities of the ^{7}Li–H$_2$O system with the ^{7}Li–(H$_2$O)$_2$ system indicates that the average interaction of the two water molecules is weaker than the interaction of a single water molecule with lithium. However, NH$_3$ interacts more strongly than H$_2$O in either case.

The Meier, Hauge, Margrave experiments and calculations, as well as calculations on the Li–NH$_3$ systems by Nicely and Dye,[70] seem to favor *partial* electron-charge movement toward Li in these complexes. However, further work regarding the exact mode of interaction is still in progress by Margrave and coworkers.

$$\overset{\delta^-}{\text{Li}}\text{-----}\overset{\delta^+}{\text{O}}\overset{\displaystyle H}{\underset{\displaystyle H}{}} \qquad \overset{\delta^-}{\text{Li}}\text{-----}\overset{\delta^+}{\text{N}}\overset{\displaystyle H}{\underset{\displaystyle H}{-H}}$$

In this work another example of low-energy light activation in a matrix was observed.[69] Visible light caused both the Li–H$_2$O and Li–NH$_3$ complexes to decompose to salts and H atoms. Likely reaction schemes were proposed.

$$\text{Li} - \text{H}_2\text{O} \xrightarrow{h\nu} \text{LiOH} + \text{H} \cdot$$

$$\text{Li} - \text{NH}_3 \xrightarrow{h\nu} \text{LiNH}_2 + \text{H} \cdot$$

$$\downarrow$$

$$\text{LiNH} + \text{H}_2$$

As with most matrix-isolation studies, noble gases are employed as inert matrices. Generally, it is assumed there is no binding interaction of M with Ar or Xe. However, transient M–Ar(Xe) Vanderwaals molecules have been detected spectroscopically during M–noble gas codepositions, while being laser irradiated. Ault, Tevault, and Andrews[70a] believe the interaction occurs in the dense gas layer just prior to condensation, and so the M–noble gas "molecule" may be considered a gaseous transient species.

5. SUBSTITUTION PROCESSES

Of all the gas-phase alkali metal atom–organic halide reactions studied, perhaps the most useful reactions have been those recently described by

Lagow and Chung.[70b] In these studies perchlorocarbons were allowed to react with excess Li vapor at $800°-1000°C$. Lithium chloride and perlithiohydrocarbons were produced in large quantities. Perlithiosilane and -germane were also prepared in this way:[70c]

$$\text{Li gas} + CCl_4 \xrightarrow{-LiCl} CLi_4 + C_2Li_4 + C_2Li_2$$
$$\text{excess} \qquad\qquad\qquad 15\% \quad 60\% \quad 20\%$$

$$\text{Li} + C_2Cl_6 \xrightarrow{-LiCl} C_2Li_6$$
$$\text{excess} \qquad\qquad\qquad 80\%$$

$$\text{Li} + SiCl_4 \xrightarrow{-LiCl} SiLi_4$$
$$\text{excess}$$

$$\text{Li} + GeCl_4 \longrightarrow GeLi_4$$
$$\text{excess}$$

It is interesting that hydrogen atoms could also be substituted by Li atoms. In fact, total replacement of H, Br, and Cl is possible in dichloro- or monobromobutanes:[70d]

$$\text{Li} + CH_2ClCHClCH_2Cl \xrightarrow[-LiH]{-LiCl} C_3Li_6 + C_3Li_8$$
$$\text{excess}$$

$$\downarrow D_2O$$

$$CD_3CD{=}CD_2 + CD_3CD_2CD_3$$

$$\text{Li} + CH_3CH_2CH_2Br \xrightarrow[-LiH]{-LiBr} C_3Li_8 + C_2Li_6$$
$$\text{excess}$$

$$\downarrow D_2O$$

$$CD_3CD_2CD_3 + CD_3CD_3$$

Likewise, H substitution was possible with alkenes as well, and in fact replacement of H was more prevalent than addition of Li to $C{=}C$.[70e]

$$\text{Li gas} + \quad \xrightarrow[700°-800°C]{-LiH} C_4Li_8$$
$$\text{excess}$$

And lastly, as one of the rare reports of high-temperature atoms being allowed to react with other high-temperature atoms, Shimp and Lagow report that Li vapor and C vapor (C_1, C_2, C_3) reacted to form mainly C_3Li_4, plus CLi_4 and C_2Li_4.[70f]

6. CLUSTER FORMATION PROCESSES

Niedermayer[71] has discussed the detailed theoretical and physical requirements for metal atom cluster growth on clean surfaces. These aspects will be discussed in more detail in Chapter 5, and have been previously summarized.[10]

Sodium atoms dispersed on clean tungsten yields one of the most ideal systems for detailed study of cluster growth. It was found that by studying the growth of Na $\rightarrow$ (Na)$_n$ on tungsten at 37°–97°C, first strained clusters grew, which exhibited heat contents far above that of the ground crystalline state of bulk sodium. The energy of desorption (E_{des}) for Na on W was at first 2.5 eV while, after four layers of Na were put down, this value for a Na atom on the surface dropped to 1.06 eV,[71] which corresponds closely to the bulk heat of vaporization of sodium. To rationalize the growth of strained small clusters, it was proposed that lattice relaxation was difficult because Na–Na interactions were so strong.

II. Alkali Metal Halide, Oxide, and Sulfide Molecules

A. Occurrence, Properties, and Techniques

With heats of vaporization ranging from 46–67 kcal/mole,[72] vaporization of alkali metal halides is quite easy. Temperatures in the range of 800°–1000°C are necessary for efficient vaporizations, and resistively heated Al_2O_3 sources are quite satisfactory. The vapors obtained are mainly but not completely monomeric.[72a] The association of MX molecules in the vapor has been the subject of a number of papers,[73–73g] and dimers and trimers have been detected in substantial amounts (0.16–0.32 mole fraction).[73d] The heats of dissociation of the dimers range from 34–48 kcal/mole, with (CsCl)$_2$ having the lowest and (LiCl)$_2$ having the highest dissociation energies.[73a,c] The (LiCl)$_3$ and (LiBr)$_3$ trimers dissociate to dimers and monomers with dissociation energies of 34 and 36 kcal/mole, respectively.[73a]

Vaporization of alkali metal oxides (M$_2$O) is also not difficult, but some decomposition accompanies vaporization. There has been controversy regarding this process as Brewer and Margrave[74] reported that monomeric Li$_2$O vaporized cleanly from Knudsen cells whereas the other alkali metal oxides decomposed to the elements. However, later work by Klemm and Scharf[75] indicated that the vaporizations were not accompanied by much decomposition except in the case of K$_2$O. In this case, slow vaporization at 450°C allowed K$_2$O to vaporize, but at higher temperatures some K atoms were evolved, leaving K$_2$O$_2$ in the residue.[75]

Table 2-4 summarizes some of the vaporization data for alkali metal halides, oxides, and sulfides.

TABLE 2-4

Vaporization Data and Spectroscopic Data for Alkali Metal Halides, Oxides, and Sulfides

Compound	mp (°C)[a]	bp (°C)[a]	Temp of vaporization (°C)[b,c]	Vapor composition	References
LiF	845	1676	1047	LiF, $(LiF)_2$ (large)	73f,g
LiCl	605	1350	783	LiCl, $(LiCl)_2$ (large)	73b
LiBr	550	1265	748	LiBr, $(LiBr)_2$ (large)	73b
LiI	449	1180	723	LiI, $(LiI)_2$ (large)	73b
NaF	993	1695	1077	NaF, $(NaF)_2$ (large)	73b
NaCl	801	1413	865	NaCl, $(NaCl)_2$	73b,f
NaBr	747	1390	806	NaBr, $(NaBr)_2$	73b
NaI	661	1304	767	NaI, $(NaI)_2$	73b
KF	858	1505	885	KF, $(KF)_2$	73b
KCl	770	1500 subl	821	KCl, $(KCl)_2$	73b,f
KBr	734	1435	795	KBr, $(KBr)_2$	73b
KI	681	1330	745	KI, $(KI)_2$	73b
RbF	795	1410	921	RbF, $(RbF)_2$	73b
RbCl	718	1390	792	RbCl, $(RbCl)_2$ (16%)	73b,d
RbBr	693	1340	781	RbBr, $(RbBr)_2$ (18%)	73b,d
RbI	647	1300	748		73b
CsF	682	1251	712		
CsCl	645	1290	744	CsCl, $(CsCl)_2$ (20%)	73d
CsBr	636	1300	748	CsBr, $(CsBr)_2$ (32%)	73d
CsI	626	1280	738		73f
Li_2O	>1700	1200	980	Li_2O	74, 75
Na_2O		1275 subl	670	Na_2O	74, 75
K_2O	d350		450	K_2O, K, O_2	74, 75
Rb_2O	d400		475	Rb_2O	74, 75
Cs_2O	d400		375	Cs_2O	74, 75
LiS	930				
Na_2S	1180				
K_2S	840				
K_2S_2	470				
Rb_2S	d530				
Rb_2S_2	420	850			
Cs_2S_2	460	>800			

[a] Reference 8c, p. B-67 through B-160.
[b] Reference 8c, p. D-183–188, temperature where compound possesses 1 torr vapor pressure.
[c] Mo, W, or Ta boats are generally satisfactory vaporization sources.

B. Chemistry

1. ABSTRACTION PROCESSES

Nothing has been published as yet.

2. ELECTRON-TRANSFER PROCESSES

Andrews and Ault[76–79] have carried out a series of matrix-isolation spectroscopic studies of the products formed from the low-temperature codeposition of MF, MCl, or MBr (M = alkali metals) with halogens or hydrogen halides. In each case electron transfer took place to yield $M^+X_3^-$ or $M^+HX_2^-$ complex salts, according to the spectra observed. In the MF–F_2 system,[76] the product $M^+F_3^-$ possessed a center of symmetry with D_{3h} point group. However, in the case of MCl + Cl_2 or Br_2, asymmetric T-shaped molecules were apparently formed. Other halogens[78] as well as mixed halogens have also been studied,[77] as have hydrogen halides.[79]

$$MF + F_2 \longrightarrow M^+F_3^-$$
$$MCl + Cl_2 \longrightarrow M^+Cl_3^-$$
$$MCl + ClF \longrightarrow M^+Cl_2^-F$$
$$MBr + HCl \longrightarrow M^+HCl^-Br$$

3. OXIDATIVE ADDITION PROCESSES

Nothing has been published—probably not applicable.

4. SIMPLE ORBITAL MIXING PROCESSES

Hauge, Gransden, and Margrave[80] have cocondensed alkali metal fluoride vapors with CO on a polished copper block, and examined by reflectance techniques the stretching frequencies for $\nu_{C\equiv O}$ and ν_{M-F} in the complexes. Both MF–CO and $(MF)_2$–CO complexes were observed. These workers

$$M-F + CO \xrightarrow[10°K]{\text{argon}} F-M-C\equiv O + (MF)_2-C\equiv O$$
$$M = Li, Na$$

found that the $\nu_{C\equiv O}$ frequency varies as a function of the reciprocal of the square (or higher order) of the metal ionic radius. Variance in $\nu_{C\equiv O}$ was not found with change in formal charge, as alkali metal MF–CO, alkaline earth MF_2–CO, and rare earth MF_3–CO complexes were compared (cf. Chapters 3, and 10) Thus, Na, Ca, and Nd have approximately the same ionic radius for the mono-, di-, and trivalent states, respectively, but a simple $\nu_{C\equiv O}$ relationship with the $+1$, $+2$, $+3$ charges on M was not found. The one general conclusion that could be made was that positive $\nu_{C\equiv O}$ shifts do serve as a qualitative probe of *molecular ionic character* for molecules with similar geometry. And it is likely that quantitative correlations will be possible once the behavior of CO in high nonlinear fields is well documented.[80] (Note that $v_{C\equiv O}$ increases upon complexation, and cf. Chapters 3 and 10 for further discussions of this work.)

Structure	$v_{C\equiv O}$ (cm^{-1})
C$\equiv$O	2140
(Li—F)—C$\equiv$O	2185.1
(Na—F)—C$\equiv$O	2172.4

Margrave and co-workers[81] have carried out further experiments with alkali metal fluoride vapors where they were codeposited at 4.2°K with alkaline earth fluoride vapors. Salt molecule combinations were prepared, such as LiMgF$_3$, and their possible molecular geometries discussed.

$$MF + M'F_2 \longrightarrow MM'F_3$$

$$M = Li, K, Cs$$
$$M' = Mg, Ca, Sr$$

5. CLUSTER FORMATION PROCESSES

See Section II,A of this chapter for a brief discussion of dimer and trimer formations for MX vapors. Akishin and Rambidi have employed electron diffraction for the study of these vapor phase associations and have discussed the probable configuration of the dimer molecules.[73b] These authors favor a symmetrical arrangement as shown below.

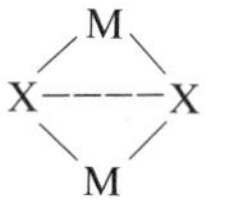

The X—X distances were determined. As examples, in (LiF)$_2$, X—X = 2.67 Å; in (LiCl)$_2$, 3.68 Å; in (NaCl)$_2$, 4.04 Å; in (NaBr)$_2$, 4.39 Å; and in (NaI)$_2$, 4.87 Å.[73b,81]

No chemistry has been reported for these vapor telomers.

References

1.   M. Gadsden and K. Salmon, *Nature (London)* **182**, 1598 (1958).
     1a. G. Kvifle, *Fys. Verden, Fra*, **21**, 253 (1959).
2.   T. M. Donahue, *Nature (London)* **183**, 1480 (1959).
     2a. E. A. Lytle and D. M. Hunton, *J. Atmos. Terr. Phys.* **16**, 236 (1959).
3.   S. N. Ghosh, *Sci. Tech. Aerosp. Rep.* **6**, 1693 (1968).
4.   J. Delannoy, *Ann. Geophys.* **16**, 236 (1960).
5.   M. Gadsden, *Ann. Geophys.* **18**, 392 (1962).
6.   P. W. Merrill, *Astrophys. J.* **105**, 360 (1947). P. S. Conti, *Astrophys. J., Suppl. Ser.* **11**, 47 (1965). P. W. Merrill and J. L. Greenstein, *Publs. Astron. Soc. Pac.* **70**, 98 (1958). O. C. Wilson, W. A. Baum, W. K. Ford, and A. Purgathofer, *ibid.* **77**, 359 (1965).

6a. G. H. Herbig, *Astrophys. J.* **140**, 702 (1964).

7. A. P. Dronov, A. G. Sviridov, and N. N. Sobolev, *Opt. Spektrosk.* **5**, 490 (1958).

8. V. Piacente, G. Bardi, and L. Malaspina, *J. Chem. Thermodyn.* **5**, 219 (1973).

8a. J. M. Walters and S. Barratt, *Proc. R. Soc. London, Ser. A* **119**, 257 (1928).

8b. L. Topor, *J. Chem. Thermodyn.* **4**, 739 (1972).

8c. "Handbook of Chemistry and Physics," 56th ed. CRC Press, Cleveland, Ohio, 1975-1976.

8d. P. L. Timms, *in* "Cryochemistry", (M. Moskovits and G. Ozin, eds.), p. 61. Wiley (Interscience), New York, 1976.

9. E. C. R. Steacie, "Atomic and Free Radical Reactions," 2nd ed. Van Nostrand-Reinhold, Princeton, New Jersey, 1954.

9a. T. Kumada, F. Kasahara, and R. Ishiguro, *J. Nucl. Sci. Technol.* **13**, 74 (1976).

10. K. J. Klabunde, *in* "Reactive Intermediates" (R. A. Abramovitch, ed.), Plenum, New York, 1979.

11. P. S. Skell and R. J. Petersen, *J. Am. Chem. Soc.* **86**, 2530 (1964).

12. R. G. Doerr and P. S. Skell, *J. Am. Chem. Soc.* **89**, 3062 and 4684 (1967).

13. von B. Mile, *Angew. Chem., Int. Ed. Engl.* **7**, 507 (1968).

14. L. Andrews, *Annu. Rev. Phys. Chem.* **22**, 109 (1971); *in* "Vibrational Spectra and Structure" (J. Durig, ed.), Elsevier, Amsterdam, 1975.

14a. E. P. Kundig, M. Moskovits, and G. A. Ozin, *J. Mol. Struct.* **14**, 137 (1972).

15. R. Ladenburg and R. Minkowski, *Ann. Phys.* (*Leipzig*) [4] **87**, 298 (1928).

16. H. Hartel and M. Polanyi, *Z. Phys. Chem., Abt. B* **11**, 97 (1930).

17. H. Hartel, N. Meer, and M. Polanyi, *Z. Phys. Chem., Abt. B* **19**, 139 (1932).

18. J. N. Harsnape, J. M. Stevels, and E. Warhurst, *Trans. Faraday Soc.* **36**, 465 (1940).

19. C. E. H. Bawn and W. J. Dunning, *Trans. Faraday Soc.* **35**, 185 (1939).

20. C. E. H. Bawn and J. Milsted, *Trans. Faraday Soc.* **35**, 889 (1939).

21. M. G. Evans and E. Warhurst, *Trans. Faraday Soc.* **35**, 593 (1939).

22. C. E. H. Bawn, *Discuss. Faraday Soc.* **2**, 145 (1947).

23. F. Fairbrother and E. Warhurst, *Trans. Faraday Soc.* **31**, 987 (1935).

24. E. Warhurst, *Trans. Faraday Soc.* **35**, 674 (1939).

25. A. G. Evans and H. Walker, *Trans. Faraday Soc.* **40**, 384 (1944).

26. E. Warhurst, *Q. Rev. Chem. Soc.* **5**, 44 (1951).

27. E. Warhurst, *Trans. Faraday Soc.* **35**, 674 (1939).

28. J. W. Hodgins and R. L. Haines, *Can. J. Chem.* **30**, 473 (1952).

29. J. Curry and M. Polanyi, *Z. Phys. Chem., B* **20**, 276 (1933).

30. W. Heller and M. Polanyi, *Trans. Faraday Soc.* **32**, 633 (1936).

31. C. E. H. Bawn and A. G. Evans, *Trans. Faraday Soc.* **33**, 1571 (1937).

32. F. Haber and H. Sachsse, *Z. Phys. Chem.*, **831** (1931).

32a. A. Fontijn, S. C. Kurzius, J. J. Houghton, and J. A. Emerson, *Rev. Sci. Instrum*, **43**, 726 (1972).

33. K. R. Wilson, G. H. Kwei, J. A. Norris, R. R. Herm, J. H. Birely, and D. R. Herschbach, *J. Chem. Phys.* **41**, 1154 (1964).

34. S. Datz and R. E. Minturn, *J. Chem. Phys.* **41**, 1153 (1964).

34a. D. L. Perry, P. F. Meier, R. H. Hauge, and J. L. Margrave, *Inorg. Chem.* **17**, 1364 (1978).

35. R. E. Minturn, S. Datz, and R. L. Becker, *J. Chem. Phys.* **44**, 1149 (1966).

36. G. H. Kwei and D. R. Herschbach, *J. Chem. Phys.* **51**, 1742 (1969).

37. R. G. Doerr and P. S. Skell, *J. Am. Chem. Soc.* **89**, 4684 (1967).

38. R. G. Doerr and P. S. Skell, *J. Am. Chem. Soc.* **89**, 3062 (1967).

39. W. T. Borden, *Tetrahedron Lett.*, p. 259 (1967).

40. J. E. Bennett, B. Mile, and A. Thomas, *Chem. Commun.* p. 265 (1965); *Proc. R. Soc. London, Ser. A* **293**, 246 (1966).

41. L. S. Andrews and G. C. Pimentel, *J. Chem. Phys.* **44**, 2527 (1966); **47**, 3637 (1967); G. C. Pimentel, *Angew. Chem., Int. Ed. Engl.* **14**, 199 (1975).
42. von B. Mile, *Angew. Chem.* **80** (13), 519 (1968); cf. von B. Mile[13]
43. D. E. Milligan and M. E. Jacox, *J. Chem. Phys.* **47**, 5146 (1967).
44. L. Y. Tan and G. C. Pimentel, *J. Chem. Phys.* **48**, 5202 (1968).
45. L. Andrews, *J. Chem. Phys.* **48**, 972 and 979 (1968).
46. L. Andrews and T. G. Carver, *J. Chem. Phys.* **49**, 896 (1968).
47. T. G. Carver and L. Andrews, *J. Chem. Phys.* **50**, 4223 and 4235 (1969).
48. T. G. Carver and L. Andrews, *J. Chem. Phys.* **50**, 5100 (1969).
49. D. W. Smith and L. Andrews, *J. Chem. Phys.* **58**, 5222 (1973).
49a. D. A. Hatzenbuhler, L. Andrews, and F. A. Carey, *J. Am. Chem. Soc.* **97**, 187 (1975).
50. L. Andrews, *in* "Cryochemistry", (M. Moskovits and G. Ozin, *eds.*), p. 195. Wiley (Interscience), New York, 1976.
51. R. C. Spiker, Jr. and L. Andrews, *J. Chem. Phys.* **58**, 702 and 713 (1973).
52. W. F. Howard, Jr. and L. Andrews, *Inorg. Chem.* **14**, 409 (1975). L. Andrews, *J. Am. Chem. Soc.* **98**, 2147 (1976).
53. P. H. Kasai and D. McLeod, Jr., *J. Chem. Phys.* **51**, 1250 (1969).
54. J. J. Turner, *Angew. Chem., Int. Ed. Engl.* **14**, 304 (1975); D. L. Perry, P. F. Meier, R. H. Hauge, and J. L. Margrave, *Inorg. Chem.* **17**, 1364 (1978).
55. R. C. Spiker, Jr., L. Andrews, and C. Trindle, *J. Am. Chem. Soc.* **94**, 2401 (1972).
56. L. Andrews, *J. Chem. Phys.* **50**, 4288 (1969); L. Andrews and R. R. Smardzewski, *ibid.* **58**, 2258 (1973).
57. L. Andrews, *J. Phys. Chem.* **73**, 3922 (1969).
58. L. Andrews, *J. Mol. Spectrosc.* **61**, 337 (1976).
59. L. Andrews, *J. Chem. Phys.* **54**, 4935 (1971).
60. D. E. Tevault and L. Andrews, *J. Phys. Chem.* **77**, 1646 (1973); also see Jacox and Milligan.[60a]
60a. M. Jacox and D. E. Milligan, *J. Mol. Spectrosc.* **52**, 363 (1974).
61. R. C. Spiker Jr. and L. Andrews, *J. Chem. Phys.* **59**, 1851 (1973); L. Andrews, *J. Chem. Phys.* **63**, 4465 (1975).
62. W. F. Howard, Jr. and L. Andrews, *J. Am. Chem. Soc.* **95**, 2056 (1973); *Inorg. Chem.* **14**, 767 (1975).
62a. C. A. Wight, B. S. Ault, and L. Andrews, *Inorg. Chem.* **15**, 2147 (1976).
62b. L. Andrews, *J. Am. Chem. Soc.* **98**, 2152 (1976).
62c. E. S. Prochaska, B. S. Ault, and L. Andrews, *Inorg. Chem.* **16**, 2021 (1977).
63. D. E. Milligan and M. E. Jacox, *J. Chem. Phys.* **55**, 3404 (1971).
64. D. E. Milligan, M. E. Jacox, and W. A. Guillory, *J. Chem. Phys.* **52**, 3864 (1970);
65. B. H. Schechtman, S. F. Lin, and W. E. Spicer, *Phys. Rev. Lett.* **34**, (11), 667 (1975).
66. E. T. Maas, Jr., *Mater. Res. Bull.* **9**, 815 (1974).
66a. P. H. Kasai, *Accts. Chem. Res.*, **4**, 329 (1971).
67. P. N. Hawker, E. P. Kundig, and P. L. Timms, *J. Chem. Soc., Chem. Commun.* p. 730 (1978).
68. C. N. Krishnan, R. H. Hauge, and J. L. Margrave, unpublished work (1974-1976).
68a. R. H. Hauge and J. L. Margrave, unpublished work, (1975-1977).
69. P. F. Meier, R. H. Hauge, and J. L. Margrave, *J. Am. Chem. Soc.* **100**, 2108 (1978).
70. V. A. Nicely and J. L. Dye, *J. Chem. Phys.* **52**, 4795 (1970).
70a. B. S. Ault, D. E. Tevault, and L. Andrews, *J. Chem. Phys.* **66**, 1383 (1977).
70b. C. Chung and R. J. Lagow, *J. Chem. Soc., Chem. Commun.* p. 1078 (1972).
70c. J. A. Morrison and R. J. Lagow, *Inorg. Chem.* **16**, 2972 (1977).
70d. L. G. Sneddon and R. J. Lagow, *J. Chem. Soc., Chem. Commun.* p. 302 (1975).

70e. J. A. Morrison, C. Chung, and R. J. Lagow, *J. Am. Chem. Soc.* **97**, 5015 (1975).

70f. L. A. Shimp and R. J. Lagow, *J. Am. Chem. Soc.* **95**, 1343 (1973).

71.   R. Niedermayer, *Angew. Chem., Int. Ed. Engl.*, **14**, 212 (1975).

72.   K. Niwa, *J. Chem. Soc. Jpn.* **59**. 637 (1938).

72a. J. E. Mayer and I. H. Wintner, *J. Chem. Phys.* **6**, 301 (1938).

73.   L. Pauling, *Proc. Natl. Acad. Sci., India, Sect. A* **25**, 1 (1956).

73a. R. C. Miller and P. Kusch, *J. Chem. Phys.* **25**, 860 (1956).

73b. P. Akischin and N. G. Rambidi, *Z. Phys. Chem. (Leipzig)* **213**, 111 (1960).

73c. S. Datz, W. T. Smith, Jr., and E. H. Taylor, *J. Chem. Phys.* **34**, 558 (1961).

73d. I. G. Murgulescu and L. Topor, *Rev. Roum. Chim.* **13**, 1109 (1968).

73e. J. Guion, D. Hengstenberg, and M. Blander, *J. Phys. Chem.* **72**, 4620 (1968).

73f. C. T. Ewing and K. H. Stern, *J. Phys. Chem.* **78**, 1998 (1974).

73g. T. T. Bykova, Y. P. Efimov, and A. M. Tyutikov, *Pis'ma. Zh. Tekh. Fiz.* **1**, 872 (1975).

74.   L. Brewer and J. L. Margrave, *U. S. A. E. C., Natl. Sci. Found.* **UCRL-1864**, 2 (1952).

75.   W. Klemm and N. J. Scharf, *Z. Anorg. Allg. Chem.* **303**, 263 (1960).

76.   B. S. Ault and L. Andrews, *J. Am. Chem. Soc.* **98**, 1591 (1976).

77.   B. S. Ault and L. Andrews, *Inorg. Chem.* **16**, 2024 (1977).

78.   B. S. Ault and L. Andrews, *J. Chem. Phys.* **64**, 4853 (1976).

79.   B. S. Ault and L. Andrews, *J. Chem. Phys.* **64**, 1986 (1976).

80.   R. H. Hauge, S. E. Gransden, and J. L. Margrave, *J. Chem. Soc., Dalton Trans.*, 745 (1979).

81.   A. S. Kanaan, R. H. Hauge, and J. L. Margrave, *J. Chem. Soc., Faraday Trans. 2* **72**, 1991 (1976).

CHAPTER 3

Alkaline Earth Metals, Metal Halides, Oxides, and Sulfides (Group IIA)

I. Alkaline Earth Metal Atoms (Be, Mg, Ca, Sr, Ba)

A. Occurrence, Properties, and Techniques

As with the alkali metals, atoms of the alkaline earth elements have been detected in the upper atmosphere. Although the more important and more fully characterized systems are Li, Mg, and K atoms (Chapter 2), in the twilight and dawn skies Be gas,[1] as well as Mg, Ca, Ba, and Sr metal vapor clouds,[2,3] have been detected. Distinguishing between atoms and ions (for example Mg^+, Ca^+) is sometimes difficult.[3] Donahue[4] believes that a layer of dust or aerosol contains the condensed metal atoms and that these atoms are volatilized in the daytime, and that as they diffuse, they are ionized by sunlight and O_3. Then, layers of Na^+, Mg^+, and Ca^+ are detected. Below this dust layer, which is presumably meteoric in origin, the metals exist mainly as oxides, but above about 100 km they exist as atoms and ions (predominantly).[4]

The occurrence of atoms and ions of alkaline earth metals in stars[5,5a] and in the sun[6] does not come as a surprise. In cool carbon stars it has been found that Sr is sometimes detected in overly abundant amounts relative to the normal ratio of Sr/Fe in other stars.[5a] Also, free atoms of Ca, Sr, and Ba in flames have been reported.[7]

In the laboratory, formation of vapors of the alkaline earth metals is quite easy, with Be being the most difficult (78 kcal/mole heat of vaporization) and Mg the least (35 kcal/mole). These metals are readily sublimed from W, Ta, or Mo boats (although Be wets these materials) or from C or BO crucibles. There have been numerous reports describing these vaporizations and reporting vapor pressures in various temperature ranges.[8] Moriya[8a] reported that ThO_2 or W were good vaporization sources for Be, Rh, and Si.

Differences in vapor composition have been noted when Be is sublimed rather than vaporized from a fused mass. Thus, some Be_2 was observed by

TABLE 3-1

Vaporization Data for the Alkaline Earth Metals

Element	mp (°C)	bp (°C)	ΔH vap[a] (kcal/mole)	Vap Temp under vac (°C)[b]	Vap method[c]	Vapor Comp	References
Be	1278	2970		1225	Knudsen cell	Be, Be_2(small)	9, 11–14a
Mg	649	1090		439	Knudsen cell, vac arc, e-beam	Mg	9, 11–15a
Ca	839	1484	43.0	597	Knudsen cell, vac arc, e-beam	Ca	9, 11–14a, 15b
Sr	769	1384	39.5	537	Knudsen cell, vac arc, e-beam	Sr	9, 11–14a, 15b
Ba	725	1640		610	Knudsen cell, vac arc, e-beam	Ba	9, 11–14a

[a] From original literature. Also see Chapter 1.
[b] Vapor pressure of the metal is approximately 10 μm at this temperature.[11,13]
[c] Al_2O_3 crucibles work well for macroscale vaporizations.

mass spectrometry upon sublimation of Be metal, but the Be_2 disappeared upon fusion.[9] (For other metals, Ca_2, Sr_2, and Ba_2 were not observed.) Actually, the Be_2 molecule is predicted to have low stability.[9,10] Drowart and Honig,[10] employing thermochemical calculations in conjunction with MS studies have predicted a dissociation energy of 0.7 eV for Be_2, quite low compared with others (Ag_2 1.63 eV, Al_2 1.7 eV, At_2 0.8 eV, Au_2 2.1 eV, C_2 6.2 eV, Cu_2 2.0 eV, Ga_2 <1.5 eV, In_2 1.0 eV, Pb_2 1.0 eV, Si_2 3.2 eV, Sn_2 2.0 eV, and Tl_2 0.6 eV).

Under Knudsen cell conditions (thermal equilibrium), the vapor compositions for the alkaline earth metals are shown in Table 3-1,[9,11–15b] along with other pertinent data.

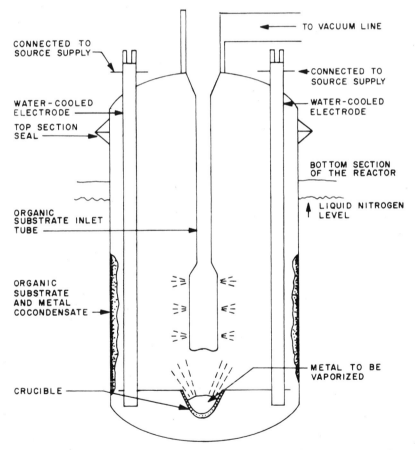

Figure 3-1. Macroscale stationary cocondensation apparatus for investigations of metal atom chemistry.

The techniques employed for study of the chemistry of alkaline earth metal atoms are identical in most respects to those used for the alkali metals.[11,16] Macroscale investigations of Mg, Ca, and Ba atom chemistry were carried out employing a stationary cocondensation apparatus first employed by Skell and Wescott for study of carbon vapor reactions (cf. Chapter 8). The general schematic for a macroscale (usually about 0.5–2 g of metal vaporized) reactor is shown in Fig. 3-1. Metal is vaporized under vacuum ($\sim 10^{-4}$ torr) and simultaneously cocondensed with incoming vapor of reactant in large excess. A matrix is formed on the cold walls (usually $-196°C$, $77°K$, liquid nitrogen). Reasonably good vacuum is a requirement so that gas-phase interactions are minimized. Low temperatures are required only so that good vacuum can be maintained. Liquid nitrogen is a convenient coolant, but it must be emphasized that higher temperatures can also be employed, some-times to advantage, as long as low vapor pressures can be maintained. The highest temperature possible and an excess of substrate are desirable in order to encourage metal atom–substrate interactions and not M–M recombina-tions (a very low activation energy process).[16]

B. Chemistry

1. ELECTRON-TRANSFER PROCESSES

The first and second ionization potentials of the alkaline earth metals indicate that electron-transfer processes should be facile, and this is found to be the case in metal atom reactions.

Employing matrix-isolation spectroscopy, a wide range of studies of alka-line earth metal atoms cocondensed with small molecules in the range of $4°–20°K$ have been carried out. Electron transfer usually takes place with N_2, O_2, O_3, N_2, and NO_2, often accompanied by formation of the metal oxide as well. Table 3-2 summarizes these microscale studies. It is interesting to compare metals, since analogous products are formed in each case. With O_2 and O_3 a good bonding approximation is that of side-on bonding with the negative charge residing on the oxygen moiety and a single positive charge on the metal.[17,18a] In the case of N_2, end-on bonding is apparently preferred, and no difference in N—N stretching frequency values could be detected upon change in metal from Ca to Sr to Ba. And finally, with NO, M—O bonded species were preferred.

$$\overset{+}{M}\!\!-\!\!\overset{\displaystyle O^-}{\underset{\displaystyle O}{\|}}\qquad \overset{+}{M}\overset{\displaystyle O}{\underset{\displaystyle O}{<}}\!\!\!\overset{-}{O}\qquad \overset{+}{M}\!\!-\!\!\overset{-}{O}\!\!=\!\!N\qquad \overset{+}{M}\!\!-\!\!\overset{-}{N}\!\!\equiv\!\!N$$

Ba > Sr   Ca > Sr > Ba

TABLE 3-2

Microscale Studies of Electron-Transfer Processes with Alkaline Earth Metal Atoms (Mg, Ca, Sr, Ba)

Reaction	Comments	References
$Ca + O_2 \rightarrow Ca^+O_2^- + Ca_2O + (CaO)_2$		17
$Sr + O_2 \rightarrow Sr^+O_2^-$	Side on bonding, C_{2v} $\nu_{O-O} = 1120$ cm^{-1}	17a,b
$Ba + O_2 \rightarrow Ba^+O_2^-$	Side on bonding, C_{2v} $\nu_{O-O} = 1115.6$ cm^{-1}	17a,b
$Mg + O_3 \rightarrow Mg^+O_3^- + MgO_2 + MgO$ in N_2 $\quad\quad Mg^+O_3^-$ (isomers) in Ar		17c,d
$Ca + O_3 \rightarrow Ca^+O_3^-$ $\quad\quad\rightarrow Ca^+O_2^-$ $\quad\quad\rightarrow CaO$	C_{2v}, M 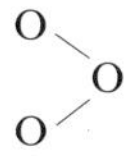	17–17b,d, 18
$Sr + O_3 \rightarrow Sr^+O_3^-$ $\quad\quad\rightarrow Sr^+O_2^-$ $\quad\quad\rightarrow SrO$	C_{2v}, M 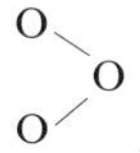	17a,b,d, 18
$Ba + O_3 \rightarrow Ba^+O_3^-$ $\quad\quad\rightarrow Ba^+O_2^-$ $\quad\quad\rightarrow BaO$	C_{2v}, M 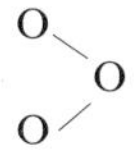	17a,b,d, 18
$Ca + NO \rightarrow Ca^+ON^-$	$\nu_{N-O} = 1357$ cm^{-1}	17a, 18a
$Sr + NO \rightarrow Sr^+ON^-$	$\nu_{N-O} = 1361$ cm^{-1}	17a
$Ba + NO \rightarrow Ba^+ON^-$	$\nu_{N-O} = 1364$ cm^{-1}	17a
$Ca + N_2 \rightarrow Ca^+N_2^-$	End-on bonding, $C_{\infty v}$ $\nu_{N-N} = 1800$ cm^{-1}	17a,b, 18
$Sr + N_2 \rightarrow Sr^+N_2^-$	End-on bonding, $C_{\infty v}$ $\nu_{N-N} = 1800$ cm^{-1}	17a,b, 18
$Ba + N_2 \rightarrow Ba^+N_2^-$	End-on bonding, $C_{\infty v}$ $\nu_{N-N} = 1800$ cm^{-1}	17a,b, 18
$Ca + NO_2 \rightarrow Ca^+NO_2^-$		18a

On macroscale, Mg has been concodensed with acetylenes, in particular 2-butyne, and the initial step appeared to be electron transfer.[19]

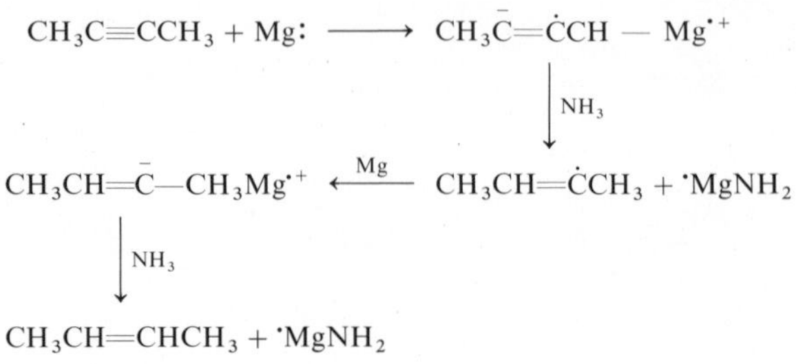

$$CH_3C{\equiv}CCH_3 + Mg: \longrightarrow CH_3\overset{-}{C}{=}\overset{\cdot}{C}CH - Mg^{\cdot\,+}$$

$$\downarrow NH_3$$

$$CH_3CH{=}\overset{-}{C}{-}CH_3Mg^{\cdot\,+} \xleftarrow{\;Mg\;} CH_3CH{=}\overset{\cdot}{C}CH_3 + {}^{\cdot}MgNH_2$$

$$\downarrow NH_3$$

$$CH_3CH{=}CHCH_3 + {}^{\cdot}MgNH_2$$

In the presence of NH_2, a Birch-type reduction process occurred, which was believed to involve two Mg atoms, although the exact fate of resultant $MgNH_2$ formed is unclear.

2. ABSTRACTION PROCESSES

Alkaline earth metal atoms are singlets in their thermal ground states and often react by oxidative addition processes (cf. next section). However, a long-lived triplet or "diradical-like" state is within reach energetically, and this electronic configuration should give rise to chemistry resembling free radicals. Skell and Girard apparently have generated ^{3P}Mg in experiments where Mg was vaporized by arcing.[19-21] In an arc, high-kinetic-energy electrons, ions, and atoms can be generated,[14a] and so higher-energy, long-lived metal atoms might be expected to be formed. Skell and Girard noted differences in chemistry of arced Mg vs thermally vaporized Mg and have attributed this to the presence of higher concentrations of ^{3P}Mg from the arc.

$$\underset{\text{(bulk)}}{Mg} \xrightarrow{\;arc\;} {}^{3P}Mg^* + {}^{1S}Mg$$

The chemistry of alkyl halides and ammonia were compared. The ^1S Mg (thermal vaporization) reacted with R—X yielding RMgX (oxidative addition) while the ^{3P}Mg (arc) reacted in radical-like processes represented below[21]:

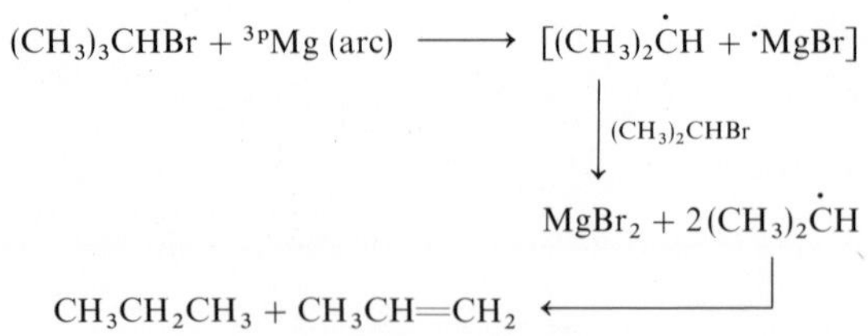

$$(CH_3)_3CHBr + {}^{3P}Mg\ (arc) \longrightarrow [(CH_3)_2\overset{\cdot}{C}H + {}^{\cdot}MgBr]$$

$$\downarrow {\scriptstyle (CH_3)_2CHBr}$$

$$MgBr_2 + 2(CH_3)_2\overset{\cdot}{C}H$$

$$CH_3CH_2CH_3 + CH_3CH{=}CH_2 \longleftarrow \rule[0.4em]{0pt}{0pt}$$

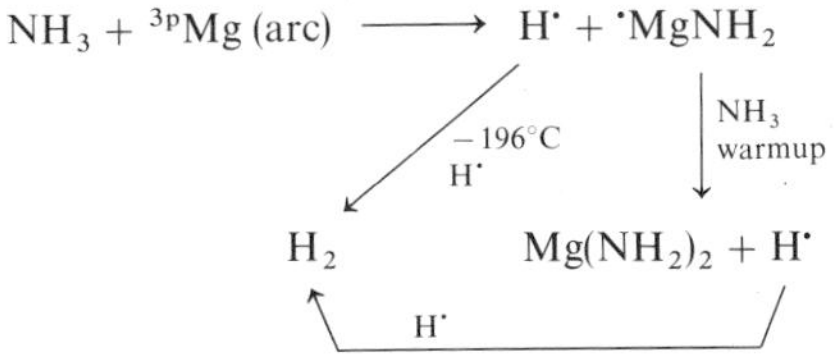

$$NH_3 + {}^{3P}Mg \text{ (arc)} \longrightarrow H^{\cdot} + {}^{\cdot}MgNH_2$$

$$H_2 \qquad\qquad Mg(NH_2)_2 + H^{\cdot}$$

The presence of inert diluents in the matrices appeared to help relax the excited states and perhaps to cage the radicals and to encourage RMgX formation from ${}^{3P}Mg$. In the case of NH_3, ${}^{1}S$ Mg initially yielded a Mg–NH_3 charge-transfer complex ($Mg^{+}NH_3{}^{-}$ perhaps), whereas ${}^{3P}Mg$ yielded $\frac{1}{2}$ mole of H_2 immediately upon codeposition and another $\frac{1}{2}$ mole of H_2 upon warmup, implying that ${}^{\cdot}MgNH_2$ was stable until warm-up.

Stereochemical experiments involving d,1- and meso-3,4-dibromobutanes with ${}^{3P}Mg$ atoms have been carried out and the resultant ratios of *trans/cis*-2-butenes determined.[19] The results implied stepwise $Br^{\cdot}$ abstraction by ${}^{3P}Mg$ with the formation of a Br-bridged free-radical intermediate which could help preserve stereochemistry.

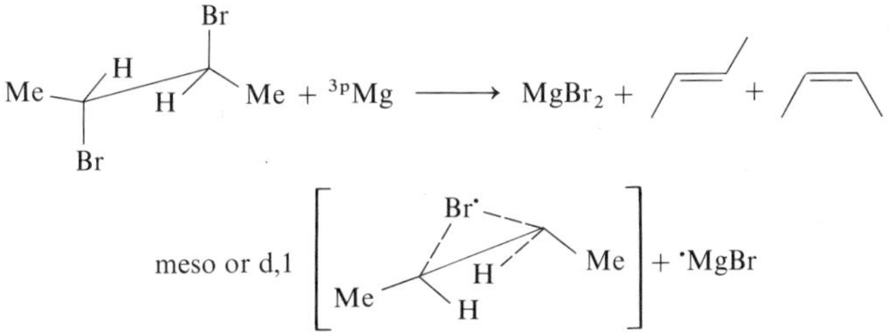

Another dihalo abstraction by ${}^{3P}Mg$ was observed with CH_2Br_2, and both $:CH_2$ and ${}^{\cdot}CH_2Br$ were generated as intermediate reactive species.[19]

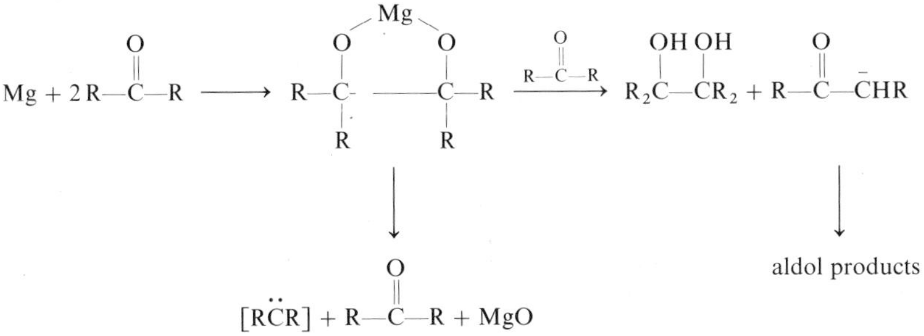

Magnesium atoms (${}^{1}S$ from thermal vaporization) have been employed as deoxygenating agents for cyclic ketones.[22] The direct abstraction of O to form MgO and a carbene ($Mg + R_2C{=}O \rightarrow MgO + R_2C{:}$) is a thermodynamically favorable process. However, Wescott and co-workers believe the carbene is only formed through a cyclic Mg coordination compound as shown below—probably best described as an electron-transfer process to give $Mg^{2+}(OCR_2)_2^{2-}$.

The true presence of a carbene or carbenoid species was confirmed by a study of cycloheptanone, where it has previously been shown that cyclo-heptenyl carbene yields cycloheptene *and* norcarane as rearrangement products.[23]

Recent gas-phase metal atom abstraction reactions have been reported by Felder, Gould, and Fontijn[24] using newly developed high-temperature fast-flow reactor techniques (cf. Chapter 7 for a more detailed discussion of this technique). Barium atoms at $180°–725°C$ and at pressures of $1–120$ torr reactant or inert bath gas reacted with N_2O to yield $BaO + N_2$. The BaO formed was in an excited state, i.e.,

$$Ba + N_2O \longrightarrow BaO^* + N_2$$
$$Ba + O_2 \longrightarrow BaO + O:$$

allowing chemiluminescence to occur, which was really the main concern of this study,[24] for the $Ba–N_2O$ reaction a rate coefficient of 4.2×10^{-10} ml/mole·sec at $600°K$ was obtained.

3. OXIDATIVE ADDITION PROCESSES

The potentially divalent alkaline earth metal atoms would be expected to undergo oxidative addition (oxidative insertion) processes fairly readily, and this has been found to be the case with alkyl and aryl halides. The mechanism of the oxidative addition is not known, however, and matrix isolation spectroscopy studies may be valuable in this area. One of the important aspects of the work is that the oxidative addition apparently always occurs upon matrix warming, and it may be that much higher temperatures than 10 or 20°K will have to be employed for the microscale studies.

According to LCAO–SCF calculations on the simplest alkaline earth system imaginable, that of Be atoms plus H_2, an unstable three-centered intermediate is first implied, followed by the formation of metastable H—Be—H.[25] A similar three-centered intermediate may be important in more complex systems such as alkyl halides. However, this has still to be proven.

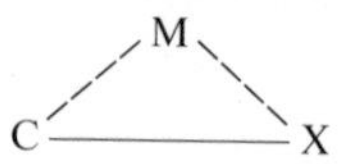

Nonsolvated Grignard reagents have been produced by codepositing on macroscale Mg atoms and alkyl halides.[26] Initially in the matrix a black complex was formed, probably a charge transfer complex $R—X^{\delta-}—Mg^{\delta+}$. Upon warming the matrix, oxidative addition to yield RMgX took place. Nonsolvated RMgX is believed to be polymeric,[19] similar to known Be complexes.[27] Their reactivities are not necessarily paralleled by solvated RMgX species since Skell and Girard found that the nonsolvated species abstracted hydrogen from acetone rather than adding to the carbonyl group, and that crotonaldehyde added 1,2 rather then 1,4 as with solvated systems.

Oxidative addition to C—F bonds has been achieved with Ca atoms.[28] Only unsaturated fluorocarbons reacted in this way, however, with saturated compounds being totally inert and unsaturated compounds reacting almost quantitatively. Upon complete warm-up of the matrices, which were usually clear colorless glasses in these reactions, defluorination took place. If water was added prior to the decomposition, R_fH species were formed ($R_fCaF + H_2O \rightarrow R_fH + HOCaF$).[28]

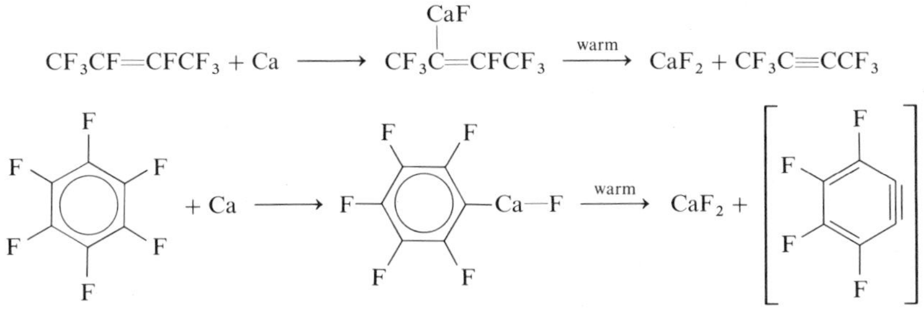

$$CF_3CF{=}CFCF_3 + Ca \longrightarrow \underset{\displaystyle |}{\overset{\displaystyle CaF}{CF_3C}}{=}CFCF_3 \xrightarrow{warm} CaF_2 + CF_3C{\equiv}CCF_3$$

4. SIMPLE ORBITAL MIXING PROCESSES

For the alkaline earth metal atoms this type of reaction has not yet been described. It is true that RX—M complexes are formed prior to oxidative addition, and these might fit under this reaction heading, but little is known about those complexes and they probably fit better under Electron Transfer Processes (RX$^-$—M$^{+\cdot}$).

5. CLUSTER FORMATION PROCESSES

Miller and Andrews[29–29d] have recently published a series of papers dealing with the formation of alkaline earth metal atom dimers in 10°K noble gas matrices. Absorption spectra and laser excitation–emission spectra have been studied in detail so that the exact bonding picture for these dimers could be elucidated. Thus, Be_2,[28a] Ca_2,[29] Mg_2,[29a] CaMg,[29b] SrMg,[29b] SrCa,[29b] and Sr_2[29c] have been investigated. These dimers were prepared by simply depositing higher concentrations of M in the matrix, such that some M_2 as well as M were trapped, usually in an argon or krypton matrix.

Further work with alkaline earth metal atoms (Group IIA) bonded to Group IIB metal atoms was reported by Miller and Andrews.[29d] Thus, at 10°K mixtures of metal atoms were deposited. In this way absorption spectra for ZnHg, CdHg, MgZn, MgCd, MgHg, Hg_2, ZnCd, MgZn, and MgCd were obtained.

II. Alkaline Earth Metal Halide, Oxide, and Sulfide Vapors

A. Occurrence, Properties, and Techniques

Although available reports are not unambiguous, the natural occurrence of vapors of alkaline earth oxides in the upper atmosphere is quite certain since the atoms themselves have been detected. Thus, at lower altitudes metal atoms could react with oxygen to form metal oxides. This reaction would be very facile, according to Diffusion Flame Studies (cf. Chapter 2) by Markstein.[29e]

Parson,[30] discussing planetary beginnings, considers atmospheric vapors of Fe, MgO, CaO, and SiO_2 as the most important substances. These have the proper vapor pressures to have been present and then to have condensed to form the genesis of a planet. Gravitational condensation does not seem as plausible in the initial stages.[30]

Table 3-3 brings together a large quantity of data on the properties, vapor compositions, geometries of molecular species, and spectroscopy of these halides, oxides, and sulfides.[12,31–52] Special note should be made of the matrix-isolation work of Weltner[34] and Margrave[42] where a large number of MX_2 molecules were isolated and studied spectroscopically. These studies indicated that $MgCl_2$ and $MgBr_2$ are linear, which agrees with the electron-diffraction studies by Klemperer and co-workers.[33] In further matrix-isolation work, Cocke and co-workers detected small amounts of Mg_2Cl_4 in organic matrices, presumably a chlorine-bridged species.[43] Also, although molecular $MgCl_2$ and $MgBr_2$ were predominant, slightly different spectra were observed due to different matrix sites (orientation of $MgCl_2$ in two different modes).[43]

Generally, the metal dihalides can be easily vaporized from C, Ni, or Pt crucibles or boats. The vapors exist almost entirely as monomeric MX_2, although there has been some controversy[31,35] regarding BeX_2, and $BeBr_2$ and BeI_2 vapors may contain substantial dimer concentrations.

It is interesting to note the change from linear to bent geometry as the metals become heavier in the dihalide series. This apparently is due to d-orbital effects.[34] Calder and co-workers,[39] by matrix isolation studies, have assigned angles to these molecules—MgF_2 158° (slight bend), CaF_2 140°, SrF_2 108°, and BaF_2 100°. Vaporization of the alkaline earth metal oxides is much more difficult than vaporization of the halides. Careful studies of the vaporization of single crystals of MgO (and Al_2O_3) have shown that the 110 face vaporizes fastest (greater than 111 face). Thus, the vaporization rate is fastest with the face that has the lowest density of surface atoms, and apparently the rate-determining step in the vaporization is diffusion range dependent (freedom of diffusion).[46] The vapor compositions of these metal

TABLE 3-3

Vaporization Data and Spectroscopic Data for Alkaline Earth Halides, Oxides & Sulfides

Compound	mp (°C)	bp (°C)	Heat of Vap (kcal/mole)	Temp of Vap (°C)	Vapor comp	Molecular species	Spectroscopic data[a] (cm⁻¹)	References
BeF_2			57 (subl)	500–1100	1% Dimer	Linear	v_2 309 (argon) v_3 1528	12, 31–36
$BeCl_2$	405	520			1% Dimer	Linear	v_2 238 (neon) v_3 1122	12, 31, 33–36
$BeBr_2$	490	520			Larger Dimer conc	Linear	v_2 207 (neon) v_3 993	12, 33–36
BeI_2	510	590			Larger Dimer conc	Linear	v_3 877 (argon)	12, 33–36
MgF_2			90	1200	99% Monomer, 1% dimer, 0.01% trimer	Linear or slightly bent, 158°	v_2 254 (neon) v_3 862	12, 31, 33, 34, 36–41
$MgCl_2$			57	650	99% Monomer, 1% dimer, 0.01% trimer	Linear	v_{asym} M–Cl 588	12, 31, 33, 38, 42, 43
$MgBr_2$	711		50	570	99% Monomer, 1% dimer, 0.01% trimer	Linear	v_{asym} M–Br 490	12, 31, 33, 42, 43
MgI_2			45	490	99% Monomer, 1% dimer, 0.01% trimer	Linear		12, 31, 33
CaF_2	1423	2500				Bent, 140°	v_1 489 (argon) v_3 561	12, 34, 37, 39
$CaCl_2$	782	>1600						12
$CaBr_2$	730	810		720				

(continued)

TABLE 3-3 (continued)

Compound	mp (°C)	bp (°C)	Heat of Vap (kcal/mole)	Temp of Vap (°C)	Vapor comp	Molecular species	Spectroscopic data[a] (cm^{-1})	References
CaI_2	784	1100						44
SrF_2	1473	2489				Bent, 108°	ν_1 447 (argon) ν_3 450	12; 12, 34, 37, 39
$SrCl_2$	875	1250						12
$SrBr_2$	643	d		770				12, 44
SrI_2	515	d						12
BaF_2	1355	2137				Bent, 100°	ν_1 421 (argon) ν_3 398	12, 34, 37, 39
$BaCl_2$	963	1560						12
$BaBr_2$	850	d						12, 44
BaI_2	740							12
BeO	2530	3900		2100				12, 45
MgO	2800	3600		1700				12, 45-48
CaO	2580	2850	91					12, 48
SrO	2430	3000	98-135	1400–1500[b]		SrO, trace Sr		12, 48-50
BaO	1923	2000	103					12, 51
BeS	d							
MgS	>2000							12
CaS	d	82					2388 Å absorp	12, 52
SrS	>2000	184					4500 Å absorp	12, 52
BaS	1200	136					4441 Å absorp	12, 52

[a] $\nu_2 = \nu_{asym}$.
[b] Vaporized from Pt source.

oxides have not been studied in any detail. In the case of SrO, mainly SrO and some Sr atoms are present in the vapor,[49,50,53] and the other oxides probably behave similarly.

B. Chemistry

Very little work has been reported dealing with the chemistry of these materials in their molecular state. Only the alkaline earth metal difluorides have received any significant attention, and these only in microscale reactions with simple molecules like CO in low-temperature matrices. A great deal of interesting work remains to be done.

1. ELECTRON-TRANSFER PROCESSES

No work has been reported.

2. ABSTRACTION PROCESSES

No work has been reported.

3. OXIDATIVE ADDITION PROCESSES

No work has been reported.

4. SIMPLE ORBITAL MIXING PROCESSES

Van Teirsburg and DeKock[54] have examined on microscale a series of MX_2–CO matrices. One alkaline earth fluoride, CaF_2, was included in the series which complexed with CO. Upon complexation, a definite bonding interaction, i.e.,

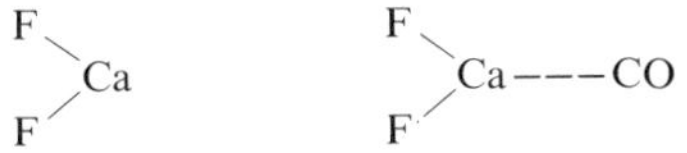

$$F\diagdown$$
$$Ca$$
$$F\diagup$$

$$F\diagdown$$
$$Ca---CO$$
$$F\diagup$$

$$\nu_{M-F} = 557.8 \text{ cm}^{-1} \qquad \nu_{M-F} = 552.2 \text{ cm}^{-1} \qquad \nu_{C=O} = 2178.0 \text{ cm}^{-1}$$
$$547.6$$

occurred as indicated by a change in ν_{C-O} and ν_{M-F}. The $\Delta\nu_{C=O}$ of about 40 cm^{-1} to high frequency indicates an interaction opposite in the normal to M–CO interactions. That is, CO is probably denoting electronic charge to CaF_2.

Hauge, Gransden, and Margrave[55] have carried out a broader scope study of MF_2–CO matrices where M = Mg, Ca, Sr, and Ba as well as a series of alkali metal, transition metal, and rare earth fluorides. For these alkaline earth fluoride–CO complexes, the following changes in vibrational frequencies were observed upon complexation (again, $\nu_{C=O}$ shifts to higher frequencies).

M	$\Delta\nu_{C\ O}\,(cm^{-1})$	$\Delta\nu_{M-F}\,(cm^{-1})$
Mg	67	23
Ca	49	4.6
Sr	43	—
Ba	35	3.7

The frequency shift of CO appeared to be a function of the metal's ionic radius. If MgF_2 is assumed to be a linear array of point charges, calculations show good agreement with experimental frequency shifts. For these calculations, it was assumed that the CO interacts with its internuclear axis perpendicular to the MgF_2 axis and intersecting at the Mg position. It was also considered that the alkaline earth fluorides behave as nearly totally ionic models; MgF_2 85% ionic, CaF_2 100%, SrF_2 100%, BaF_2 100%.

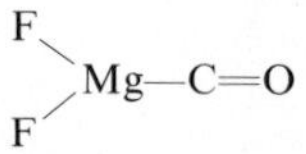

$$
\begin{array}{c}
F \\
\diagdown \\
\diagup Mg-C{=}O \\
F
\end{array}
$$

Thus, positive CO shifts do serve as a qualitative probe of molecular ionic character for molecules with similar geometry, and perhaps could serve as a quantitative measure after further data is gathered.

Vapors of MgF_2 have been codeposited with alkali metal fluorides to yield $MMgF_3$ in a cold matrix.[56] Also, MgF_2 and CaF_2 (or SrF_2) have been codeposited, yielding matrix-isolated $MgCaF_4$ and $MgSrF_4$ complex salts. These latter species are believed to exist in cyclic planar C_{2v} structures,[56] whereas Mg_2F_4 probably has D_{2h} symmetry.

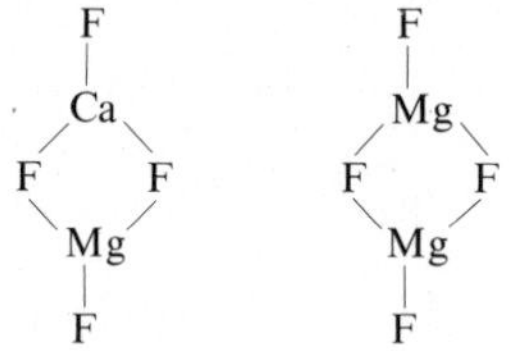

$$
\begin{array}{cc}
\begin{array}{c}
F \\
| \\
\cdot\ Ca \\
\diagup\ \ \diagdown \\
F\quad\quad F \\
\diagdown\ \ \diagup \\
Mg \\
| \\
F
\end{array}
&
\begin{array}{c}
F \\
| \\
Mg \\
\diagup\ \ \diagdown \\
F\quad\quad F \\
\diagdown\ \ \diagup \\
Mg \\
| \\
F
\end{array}
\end{array}
$$

5. CLUSTER FORMATION PROCESSES

Only the gas-phase dimerization and trimerization previously discussed (Table III-3) are relevant.

References

1. A. A. Pokhunkov, Iskusstv. Sputniki Zemli **13**, 110 (1962).
2. W. Smilga, *Z. Naturforsch., Teil A* **23**, 417 (1968).
3. A. V. Jones, *Ann. Geophys.* **22**, 189 (1966).

4. T. M. Donahue, *Space Res.* **7**, 165 (1966).
5. U. Ibrus, *Eesti NSV Tead. Akad. Toim. Fuus., Mat.* **18**, 79 (1969); P. S. Conti and S. E. Strom, *Astrophys. J.* **152**, 483 (1968).
5a. K. Utsumi, *Pub. Astron. Soc. Jpn.* **22**, 93 (1970).
6. C. Zwaan, *Bull. Astron. Inst. Neth.* **19**, 1 (1967).
7. P. J. T. Zeeger, W. P. Townsend, and J. P. Winefordner, *Spectrochim. Acta, Part B* **24**, 243 (1969).
8. I. Ansara and E. Bonnier, *Conf. Int. Metall. Beryllium [Commun.] 3rd, 1965* p. 17 (1966); M. M. Spivak and A. K. Yudina, *Tr. Inst. Metall. Obogashch., Akad. Nauk. Kaz. SSR* **26**, 42 (1967); J. Bohdansky and H. E. J. Schins, *J. Phys. Chem.* **71**, 215 (1967); A. V. Grosse, *J. Inorg. Nucl. Chem.* **26**, 1349 (1964); G. P. Kovtun, A. A. Kruglykh, and V. S. Pavlov, *Izv. Akad. Nauk SSSR, Met. Gorn. Delo,* p. 177 (1964).
8a. Y. Moriya, *Electrotech. J.* **2**, 219 (1938).
9. O. T. Nikitin and L. N. Goroknov, *Zh. Neorg. Khim.* **6**, 224 (1961).
10. J. Drowart and R. E. Honig, *J. Phys. Chem.* **61**, 980 (1957).
11. P. L. Timms, *in* "Cryochemistry", (M. Moskovits and G. Ozin, eds.), p. 61. Wiley (Interscience), New York, 1976.
12. "Handbook of Chemistry and Physics," 56th ed. CRC Press, Cleveland, Ohio, 1975–1976.
13. S. Dushman, "Vacuum Technique," p. 745. Wiley, New York, 1949; S. Dushman, *in* "Scientific Foundations of Vacuum Technique", (J. M. Lafferty, ed.), p. 691. New York, 1962.
14. B. Siegel, *Q. Rev., Chem. Soc.* **19**, 77 (1965).
14a. D. Stuewer, *Adv. Mass Spectrom.* **6**, 665 (1974); D. Cain and P. R. Barnett, *Appl. Spectrosc.* **31**, 321 (1977); T. Makita, H. Kishi, and K. Kodera, *Shitsuryo Bunseki* **21**, 293 (1973).
15. M. Staerk, *Optik (Stuttgart)* **36**, 139 (1972).
15a. J. M. Freese, A. W. Lynch, and R. T. Meyer, *Anal. Chem.* **45**, 1438 (1973).
15b. G. DeMaria and V. Piacente, *J. Chem. Thermodyn.* **6**, (1974).
16. K. J. Klabunde, *in* "Reactive Intermediates", (R. Abramovitch, ed.), Plenum, New York, 1979.
17. L. Andrews and B. S. Ault, *J. Mol. Spectrosc.* **68**, 114 (1977).
17a. L. Andrews, *in* "Cryochemistry", (M. Moskovits and G. Ozin, eds.), p. 195. Wiley (Interscience), New York, 1976.
17b. D. M. Thomas and L. Andrews, *J. Mol. Spectrosc.* **50**, 220 (1974).
17c. L. Andrews, E. S. Prochaska, and B. S. Ault, *J. Chem. Phys.* **69**, 556 (1978).
17d. B. S. Ault and L. Andrews, *J. Mol. Spectrosc.* **65**, 437 (1977).
18. B. S. Ault and L. Andrews, *J. Chem. Phys.* **62**, 2312, and 2320 (1975).
18a. D. E. Tevault and L. Andrews, *Chem. Phys. Lett.* **48**, 103 (1977).
19. M. J. McGlinchey and P. S. Skell, *in* "Cryochemistry" (M. Moskovits and G. Ozin, eds.), p. 137. Wiley (Interscience), New York, 1976.
20. P. S. Skell and M. J. McGlinchey, *in* "Cryochemistry", (M. Moskovits and G. Ozin, eds.), p. 195. Wiley (Interscience), New York, 1976.
21. Private communications with J. E. Girard and P. S. Skell.
22. L. D. Wescott, Jr., C. Williford, F. Parks, M. Dowling, S. Sublett, and K. J. Klabunde, *J. Am. Chem. Soc.* **98**, 7853 (1976).
23. L. Friedman and H. Schechter, *J. Am. Chem. Soc.* **83**, 3159 (1961).
24. W. Felder, R. K. Gould, and A. Fontijn, *J. Chem. Phys.* **66**, 3256 (1977).
25. V. Griffing, J. P. Hoare, and J. L. Vanderslice, *J. Chem. Phys.* **24**, 71 (1956).
26. P. S. Skell and J. E. Girard, *J. Am. Chem. Soc.* **94**, 5518 (1972).
27. A. I. Snow and R. E. Rundle, *Acta Crystallogr.* **4**, 348 (1951); H. Kleinfeller, *Ber. Dtsch. Chem. Ges. B* **62**, 2736 (1929).
28. K. J. Klabunde, J. Y. F. Low, and M. S. Key, *J. Fluorine Chem.* **2**, 207 (1972).

28a. J. M. Brom, Jr., W. D. Hewett, Jr., and W. Weltner, Jr., *J. Chem. Phys.*, **62**, 3122 (1975).

29. J. C. Miller and L. Andrews, *Chem. Phys. Lett.* **50**, 315 (1977); J. C. Miller, B. S. Ault, and L. Andrews, *J. Chem. Phys*, **67**, 2478 (1977); J. C. Miller and L. Andrews, *J. Chem. Phys.* **68**, 1701 (1978); **69**, 2054 (1978).

29a. J. C. Miller and L. Andrews, *J. Am. Chem. Soc.* **100**, 2966 (1978).

29b. J. C. Miller and L. Andrews, *J. Am. Chem. Soc.* **100**, 6956 (1978).

29c. J. C. Miller and L. Andrews, *J. Chem. Phys.* **69**, 936 (1978).

29d. J. C. Miller and L. Andrews, *J. Chem. Phys.* **69**, 3034 (1978).

29e. G. H. Markstein, *Symp. (Int.) Combust.* [Proc.], *9th, 1962* p. 137 (1962).

30. A. L. Parson, *Mon. Not. R. Astron. Soc.* **105**, 244 (1945); *Nature (London)* **154**, 707 (1944).

31. J. Berkowitz and J. R. Marquart, *J. Chem. Phys.* **37**, 1853 (1962).

32. E. P. Ozhigov and A. I. Zatsarin, *Tr. Dal'nevost. Fil. Akad. Nauk. SSSR, Ser. Khim.* **5**, 24 (1961).

33. A. Buechler, J. L. Stauffer, and W. Klemperer, *J. Am. Chem. Soc.* **86**, 4544 (1964).

34. W. Weltner, Jr., *Adv. High Temp. Chem.* **2**, 85 (1969).

35. O. Rahlfs and W. Fischer, *Z. Anorg. Allg. Chem.* **211**, 349 (1933); J. Berkowitz and W. A. Chupka, *Ann. N. Y. Acad. Sci.* **79**, 1073 (1960); L. Brewer, *in* "The Chemistry and Metallurgy of Miscellaneous Materials", (L. L. Quill, ed.), p. 215. McGraw Hill, New York; 1950; J. A. Blauer, M. A. Greenbaum, and M. Farber, *J. Phys. Chem.* **69**, 1069 (1965); W. Fischer, T. Petzel, and S. Lauter, *Z. Anorg. Allg. Chem.* **333**, 226 (1964); D. L. Hildenbrand and N. D. Potter, *J. Phys. Chem.* **67**, 2231 (1963). D. T. Peterson and J. F. Hutchinson, *J. Chem. Eng. Data* **15**, 320 (1970).

36. A. Snelson, *J. Phys. Chem.* **70**, 3208 (1966); *J. Chem. Phys.* **46**, 3652 (1967).

37. H. V. Wartenberg and O. Fitzer, *Z. Anorg. Allg. Chem.* **151**, 313 (1926).

38. D. L. Hildenbrand, W. F. Hall, F. Ju, and N. D. Potter, *J. Chem. Phys.* **40**, 2882 (1964).

39. V. Calder, D. E. Mann, K. S. Seshadri, M. Allavera, and D. White, *J. Chem. Phys.* **51**, 2093 (1969).

40. R. R. Hammer and J. A. Pask, *J. Am. Ceram. Soc.* **47**, 264 (1964).

41. R. H. Hauge, A. S. Kanaan, J. L. Margrave, *J. Chem. Soc., Faraday Trans. 2,* **71**, 1082 (1975).

42. S. P. Randall, F. T. Greene, and J. L. Margrave, *J. Phys. Chem.* **63**, 758 (1959).

43. D. L. Cocke, C. A. Chana, and K. A. Gingerich, *Appl. Spectrosc.* **27**, 260 (1973).

44. A. Stock and H. Heynemann, *Ber. Dtsch. Chem. Ges.* **42**, 4088 (1909).

45. E. S. Lukin and D. N. Poluboyarina, *Ogneupory* **29**, 418 (1964).

46. M. Peleg and C. B. Alcock, *High. Temp. Sci.* **6**, 52 (1974).

47. W. Kroenert and A. Boehm, *Glas-Email-Keramo-Tech.* **23**, 319 (1972).

48. M. Farber and R. D. Srivastava, *High Temp. Sci.* **8**, 73 (1976).

49. N. D. Morgulis, V. M. Gavrilyuk, and A. E. Kulik, *Dok. Akad Nauk SSSR* **101**, 479 (1955).

50. M. Asano, Y. Yamamoto, N. Sasaki, and K. Kubo, *Kyoto Daigaku Kogaku Kenkyusho Iho* **40**, 44 (1971).

51. M. S. Chandrasekharaiah and L. B. Brewer, *J. Karnatak Univ.* **4**, 16 (1960).

52. L. S. Mathur, *Proc. R. Soc. London, Ser. A* **162**, 83 (1937).

53. M. Asano, Y. Yamamoto, N. Sasaki, and K. Kubo, *Bull. Chem. Soc. Jpn.* **45**, 82 (1972).

54. D. A. Van Teirsburg and C. W. DeKock, *J. Phys. Chem.* **78**, 134 (1974).

55. R. H. Hauge, S. E. Gransden, and J. L. Margrave, *J. Chem. Soc., Dalton Trans.* p. 745 (1979).

56. A. S. Kanaan, R. H. Hauge, and J. L. Margrave, *J. Chem. Soc., Faraday Trans.* p. 272, 1991 (1976).

CHAPTER 4

Early Transition Metals, Metal Halides, Metal Oxides, and Metal Sulfides Groups IIIB–VIIB)

I. Early Transition Metal Atoms
(Sc, Ti, V, Cr, Mn, Y, Zr, Nb, Mo, Tc, Hf, Ta, W, Re)

A. Occurrence, Properties, and Techniques

Most of these metals are quite refractory and thus occur in nature as atoms only under extreme conditions. Spectral lines for Sc, Ti, V, Cr, Mn, Y, Zr and other metals have been detected in cool carbon stars.[1–4] These metals can exist as neutral or ionized atoms,[3] and abundances have been tabulated for some of these elements in stars.[4]

Vaporization of many of these elements in the laboratory can be quite difficult, requiring specialized procedures. Generally, their heats of vaporization are very high (cf. Chapter 1, Fig. 1-1) and the problem that immediately presents itself is the choice of vaporization source material. The answer for some of these metals, notably Mo and W, is "none"—no crucible material is satisfactory. Fortunately, however, Mo and W sublime and so it is possible simply to heat resistively, Mo or W wires directly.

Because of the difficulty in vaporizing some of these elements, some new vaporization techniques will be introduced in this chapter. These are electron-beam, laser, arc, and induction heating. These methods have been employed extensively for production of thin films and for shadowing electron microscope specimens.

Pulsed laser evaporation of metals employing Nd glass lasers has been studied photographically.[5] Heating of 10^{10} deg/sec has been observed, followed by formation of a vapor cloud, a crater in the metal surface, and then by ejection of liquid droplets of metals along with the generation of a powerful supersonic jet vapor.[6] Shock waves related to these supersonic

TABLE 4-1

Vaporization Data for the Early Transition Metals (Before Vapor Synthesis Chemistry)

Element	mp (°C)[a]	bp (°C)[a]	Heat of vap (kcal/mole)[b]	Vaporization method[c]	Vapor comp[d]	References
Sc	1541	2831	78, 91	Arc	Sc	20, 21, 30, 31
Ti	1660	3287		Laser, e-beam, arc	Ti	8, 11–13, 18–21
V	1890	3380		e-Beam, arc	V	14, 15, 20, 21
Cr	1857	2672		Laser, e-beam, arc	Cr	6, 11, 14, 15, 19–21, 23, 24
Mn	1244	1962		Arc, levitation	Mn	18–21
Y	1552	3338	91	Arc	Y	20, 21, 30
Zr	1852	4377	148	Laser, e-beam, arc	Zr	8, 12, 16, 20, 21
Nb	2468	4742	175	e-Beam, arc	Nb	14, 15, 20, 21, 32
Mo (Tc)	2617	4612	158	Laser, e-beam, arc	Mo	5, 7, 10, 16–18, 20–22
Hf	2227	4602	146, 148	e-Beam	Hf	32–34
Ta	2996	5425		Laser, e-beam, arc	Ta	8, 10, 18, 34
W	3410	5660		Laser, e-beam, arc	W	5, 7, 10, 11, 17, 18, 22, 23, 34
Re	3180	5627		e-Beam, arc	Re	20, 21, 34

[a] From Handbook of Chemistry and Physics,[26] p. B-67.
[b] From the original literature; also see Fig. 1-1 of Chapter 1.
[c] It is possible to vaporize all of these metals in small amounts by resistive heating.
[d] Monoatomic vapors.[27–29]

jets have been observed.[7] Focussed ruby lasers have also been observed to form craters rapidly in the metal surface.[8]

Electron-beam vaporizations have been studied extensively.[9-17] Generally, 5–6 Kv electron guns have been employed, and often work-accelerated beams, with the evaporant itself serving as the anode, are used.[9] Many materials can be vaporized, including metal oxides, carbides, and alloys, as well as metals.[11] Short, very energetic electron-beam bursts can also be employed if desired.[13]

Arc vaporization of early transition metals has also been examined.[18-25] X-ray photographs of electric arcs between carbon electrodes with the metal in a crater of the lower carbon electrode have shown that many metals can be vaporized in this way, but often the metal condenses on the upper electrode and forms nonvolatile compounds.[18] Also, some metals diffuse into the carbon electrode. Vaporization rates decrease in the order Bi, Tl, Pb, Sn, Ni, Fe, Mn, Ag, Hg, As, Te, Cd, Zn, Sb, Cu, In, Ga, Ge, Ti, Th, Ta, Mo, W.[18] Direct current arcs have also been used successfully for metal vaporizations.[19,20] Interrupted arcs with vibrating electrodes also have been used,[21,22] as has inductive heating.[23]

Table 4-1[5-8,10-34] summarizes vaporization procedures employed before metal atom chemistry investigations were begun for these metals.

1. ELECTRON-BEAM VAPORIZATION AND METAL ATOM CHEMISTRY (VAPOR SYNTHESIS)

The electron beam (e-beam) can be focussed on the center portion of the metal sample and this center portion rapidly heated, melted, and vaporized without melting the outer portion of the metal sample. Thus, vaporization can take place by a "containerless method,"[35] a great advantage for metals that corrode crucible materials when molten. Another advantage of e-beam techniques is that the technology is far advanced and large-scale vaporizations (Kg/hr) are possible.[34] Also, many types of substances can be vaporized in this way, including metals, metal halides, metal oxides, metal sulfides, and alloys, to date mainly for the production of thin films for decorative, protective, reflective, or electronic uses.[36,37]

The use of electron-beam vaporization for generation of atoms or molecules for carrying out vapor synthesis is also not new. The first reports, in the late 1960s, appeared at the very inception of vapor synthesis chemistry. However, problems plagued the method for years. These problems are (1) the fact that extremely good vacuum ($<10^{-5}$ torr) is required in order for the emitting high voltage filament to avoid arcing, (2) that stray electrons and X rays may damage sensitive organometallic products, and (3) substantial cost. These problems still exist, but recent results have indicated that special shielding, electrostatic focussing, use of very high power apparatus, and reversing of polarity (i.e., have filament at ground potential and sample at high voltage) can serve greatly to alleviate problem (2), although problem (3), cost, becomes more severe.

Figure 4-1 illustrates schematically an electron-beam apparatus.[38] A beam of electrons emitted from a hot tungsten wire is focussed, either electrically or magnetically, on the metal sample, which is usually held in a carbon block or shallow crucible. The carbon block rests in turn on a copper hearth that can be continuously water cooled. This cooling leads to high energy losses, however, and so the energy efficiency of this apparatus is not high, about 35%, for large-scale Ni vaporization.[34] Electron backscattering energy loss is only about 18% for Ni, but as high as 38% for Hf, Ta, W, and Re.[34]

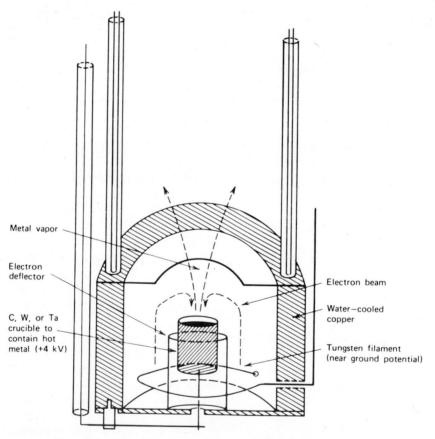

Figure 4-1.   Schematic for electron beam vaporization (after Timms).[35]

From 1968 to 1973, the infancy of vapor synthesis chemistry, electron beams were used to vaporize boron,[39,40] silicon,[41,42] and titanium.[43,44] However, the problems previously mentioned plagued these studies, and only recently has an electron-beam method been used successfully to vaporize tungsten for the preparation of bis(benzene)tungsten(0).[45]

In brief summary, these studies have shown that for vapor synthesis chemistry it is best to (1) minimize electron and X-ray scatter by the inclusion of supplementary magnetic fields or negatively charged wire mesh, (2) choose the polarity such that the crucible is at a positive potential relative to the cocondensation surface (have the crucible at high positive potential and the filament at ground potential), and (3) keep the accelerating voltage low (<20 Kev) to minimize X rays (this causes a need for increased currents).

Further work on perfecting electron-beam apparatus designs for vapor synthesis chemistry are definitely needed, but costs must also be held down. This method has the greatest potential for large-scale metal atom chemistry.

2. LASER EVAPORATION METHODS IN METAL ATOM CHEMISTRY

For vapor synthesis chemistry, only Koerner von Gustorf and co-workers have employed lasers to any great extent.[46–48] These workers have used YAG (Yttrium–Aluminum–Garrett doped with Nd^{3+}) lasers most effectively, although CO_2 infrared lasers were also used but with less efficiency. Figure 4-2 shows a schematic for the Koerner von Gustorf apparatus.

The greatest difficulty with the laser method is that much of the light energy is reflected by the molten metals, which act much like a mirror, especially for IR lasers. Another difficulty is the "window problem," which exists because a window is needed for the laser light to enter the vacuum chamber wherein the metal is evaporated. Once metal begins to evaporate, however, the window becomes coated, thus shutting out the light. In spite of these serious difficulties, Koerner Von Gustorf has been able to put the laser method to good use.[46] The window problem was solved by use of a slight gas pressure around the window called a "gas window." A slight gas stream flows out of the slit of the gas window and is continually pumped away, keeping the window clear of metal deposits; and with the use of YAG lasers instead of CO_2 lasers, better light-absorption efficiency can be obtained, although energy loss is still a serious problem.

One advantage of the laser application is that no other evaporation method allows the production of comparable energy flux densities. Very small areas can be heated rapidly. A wire tip can be evaporated continuously by careful wire feed. Also, metals can be evaporated intermittantly with laser bursts. Other advantages include (1) the possible use of different types of vaporization apparatus with the same laser and (2) the fact that stray

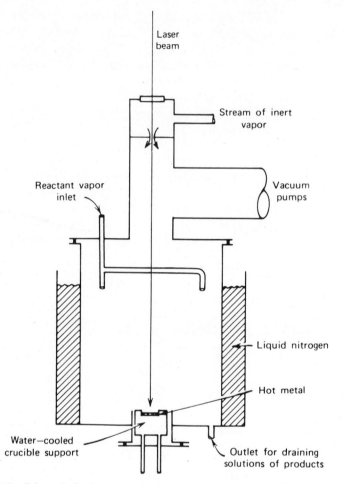

Figure 4-2. Schematic for laser vaporization and cocondensation of vapor with chemical substrates (after Timms).[35]

radiation, especially IR radiation, is less damaging to products than stray electrons and X rays from electron-beam setups.

Koerner von Gustorf and co-workers have used their apparatus for vaporization of Cr, Mn, Fe, Ni, Cu, Er, Dy, Ho, Zn, Sn, Pb, and Al.[46]

Laser evaporation methods should be investigated more thoroughly for refractory metal vaporizations and perhaps for the generation of electronically excited metal atoms.

Neither electron-beam or laser methods have been employed for matrix-isolation spectroscopy investigations. Only the resistive heating method described in Chapter 5 has been used, since this method apparently yields lower concentrations of metal atom dimers, trimers, oligomers, and metal particles.

B. Chemistry

1. ELECTRON-TRANSFER PROCESSES

There has been no work reported on electron-transfer reactions with these metal atoms.

2. ABSTRACTION PROCESSES

Several macroscale cocondensation studies on deoxygenation have been reported. Gladysz, Fulcher, and Tagashi compared the effectiveness of Ti, V, Cr, Co, and Ni for deoxygenation of cyclohexeneoxide, as determined by the amount of cyclohexene produced/metal atom.[49] The equivalents of oxygen removed/metal atom were found to be Ti = 0.9, V = 2.8, Cr = 2.7, Co = 1.2, and Ni = 0.6. Chromium was studied further with 2,6-dimethyl-pyridine oxide, triethylphosphine oxide, dimethyl sulfoxide, and nitro- and nitrosoarenes,[49–53] and in each case low-yield deoxygenation reactions were observed. For the nitro- and nitrosoarene work, nitrene or nitrenoid species were believed to be formed as intermediate species.

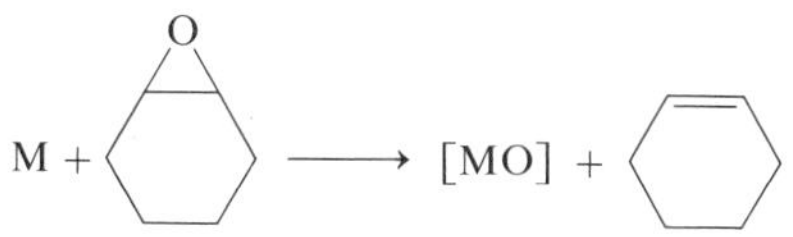

$$M + \text{[epoxide]} \longrightarrow [MO] + \text{[cyclohexene]}$$

In similar deoxygenation work, we have found that Ti atoms deoxygenate a variety of substrates, including ketones and even ethers such as THF.[52] V and Cr atoms behave similarly. These abstraction reactions are not high yield reactions (5–30% based on metal vaporized) and other reactions also take place.

Desulfurization processes have also been observed when Cr atoms were codeposited with thiophene.[53]

3. OXIDATIVE ADDITION PROCESSES

Although we have studied oxidative addition reactions extensively for the later transition metals (Group VIII) and have attempted similar studies with the early transition metals, very little successful work has been accomplished. A very useful substrate for the study of oxidative additions with Co, Ni, Pd, and Pt is C_6F_5Br (cf. Chapter 5). However, when macroscale cocondensations of C_6F_5Br with Ti, V, and Cr were carried out only very unstable organometallic species were produced. We have not been successful at isolating or trapping these unstable materials. However, in the case of C_6F_5Br–Ti the product (possibly C_6F_5TiBr) served as a tremendously active butadiene polymerization catalyst at $-78°C$.[54] A thin, rubbery polymer sheet formed almost explosively when 1,3-butadiene was added to the cold C_6F_5Br–Ti matrix.

In similar work we found that benzyl chloride–V,Cr, Mn, and Fe depositions yielded varying quantities of catalytic self alkylation products plus HCl. This observation indicates that the benzyl metal halides or metal halides formed served as catalysts for Friedel–Crafts-type self alkylation by benzyl chloride.[54]

$$C_6H_5—CH_2Cl + M \longrightarrow C_6H_5—CH_2MCl + MCl_n$$

$$n\,HCl + C_6H_5{+}CH_2—C_6H_4—CH_2—C_6H_4{)}_n Cl \longleftarrow \Big| \quad C_6H_5CH_2Cl$$

It appears that much more work needs to be carried out in this area of metal atom chemistry. Extremely active catalysts and a new series of low valent early transition metal organometallics may result.

4. SIMPLE ORBITAL MIXING PROCESSES

This process, the mixing of atomic orbitals of the metal atom with ligand orbitals to form a strong covalent-like bond, has been most extensively investigated for the early transition metals.

$$M + L \longrightarrow M(L)_n$$

a. Alkene Reactions.   Macroscale cocondensations of Cr with ethylene caused polymerization to give polyethylene. Styrene gave polystyrene [containing some $(arene)_2Cr$] in a similar way on matrix warm-up. However, if PF_3 or CO were added to the styrene–Cr mixture at low temperature, $C_8H_8Cr(L)_3$ complexes could be prepared.[55]

b. Diene Reactions.   Cyclopentadiene, on cocondensation with metal atoms, reacts by a combination of C—H oxidative addition and π-bond simple orbital mixing. Thus, Cr yields chromocene and H_2,[56,57] presumably through a bis-cyclopentadienylchromium dihydride intermediate. In the cases of Mo and W, the dihydride species is stable and can be isolated.[58,59]

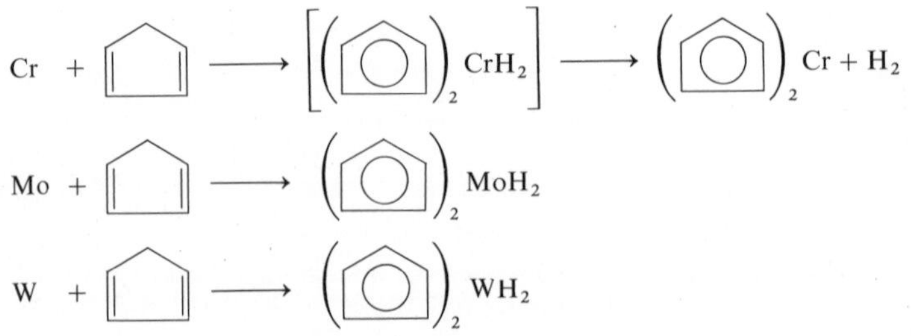

Manganese atoms codeposited with cyclopentadiene–benzene mixtures give a low yield of the mixed complex[46]:

Mn + ⬡ + ⬡ ⟶ ⬡—Mn—⬡ + ½H₂

$Mn + \text{(cyclopentadiene)} + \text{(benzene)} \longrightarrow \text{(benzene)}-Mn-\text{(cyclopentadiene)} + \tfrac{1}{2}H_2$

1,5-Cyclooctadiene (COD) is an excellent ligand both in metal atom chemistry and in organometallic chemistry in general. Many transition metal atoms have been cocondensed with (COD) yielding stable complexes. In the case of Cr both 1,3-COD and 1,5-COD have been studied in codeposition schemes. The $(COD)_n Cr$ complexes were found to be unstable, but addition of trapping ligands at low temperature yielded stable $(1,5\text{-COD})-CrL_4$ complexes. Note that both 1,3-COD and 1,5-COD yielded the same 1,5-COD complex, indicating that rapid low-temperature isomerizations were possible and that 1,5-COD is the preferred ligand. One additional compound, a chromium hydride species $C_8H_{11}Cr(PF_3)_3H$, was also formed when PF_3 was the trapping ligand[46]: Hydrogen transfers are also quite facile with 1,3- or 1,4-cyclohexadiene in the presence of Cr atoms (in a warming matrix).[60] Disproportionation was induced catalytically to yield benzene and cyclohexane.[61] Although little is known about these hydrogen-transfer processes, it seems likely that R—Cr—H species (R = allyl) are intermediates, which could be formed by oxidative addition of allyl C—H to Cr atoms. When the 1,3-cyclohexadiene–Cr complex is trapped with PF_3, $(\eta^4\text{-}C_6H_8)_2Cr(PF_3)_2$ can be isolated.[60]

With $Mn-C_4H_6$, reaction of Mn atoms with C_4H_6 followed by CO trapping yielded bis(butadiene)Mn–CO complex.[62]

1,3-Butadiene codeposited at $-196°C$ with Cr atoms yields an unstable $Cr(diene)_n$ complex, and upon low-temperature addition of trapping ligands CO or PF_3, $(diene)_n CrL$ complexes can be prepared.[60,63,64] The CO derivative is believed to be an important catalysis intermediate in $Cr(CO)_6$-catalyzed oligomerization of butadiene. It is a very labile complex above $0°C$, which also can be generated by $Cr(CO)_6$ photolysis in the presence of 1,3-butadiene.

1,3-Butadiene–Mo or W macroscale depositions yielded unique, stable $(diene)_3M$ complexes.[65] The diene molecules, as shown by X-ray crystallography, are wrapped about the Mo and W cis-oid so that all 12 carbons are equidistant from the metal.[66]

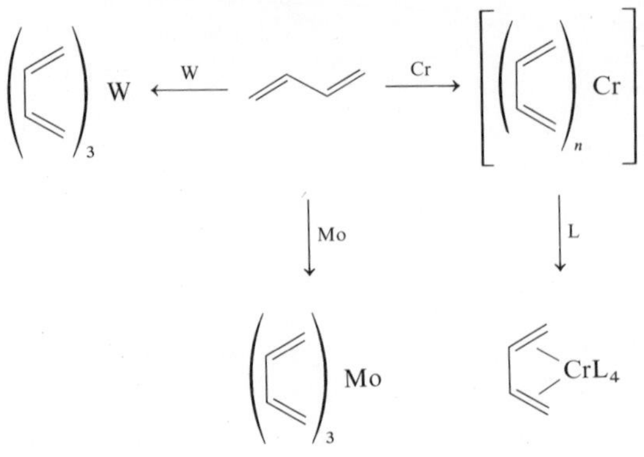

M. L. H. Green and co-workers have carried out a wide variety of butadiene oligomerization experiments using metal atoms or products from metal atom reactions as catalysts.[67,68] A summary of these studies[36,37] indicates that in the presence of benzene [which can complex the metal atoms to yield $(C_6H_6)_2M$], Ti yielded polymer with butadiene, V and Cr yielded oligomers plus triene, and Mn yielded trimer. The addition of the alkylating agent Et_2AlCl caused some product distribution changes, but mainly a great increase in % diene converted. Several other catalytic systems were also investigated, including M atom–Al atom, M atom–Al atom–EtCl, Al atom–$TiCl_4$, M atom–EtCl, M atom–$P(C_6H_5)_3$, TiO_2, FeO, and NiO.[67,68]

Overall, these studies add to the weight of evidence that metal atom vapor synthesis techniques may someday be useful for low-temperature catalysis processes. One possible application would be low-temperature polymerization on surfaces.

c. Triene (Nonaromatic) and Tetraene Reactions. Cycloheptatriene has been codeposited on a macroscale with Cr, Mo, and W. As with dienes, Cr yielded the most unstable species.[60,63,64,69] Two products formed in low yield, $C_{14}H_{16}Cr$ and $C_{14}H_{17}Cr$, were thought to be derived from bis-cycloheptatriene Cr(O) with subsequent hydrogen disproportionations.[46,60] Deposition of triene–PF_3 resulted in a low yield of (cycloheptatriene)-$Cr(PF_3)_3$.[46,60] With Mo and W, hydrogen-transfer processes also took place, yielding products more amenable to full characterization. Thus, $\eta^7-\eta^5$ sandwich complexes were formed by H-transfer processes.[69] These complexes had been previously prepared by classical synthetic methods.

Timms and co-workers have carried out more detailed cycloheptatriene studies with Ti, V, Cr, Fe, and Co.[70] In the case of Ti, hydrogen transfers took place yielding the $\eta^7-\eta^5$ sandwich compound shown below. With Cr, triene, and PF_3 a η^6-ring complex was isolated.

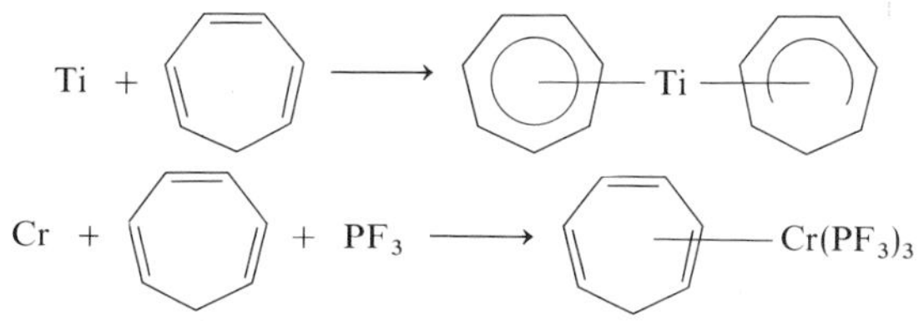

With cyclooctatetraene (COT) and Ti, Fe, or Co, polymers were formed. However, with Cr a Cr–Cr bonded $Cr_2(C_8H_8)_3$ complex was formed.[70,71] In the case of Ti–COT, a novel "triple decker sandwich" was obtained.[72]

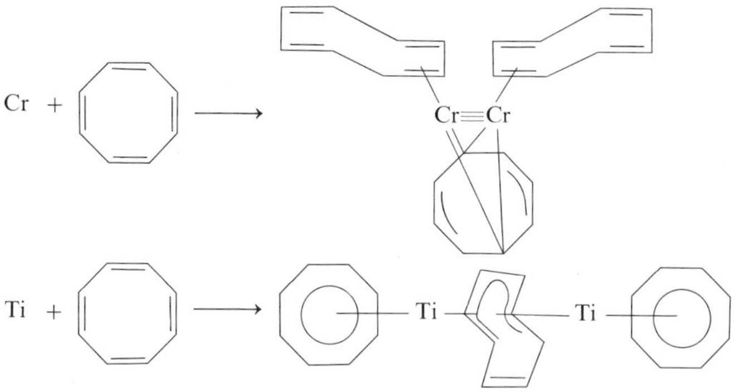

d. Alkyne Reactions. Alkyne reactions with the early transition metals have been studied sparingly. One brief report of trimerization of 1-butyne, 2-butyne, and 1-pentyne with Cr atoms is available.[63] The expected arenes were formed, and no evidence for the anticipated bis(arene)Cr(0) was found. It seems likely that mono(arene)Cr(0) could be formed, the arene dissociate, and Cr(0) go on to trimerize alkyne again. It is not expected that the mono-(arene)Cr(0) would have much stability since recent matrix isolation studies imply that the formation of bis(arene)Cr(0) complexes may be a ternary process (two arenes plus Cr atom must collide simultaneously).[73]

Gladysz and co-workers,[64] have found that cyclic diynes can be trimerized by Cr atoms, to yield a unique arene–triyne derivative. Thus, 1,7-cyclodo-decadiyne with Cr caused the transformation shown below. Further Cr atom "template" action to close the remaining -yne bonds did not take place:

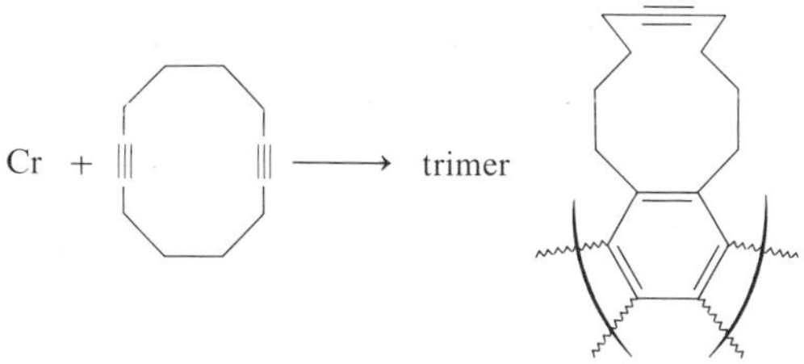

A similar conversion of 1,8-cyclotetradecadiyne was also carried out by Cr atoms.

e. Arene Reactions. (*i*) *Stable Symmetrical Bis(Arene) Sandwich Complexes.*   Probably the most impo tant aspect of the chemistry of the early transition metal atoms is the formation of bis(arene)M(0) complexes. The metal atom method has provided an important breakthrough[74] in synthesis, both in terms of speed and convenience of preparation of already known sandwich compounds and for preparation of many new sandwich compounds not preparable by classical means.[74a] A host of variously substituted bis(arene)chromium(0) compounds have been made simply by cocondensing Cr atoms with the appropriate arene ligand (cf. Table 4-2).[43,45,57,63,64, 71,75–85] The products are generally very easy to purify by crystallization or sublimation. And since Cr is quite easy to vaporize by resistive heating, it is not difficult to prepare these compounds in synthetically useful amounts (grams). Also, theoretical economic studies indicate that such compounds could be prepared industrially with continuous operation metal vapor reactions, on a tons/year basis for about \$20/pound.[86]

Large laboratory or industrial scale syntheses of Mo and W sandwich compounds would not be feasible by resistive heating, but are quite feasible by electron-beam vaporization. Either method if carried out correctly can yield many new bis(arene)Mo(0) and W(0) compounds.[45,80] Recent results of Green and co-workers[45,89] have shown that by employing a 2–4 KW electron gun with reverse polarization and proper shielding, gram quantities of bis(arene)W(0) compounds are readily prepared in a few hours. This is a

TABLE 4-2

Stable Bis(Arene)M(0) Complexes of the Early Transition Metals Prepared by Metal Atom Techniques

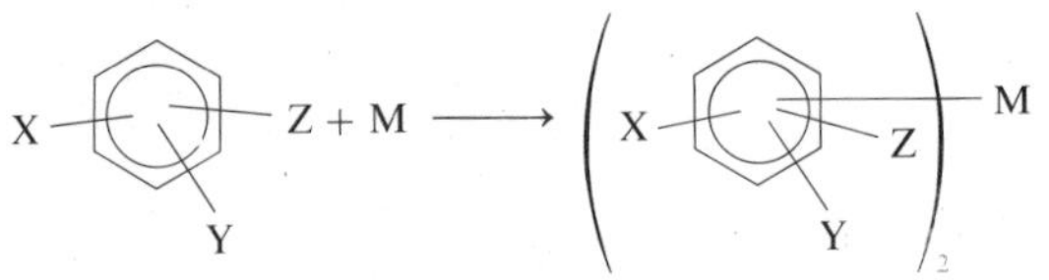

Metal[a]	X	Y	Z	References
Cr	H	H	H	57, 75
Cr	F	H	H	63, 64, 76
Cr	Cl	H	H	63, 64
Cr	CF_3	H	H	76
Cr	$COOCH_3$	H	H	63, 64
Cr	CH_3	H	H	63, 64
Cr	C_2H_5	H	H	63, 64, 77

TABLE 4-2 (*continued*)

Metal[a]	X	Y	Z	References
Cr	CH(CH$_3$)$_2$	H		77
	F	F(*ortho*)[b]		63, 64, 78
	F	F(*meta*)[b]		76, 78
	F	F(*para*)		63, 64, 78
	Cl	CF$_3$(*ortho*)[b]		76
	Cl	CF$_3$(*meta*)[b]		76
	Cl	CF$_3$(*para*)[b]		76
	CF$_3$	CF$_3$(*meta*)[b]		76
	CF$_3$	CF$_3$(*para*)[b]		76
	CH(CH$_3$)$_2$	H		77
	CH(CH$_3$)$_2$	CH(CH$_3$)$_2$		77
	CH$_2$CH$_2$CH$_2$CH$_2$[c]			79
Mo	H			80
	F			80
	Cl			80
	CH$_3$			80
	OCH$_3$			80
	N(CH$_3$)$_2$			80
	COOCH$_3$			80
W	H			45, 80
	F			80
	CH$_3$			45, 80
	OCH$_3$			80
W	CH$_3$	CH$_3$(*ortho*)		80
	CH$_3$	CH$_3$	CH$_3$	45
V	H			76
	F			76
	Cl			76
	CF$_3$			76
	F	F(*ortho*)		76
	F	F(*meta*)		76
	F	F(*para*)[b]		76
Nb	H			81
	CH$_3$			81
	CH$_3$	CH$_3$	CH$_3$	81
Ti	H			43, 44
	CH$_3$			43, 44
	CH$_3$	CH$_3$	CH$_3$(1,3,5)	43, 44

[a] Also see Timms and Turney[71] for a review of these types of complexes.

[b] X-ray structure completed.[82-85]

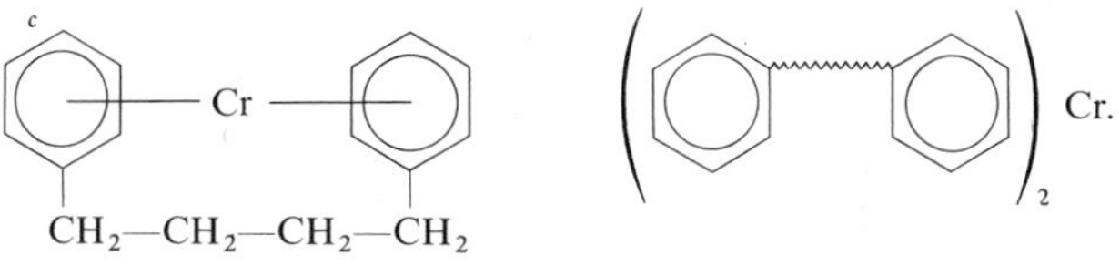

very significant breakthrough since stray electron damage to products, a problem that has plagued e-beam methods for years, has been all but eliminated, and opens the way for studying much more chemistry of the early refractory transition metals. In fact, further work of Cloke, Green, and Price has shown that similar e-beam methods can be used to prepare bis(arene)-Nb(0) complexes for the first time.[81] Good yield (gram quantities) of the very electron-rich paramagnetic benzene, toluene, and mesitylene sandwich compounds were prepared.

A series of new bis(arene)vanadium(0) complexes have also been prepared by metal atom methods.[76] Resistive heating of V in W boats was employed where the V was held just at its softening point during vaporization. A variety of new F- and CF_3-substituted derivatives were readily synthesized.

Green and co-workers[43,44] have also prepared a series of bis(arene)titanium(0) complexes, which were totally unknown prior to their work. Electron-beam methods were employed for this work, although resistive heating methods have also been used successfully for their preparation.[88] These Ti(0) complexes must be isolated under argon, and the presence of Ti particles must be avoided as much as possible since these particles will autocatalytically decompose the sandwich complex.

In view of this successful synthetic work with Ti, V, Cr, Mo, W and Nb, it appears likely that stable sandwich complexes of some of the metals Sc, Y, Zr, Hf, and Ta will soon be prepared by metal atom methods. The Mn, Tc, Re, series appears less promising because Mn–arene complexes have already been shown to be unstable,[89] Tc is radioactive and short-lived. However, Re must still be investigated.

(ii) *Mixed Arene–Arene; Arene–Ligand, and Pyridine Sandwich Complexes.* Codeposition of two different arenes simultaneously with metal atoms yields three complexes, $(arene)_2M$, $(arene')_2M$, and $(arene)(arene')M$. Sometimes these can be separated chromatographically or chemically. For example, $(C_6H_6)Cr(C_6F_6)$ is air stable whereas $(C_6H_6)_2Cr$ is not, and $(C_6F_6)_2Cr$ is thermally unstable. Therefore, McGlinchey and co-workers[90–95] isolated $(C_6H_6)Cr(C_6F_6)$ simply by exposing the solid mixture to air, and then extracting the remaining product.

Further work by the McGlinchey group has shown that codeposition of C_6H_6—C_6F_5H with Cr yielded $(C_6H_6)Cr(C_6F_5H)$. The H of the C_6F_5H group is readily abstracted by base, and therefore the lithiated compound is readily available. This material has been derivatized in a number of ways, as outlined below.[93]

$$(LiC_6F_5)Cr(C_6H_6) + XY \rightarrow (XC_6F_5)Cr(C_6H_6)$$

$X = SnMe_3, Re(CO)_5, SiMe_3, CO_2Li, C(Me)_2OH, CHOH$

TABLE 4-3

Mixed Arene and Arene–Ligand Complexes of the Early Transition Metals Prepared by Metal Atom Techniques

Metal	Arene	Arene'	L	Products	References
Cr	C_6H_6	C_6F_6		$(C_6H_6)Cr(C_6F_6)$	36, 37, 95
Cr	C_6H_6			$C_6F_5H^{c,d}$	93
	C_6H_6		PF_3	$(C_6H_6)Cr(PF_3)_3$	77
	C_6F_6		PF_3	$(C_6F_6)Cr(PF_3)_3$	77
Cr	$C_6H_5CH(CH_3)_2$		PF_3	$C_6H_5CH(CH_3)_2Cr(PF_3)_3$	77
Cr	$C_6H_3(CH_3)_3$		PF_3	$C_6H_3(CH_3)_3Cr(PF_3)_3$	77
	$C_5H_5N^a$		PF_3	$(C_5H_5N)Cr(PF_3)_3$	35
				$(C_5H_5N)Cr(PF_3)_5$	
	$2,6(CH_3)_2C_5H_3N^b$			$[(CH_3)_2C_5H_3N]_2Cr$	96, 97
W	C_6H_6		C_5H_6	C_6H_6–$W(H)(C_5H_5)$	80
Mn	C_6H_6		C_5H_6	C_6H_6–MnC_5H_5	9, 18, 36, 37, 46

[a] Pyridine π-complexed $(C_5H_5N)Cr(PF_3)_3$ and Pyridine σ-complexed $(C_5H_5N)Cr(PF_3)_5$.

[b] 2,6-dimethylpyridine with Cr gave two crystalline forms of π-complexes. The methyl groups apparently protect the nonbonding electrons on nitrogen and disallow Cr complexation to that site.

[c] The $(C_6H_6)Cr(C_6F_5H)$ compound can be readily lithiated to yield $(C_6H_6)Cr(C_6F_5Li)$, and compound can be readily derivatized (see text).[93]

[d]

has also been prepared.[93]

Table 4-3 summarizes the mixed systems that have been prepared by codeposition of the ligand mixture with metal atoms.[35–37,46,77,80,93,95–97]

   (iii) *Effects of Substituents on* $(Arene)_2M(0)$ *Stability and Chemistry.* Highly electron-demanding arenes such as C_6F_6 and $C_6(CF_3)_6$ form thermally unstable complexes with the early transition metals, and with C_6F_6, and sometimes explosions result.[89] However, a few electron-demanding substituents can be tolerated, such as with $C_6H_4F_2$ or $C_6H_4(CF_3)_2$.[76] Actually, the presence of one or two (but not more) F or CF_3 allows easier isolation of the complexes because their air stability becomes much greater and yet they still have good thermal stability. There is apparently a delicate balance; whereas more electron-rich arenes yield more strongly bonded complexes, they are in turn more readily oxidized.[76] Indeed, electrochemical studies on a series of variously substituted bis(arene)Cr(0) complexes showed remarkably large variations in $E_{1/2}[(Ar)_2Cr(0) \rightarrow Ar_2Cr^+]$, ranging from -0.2 volts for $(C_6H_5OCH_3)_2Cr$ to $+1.2$ volts for $[(CF_3)_2C_6H_4]_2$ Cr.[98–100] Calculations in conjunction with Hammett substituent correlations indicated that the bonding in these systems is greatly affected by substituents because the σ-framework of the arene ring is greatly involved as well as is the π-framework in the M-ring bond.[99,100]

Spectroscopic and X-ray studies on these complexes have shown that substituents can affect the chemistry[90–93] and structures[82–85] greatly. Structurally, for example, bis(1,4-difluorobenzene)V shows a ring deformation, with the carbons bearing the F atoms moved slightly up out of the plane.[82] Also, it is interesting that this structure, as well as a variety of other substituted bis(arene)M complexes, shows the carbons of the rings, and in some cases even F atoms, eclipsed as shown below:

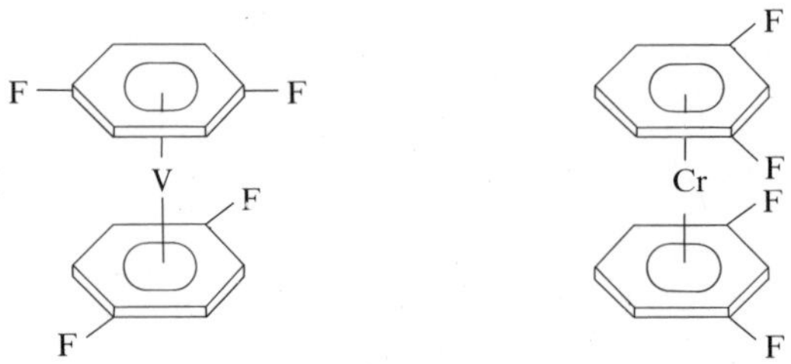

F—⬡(V)—F   (Cr)

Lagowski and co-workers have carried out some detailed ^{1}H- and ^{13}C-NMR studies on a series of bis(arene)Cr complexes.[84,85,101] Generally, they concluded that complexation decreases the aromaticity of the arene significantly, and that substituent effects are not transmitted through the Cr atom. On the other hand, Hao and McGlinchey[93,94] have also carried out very careful spectroscopic and chemical studies on these systems and have concluded that a variety of correlations of chemistry and oxidative stability with σ_m-substituent constants can be made. These workers propose that a π-Cr—C_6H_6 moiety is actually an electron-releasing group. Furthermore, they propose that unsymmetrical "chromarenes"[93] [bis(arene)Cr(0) complexes] exhibit an internally compensating stabilization effect, analogous to that of the $Cr(CO)_3$ moiety.

(iv) Mechanism of Bis(benzene)Cr(0) Formation from Cr atoms and Benzene. Benzene itself consistently gives lower yields of sandwich compounds than substituted benzenes. There appear to be two possible reasons for this. (1) The high freezing point of benzene, which causes the C_6H_6–M matrix to be immediately highly crystalline, may hinder the exact approach of the π-orbitals to the metal atom. This seems more believable after considering the recent matrix-isolation work of Boyd, Lavoie, and Gruen[102] and Efner, Smardzewski, Tevault, and Fox[103,104] that indicates the formation of $(C_6H_6)_2Cr$ may require a ternary process (all three colliding at once).

(2) Substituted arenes have their π-bonding perturbed enough that the π-system is more polarizable and can complex more readily with the approaching metal atom. Timms[86] has likened p-substituted arenes to cyclic diene molecules in metal atom reactions because of the polarization in the arene, and believes this explains why such high yields of bis(arene)M(0) complexes are generally obtained with p-$C_6H_4X_2$ derivatives.

f. Phosphine and Isonitrile Reactions. As exhibited in previous sections in this chapter, phosphines, particularly PF_3, have been employed for trapping or stabilizing organometallic species generated in low-temperature metal atom matrices. However, pure phosphine–metal atom studies with the early transition metals have been lacking. The only complex prepared this way prior to 1978 is $Cr(PF_3)_6$.[105] However, recent work by King and Chang,[106] with the unique aminophosphine systems $(CH_3)_2NPF_2$ and $(CH_3)N(PF_2)_2$, has shown that a variety of new homoleptic M–$(L)_n$ systems are now available. For example, codeposition of Cr vapor with these ligands yielded the complexes shown below (Mn yielded no stable products):

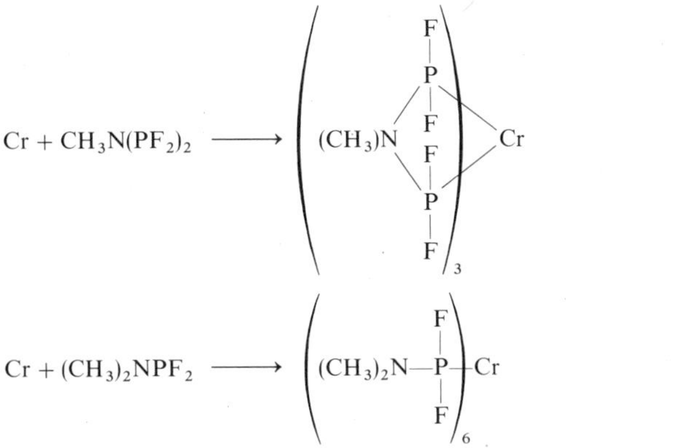

$$Cr + CH_3N(PF_2)_2 \longrightarrow \left((CH_3)N \underset{P}{\overset{P}{\left\langle \begin{matrix} F \\ F \end{matrix} \right\rangle}} Cr \right)_3$$

$$Cr + (CH_3)_2NPF_2 \longrightarrow \left((CH_3)_2N-P-Cr \right)_6$$

$$Cr + (CH_3)_2NPF_2 + (CH_3)N(PF_2)_2 \longrightarrow Cr[PF_2N(CH_3)]_4[(PF_2)_2N(CH_3)]$$
$$4 \quad : \quad 1$$

These materials exhibited impressive thermal and oxidative stabilities.[106]

Similarly, CrL_6 complexes where L = *tert*-butylisocyanide, methyl-isocyanide, cyclohexylisocyanide, and vinylisocyanide can be prepared by metal atom methods.[107] Codeposition of PF_3–isocyanide mixtures have yielded mixed complexes.

g. Carbon Monoxide Reactions. In transition metal chemistry PF_3 and CO behave similarly because of their similar π-acid ligand characteristics. For macroscale reactions only PF_3 can be used because CO is too volatile for depositions at $-196°C$, liquid nitrogen temperature. Therefore, no macroscale M–CO reactions have yet been investigated. However, on a microscale, lower temperatures can be used, and almost every metal in the periodic chart has been codeposited with CO for matrix isolation studies. The v_{CO} stretch in the IR is generally very sensitive to the M—CO bond, and thus IR has been used extensively for study of these systems.

Pure CO or CO diluted with Ar or Kr have been deposited with metal atoms, usually at about 10°K. Employing these dilution techniques, $M(CO)_n$ (where $n = 1, 2, 3, 4, 5$, and/or 6) have been prepared for some of the early transition metals.

$$M + CO \rightarrow M-CO \rightarrow M(CO)_2 \rightarrow M(CO)_3 \rightarrow \rightarrow \rightarrow M(CO)_6$$

For metals such as Cr, Mo, or W the final $M(CO)_6$ species is thermally stable, but the $M(CO)_{1-5}$ intermediates are not, and are only observable under the correct dilution and temperature conditions in the matrix. Actually determining which IR absorptions belong to which species is not trivial. One important tool for this, in addition to dilution and temperature variations, is isotope labeling (^{13}CO mixed with ^{12}CO). For a detailed account of these techniques, the reader is referred to several excellent reviews.[108–115]

The most important aspects of the M atom–CO work are related to the determination of which metal–carbonyl stoichiometry is favored, the bonding geometry of the complex, and how the observed v_{CO} values compare with theoretically predicted v_{CO} values for such simple species.

Table 4-4 summarizes the M–CO work for the early transition metals.[109,110,116–125] All of these complexes possess covalent-like M–CO bonds. The apparent geometries of the complexes and other comments are also included.

Comparisons of early transition metals within a family indicate close similarities for the geometries of $M(CO)_3$, $M(CO)_4$, $M(CO)_5$, and $M(CO)_6$ respectively; thus, Cr vs Mo vs W follow the same geometrical configurations, and v_{CO} values are very close, with Mo apparently bonding just slightly less strongly than Cr or W (cf. $M(CO)_3$ and $M(CO)_6$).

The unsaturated $M(CO)_n$ species are apparently very reactive, as illustrated by the elegant work of Turner and co-workers concerning the interactions of $Cr(CO)_5$, $Mo(CO)_5$, and $W(CO)_5$ with atoms of relatively unreactive gases, CH_4, and SF_6.[122–124]

Comparisons among neighboring families of these metals indicate some striking geometrical differences. For example, V vs Cr in $M(CO)_3$, $M(CO)_4$, and $M(CO)_5$ show differences in each case, although the final product $M(CO)_6$ is of course octahedral in both systems.

h. Dinitrogen and Dioxygen Reactions. Little matrix-isolation work with dinitrogen and dioxygen and the early transition metal atoms has been reported. A $Cr(O_2)_2$ system has been reported[126,127] as well as a series of

TABLE 4-4

Metal Carbonyl Complexes of the Early Transition Metals Prepared by Matrix Isolation Metal Atom Techniques

Complex	$\nu_{C=O}$ (cm^{-1})	Probable geometries	Comments	References
$Ti(CO)_6$	1945, 1953, 1985	Octahedral-low spin		116, 117
V–CO	1904 (argon)	Non-linear		109, 110, 118
$V(CO)_2$	1974, 1882, 1880	$C_{3h}, C_{2h}, D_{\infty h}$		109, 110, 118
	1723, 1719	Three forms		
$V(CO)_3$	1920	D_{3h}		109, 110, 118
$V(CO)_4$	1893	D_{4h} or T_d		109, 110, 118
$V(CO)_5$	1952, 1943	D_{3h}		109, 110, 118
$V(CO)_6$	1971	O_h, stable		109, 110, 116, 118, 119
Ta–CO	1819, 1831			109, 110, 120
$Ta(CO)_2$	1891, 1897			109, 110, 120
$Ta(CO)_3$	1916			109, 110, 120
$Ta(CO)_4$	1943			109, 110, 120
$Ta(CO)_5$	1953			109, 110, 120
$Ta(CO)_6$	1967			109, 110, 120
$Cr(CO)_3$	1867	C_{3v}		121, 122
$Cr(CO)_4$	1940, 1934, 1896	C_{2v}, low spin		121, 122
$Cr(CO)_5$	2093, 1966, 1936	C_{4v} or D_{3h} interacts with rare gases, CH_4, SF_6		121–124
$Cr(CO)_6$	1990 (argon)	O_h, stable		116, 121, 122
$Mo(CO)_2$	1915, 1911	$D_{\infty h}$		121, 122
$Mo(CO)_3$	1869	C_{3v}		121, 122
$Mo(CO)_4$	1951, 1895	C_{2v}		121, 122
$Mo(CO)_5$	2098, 1973, 1933	C_{4v}, interacts with rare gases, CH_4, SF_6		121, 122
$Mo(CO)_6$	1993	O_h, Stable		121, 122
$W(CO)_3$	1865	C_{3v}		121, 122
$W(CO)_4$	1939, 1894	C_{2v}		121, 122
$W(CO)_5$	2097, 1963, 1932	C_{4v}, interacts with rare gases, CH_4, SF_6		121, 122
$W(CO)_6$	1987	O_h, stable		121, 122
$Mn(CO)_5$	1910, 1933, 1940	C_{4v}		124
$Re(CO)_5$	1995, 1977	D_{3h}		125

N_2 complexes of Ti,[117] V,[128] Nb.[129] In these cases it is believed that $Nb(N_2)_4$ is formed in D_{4h} geometry, and $Ti(N_2)_6$ and $V(N_2)_6$ in octahedral geometries.

i. Polymer Reactions. Francis and Timms[130] have employed solution-phase metal atom reactor techniques (cf. Chapter 5) so that M atoms could be allowed to react with liquid methylphenylsiloxane polymers. It will be recalled that Ti, V, Cr, and Mo atoms complex readily with arenes, and the same type of complexation takes place when these metal atoms interacted with the polymer. Thus, π-arene complexation took place with very high efficiency, and since the resulting colored oils were not significantly more viscous than the starting polymer, it is likely that intramolecular bis(arene) complexes were predominantly formed. Some of the chemical and spectroscopic properties of these fluid materials were also reported.[130]

5. CLUSTER FORMATION PROCESSES

The tremendous importance of small metal clusters in catalysis is certainly well-known, and this has served as an impetus for broad ranging theoretical and experimental studies of cluster growth on clean surfaces.[131] These studies have been dependent on ultra high vacuum techniques and on maintaining the surface temperatures such that cluster growth and cluster shrinkage by surface atom migration are often in equilibrium. This, of course, is not the case in the low-temperature matrix that metal atom chemists employ. At low temperatures, cluster growth by M—M bond formation is not reversible. The size of the cluster produced will depend on several factors, including temperature, concentration of M, and strength of interaction of the ligand present (or diluent such as Ar) with M, M_2, M_3, etc.

a. Matrix Isolation Spectroscopic Studies of Dimers and Small Clusters. The first step in the embryonic clustering of dispersed metal atoms is the formation of dimers. Kundig, Moskovits, and Ozin have studied the dimerization of some early transition metal atoms by matrix isolation spectroscopy.[132] From their work as well as others it has become generally well accepted that dimerization is less efficient (usually an unwanted process) when the metal and matrix (inert gas) are high in atomic weight, and that dimerization efficiency increases with increase in M concentration and temperature. Thus, Li and Be readily diffuse below $40°K$, but Mg, Ca, Pb, Cu, Ag, and Au have a lesser tendency to do so. Dimerization is likewise more efficient with matrix material Ne > Ar > Kr > Xe.

It is quite possible, however, for heavy metal atoms to diffuse and dimerize during metal atom cocondensation experiments, because a dense liquid-like layer is continually formed as the matrix freezes. It is in this layer that even heavy metal atoms are mobile.

In spite of previous discussion, some metal atoms, for reasons not presently understood, diffuse surprisingly well during matrix deposition. Very high concentrations of M_2 have been observed with these metals. Manganese is a good example. For Mn–CO depositions $Mn_2(CO)_{10}$ is believed to be the main product even at Mn/CO ratios of 1/1000. Evidently Mn has some very unusual properties in terms of mobility in the quasi-liquid layer. However, the dimer product is apparently a CO-bridged species in the matrix, whereas normally $Mn_2(CO)_{10}$ is M–M bonded.[124,132] Other species such as Mn_2–CO and $Mn_2(CO)_2$ were also believed to be formed.[124,132]

Dimerization occurred, but to a lesser extent with Re. Here an M–M bonded dimer $Re_2(CO)_{10}$ could be produced.[132] In the case of V and Cr, M–M bonded dimers $V_2(CO)_{12}$ and $Cr_2(CO)_{10}$ were produced, although it was observed that another dimer possibly bridging $V_2(CO)_{12}$, was also present in the matrix.[124,132,133]

Kinetic analyses of these dimerization processes have been discussed.[132,134] Both a statistical frozen-matrix approach (calculable probability that M and M are neighbors and react to give M_2) and a highly mobile metal atom approach (diffusion is rapid in quasi-liquid layer) have been used.[132,134] It was found, not unexpectedly, that the diffusion mechanism approach appears to be corroborated best by experimental results, and that the eventual M_2 concentration is proportional to the square of the M/substrate ratio. Concentrations of higher metal aggregates vary as some higher power of the M/substrate ratio (in these analyses "reactive" matrices are assumed). Kinetic analysis of the nonreactive materials is somewhat more involved (e.g., M + Ar + CO), and any further conclusions are not warranted at this time.

Clustering of V atoms in alkane matrices has been studied.[135] No signs of actual reaction with the alkanes were detected in the temperature range of $10°$–$40°K$, but change in the alkane did affect the rate with which V atoms dimerized and telomerized. A relationship between alkane chain length and metal atom mobility was derived. It was found that under similar conditions, alkane matrices were more efficient than argon matrices for isolation of metal atoms (discourage M_2 and M_n formation), and this isolation effect gets better with increasing chain length in the n-alkane.

Recently, a new dimension in cluster synthesis in matrices was developed by Ozin and co-workers. They found that selective dimerization and tri-merization processes could be carried out by photolysis at the wavelength where the isolated metal atom absorbs light. This absorbed energy is transferred to the surrounding inert matrix, warming it locally and allowing the metal atom to migrate. For example, Cr atoms photolyzed with 335-nm wavelength light caused Cr atoms to move more rapidly in an Ar or Kr matrix than Mo atoms in the same matrix.[135a] Clusters of $(Cr)_n(Mo)_m$ (where $n = 1$–3 and $m = 1$–3) could be prepared with selectivity. Similarly, if 295-nm light was used, Mo atoms were forced to migrate more rapidly.[135a]

Thus, using either matrix-warming or photolysis techniques, a wide variety of new M_2 molecules have been prepared in a low-temperature matrix. Recent examples include Nb_2, Mo_2,[135b] Cr_2, Mo_2, $CrMo$,[135c] V_2,[135d] Ti_2, and Sc_2.[135e] In addition, by use of reactive matrices (i.e., CO/Ar or N_2/Ar) some new M_2–carbonyls and M_2–dinitrogen complexes have been prepared in low-temperature matrices [for example, $V_2(CO)_{12}$ (believed to be CO bridged)[110,133] and $V_2(N_2)_n$[135f]].

The importance of these studies relates to understanding the bonding, electronic structure, and reactivity of tiny metal clusters as they relate to larger clusters, which of course are of utmost importance in heterogeneous catalysis. Already some chemisorption studies have been clarified by these studies, (cf. Chapter 5),[128,134] and it appears that further useful knowledge will be gained by this type of work. However, among heterogeneous catalysis chemists, there is disagreement as to whether these low-temperature matrices are well characterized or not.[136]

b. Macroscale Metal Atom Cluster Studies.   Macroscale production of discrete organometallic clusters and reactive metal particles by metal atom methods has been very important in studies of later transition metals (cf. Chapter 5) and main group metals (cf. Chapters 7 and 8). However, studies have not yet been reported on any of the early transition metals.

6. DISPROPORTION AND LIGAND TRANSFER PROCESSES

A brief report of the reaction of $(allyl)_2Sn$ with Cr atoms was given by Timms and Turney.[71] Here a $Cr_2(allyl)_6$ complex was obtained, possibly of the following formulation:

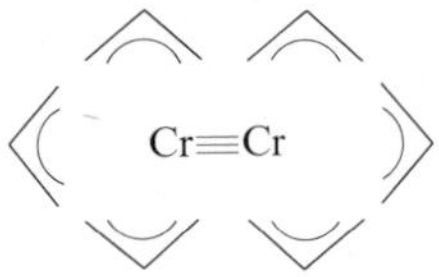

II. Early Transition Metal Halide, Oxide, and Sulfide Vapors

A. Occurrence, Properties, and Techniques

Natural occurrence of the vapors of these species is rare, necessitating very unusual conditions such as in the atmosphere of stars. The bands for TiO, ScO, YO, and ZrO have been found to be very strong in the spectrum of some stars.[137,138]

Table 4-5 summarizes available vaporization data in the literature and includes those few studies found on the molecular composition of the vapors.[26,139–186] The ease of vaporization varies remarkably throughout this series, with some of the halides being volatile liquids or gases while some of the oxides are extremely refractory.

TABLE 4-5

Vaporization Data for the Early Transition Metal Halides, Oxides, and Sulfides

Compound	mp (°C)[a]	bp (°C)[a]	Heat of Vap (kcal/mole)	Vapor composition or comments	References
ScBr$_3$	subl >1000				139, 140
ScCl$_3$	939	subl 800–850	66	ScCl$_3$, (ScCl$_3$)$_2$	141
ScF$_3$			94	D$_{3h}$ geometry	
Sc$_2$O$_3$	—	—			
TiBr$_2$	d >500				
TiBr$_4$	39	230			
TiCl$_2$		d 475			
TiCl$_3$	d 440	660			
TiCl$_4$	−25	136		TiCl$_4$, (TiCl$_4$)$_2$	139
TiF$_3$	1200	1400			
TiF$_4$	>400	subl 284			
TiI$_2$	600	1000			
TiI$_4$	150	377			
TiO$_2$	1825				
TiO$_2$(rutile)	1830	2500–3000	138	Ti$_3$O$_5$, Ti$_4$O$_7$, TiO, TiO$_2$	142, 143
TiO	1750	>3000		Ti$_2$O, TiO, Ti$_2$O$_3$, Ti$_3$O$_5$, Ti$_4$O$_7$, Ti$_5$O$_9$, Ti$_{10}$O$_{19}$, TiO$_2$, Ti	144–146
Ti$_2$O$_3$	d 2130				
TiS		vap 1900	140.7	TiS, Ti, S$_2$, S	147
Ti$_2$S$_3$		—			
VBr$_3$	d	—			
VCl$_2$	—	—			
VCl$_3$	d	—			
VCl$_4$	−28	149	9.9	VCl$_4$	148
VF$_3$	>800	subl			
VF$_4$	d 325	—			

(continued)

TABLE 4-5 (*continued*)

Compound	mp (°C)^a	bp (°C)^a	Heat of Vap (kcal/mole)	Vapor composition or comments	References
VF_5	—	111		VF_5	149
VF_2	subl 750–800	—			
VI_2	—	—			
VO	1967				149a
VO_2	1970				
V_2O_3	690				
V_2O_5	d	d 1750	13.5		150
VS	d	—		VS, V, S	149a, 151
V_2S_5		—			
$CrBr_2$	842	—			
$CrBr_3$	subl	—			
$CrCl_2$	824				
$CrCl_3$	1150	1300			
CrF_2	1100	>1300			
CrF_3	>1000	subl 1100			
CrF_5				CrF_5	149
CrI_2	856	subl 800			
CrI_3	600	d			
CrO	d				
CrO_2	d				
Cr_2O_3	2435	4000			
CrS	1550	—			
Cr_2S_3	d				
$MnBr_2$	d	—			
$MnCl_2$	650	1190			
$MnCl_3$	d				
MnF_2	856	—		MnF_2, Mn_2F_5	142

MnF₃	d	—		MnF₃, Mn₂F₅	142
MnI₂	638	subl 500		telomers	152
Mn₃O₄	1705	—			
MnO₂	d	—			
Mn₂O₇	5.9	d			
MnO	—	—			
MnO₃	d	—			
YBr₃	904	—			
YCl₃	721	1507(1488)	72		153
YF₃	1387	650(0.02 torr)	109	YF₃(C₃ᵥ geometry), YF₂(C₂ᵥ geometry)	141, 154
YI₃	1004	—			
Y₂O₃	2410	—		YO	155, 156
Y₂S₃	3000	—			
ZrBr₂	—	—			
ZrBr₄	450	subl 357	76.4		157
ZrCl₂	d 350	—			
ZrCl₃	d 350	—			
ZrCl₄	437	subl 331	23.7		158
ZrF₄	subl 600	subl 903	56.6	ZrF₂, ZrF₃, ZrF₄	159–161
ZrI₄	499	d 600			
ZrO₂	2700(2687)	5000		ZrO, ZrO₂	162, 163
ZrS₂	1550	vap 1400		Zr₂S₅, S₂	164
NbBr₅	265(268)	362		NbBr₅	160, 165
NbCl₅	205(209)	254	31.4		151, 165
NbF₅	72(80)	236(235)	12.9	Nb₃F₁₅, NbF₅	166–168
NbO	—	vap 1630	140	Nb, NbO, NbO₂ mainly	169
NbO₂	—	—			
Nb₂O₃	1780	—			
Nb₂O₅	1460	—			
MoBr₂	—	—			
MoBr₃	d	—			

(continued)

73

TABLE 4-5 (*continued*)

Compound	mp (°C)[a]	bp (°C)[a]	Heat of Vap (kcal/mole)	Vapor composition or comments	References
$MoBr_4$	d	—			
$MoCl_2$	d	—			
$MoCl_4$	d	—			
$MoCl_5$	194	268			
MoF_6	17	35			
MoF_5	—	—			
MoI_2	—	—			
MoI_4	d 100	—			160
MoO_2	—	—	134	Mo_3O_9	170
MoO_3	795	subl 1155	79	Mo_3O_9, Mo_4O_{12}, Mo_5O_{15}	170–173
Mo_2O_3	—	—			
Mo_2O_5	—	—			
MoS_2	1185	subl 450		S_2	174
Mo_2S_3	d 100	—			
MoS_4	d	—			
MoS_3	d	—			
Tc					
none					
$HfBr_4$	subl 420	—			
$HfCl_4$	subl 319	—	23.5		158
HfF_4	—	—			
HfI_4	—	subl 400 (vac)	34		175
HfO_2	2812	5400			162
HfS_2		vap 1500		Hf_2S_5, S_2	164
$TaBr_5$	265(280)	349		$TaBr_5$	160, 165
$TaCl_5$	216(220)	242(239)			165
TaF_5	97(95)	230(229)	13	Ta_3F_{15}, TaF_5	166, 168
TaO_2	d	—			

Ta$_2$O$_5$	1800	—			
TaS$_2$	>1300	—			
WBr$_2$	d 400	—			
WBr$_5$	276(286)	333(392)		telomers	176
WBr$_6$	232	—			
WCl$_2$	d	—			
WCl$_4$		—			
WCl$_5$	248	276			175, 177
WCl$_6$	275(330)	347	14.9(13.7)		149
WF$_5$		—		WF$_4$, WF$_6$	
WF$_6$	2.5	18			
WI$_2$	d	—			
WI$_4$	d	—			
WO$_2$[b]	1500	1430	90	W$_3$O$_9$	170, 172
WO$_3$[b]	1473		108	W$_3$O$_9$	170
W$_2$O$_5$[b]	800	1530			
WS$_2$	1250				
WS$_3$	subl 500	—			
ReBr$_3$	—	—	48	Re$_3$Br$_9$	160, 161, 178
ReCl$_3$	—	>550		Re$_3$Cl$_9$	161, 178, 179
ReCl$_4$	d	500			
ReCl$_5$		—			
ReF$_4$	124	d 500			
ReF$_5$		—			
ReF$_6$	19	48			165
ReO$_2$	d 1000	vap 350	65	Re$_2$O$_7$, ReO$_3$, HReO$_4$	180–184
ReO$_3$	—	vap 750		Re$_2$O$_7$, ReO$_3$, HReO$_4$	172, 181–183
Re$_2$O$_7$	297	subl 250	17.7		184, 185
ReS$_2$	—	d	22.6		180
Re$_2$S$_7$	—	d			

[a] Taken from "Handbook of Chemistry and Physics,"[26] or in reference shown.

[b] The W—O system yields W$_4$O$_{12}$, W$_3$O$_9$, W$_3$O$_8$, and W$_2$O$_6$ vapor species about 1350°C.[186]

Alcock and co-workers[187,188] have determined that the free evaporation rate of certain oxides, sulfides, borides, or carbides is sometimes lower than would be predicted by thermodynamic data from Knudsen cell experiments. In these instances the vaporization coefficient they define would be less than unity. The oxide Al_2O_3 vaporizes in this way, whereas ZrO_2, ThO_2, TiB_2, TiC, Si_3N_4 have a coefficient of unity. These considerations are of importance.[189] It may be that many compounds as well as elements do not vaporize off a hot surface or from a molten drop the same as they do in thermal equilibrium in a Knudsen cell. Vapor compositions may vary considerably, and much more study and comparisons are needed.

The structures of gaseous metal dihalides have been discussed by Charkin and Dyatkina.[190] Generally, it is believed that the dihalides of the following metals are bent in the gaseous state: Sc, Ti, V, Cr, Y, Zr, Nb, Mo, La, Hf, Ta, W, Ru, Rh, Pd, Re, Os, Ir, and Pt. However, linear MX_2 structures are predicted for Mn, Fe, Co, and Ni.

B. Chemistry

Nothing has been reported concerning the chemistry of these species other than film-forming properties or chemical vapor deposition of the vapors to metal film (e.g., WF_6 vapor $\rightarrow$ W film). Some of these processes are discussed by Packard.[191] These are not considered chemical reaction studies for purposes herein.

Potentially, the chemical reactions of many of these species with organic compounds could be quite interesting since in the condensed crystalline phase many of these materials have catalytic effects on organic molecules. For example, MoO_2, MoO_3, WO_3, etc. are extremely important in hydrotreating catalysts. How do they interact on a molecular level with organics? (In this regard, perhaps metal carbides should also be included in the list in Table IV-5.)

On the other hand, perhaps many of the molecular species listed would be good ligands for coordination with metal atoms to form M—M′ bonded species. Possibly a new series of non-organometallic ligands could be uncovered, for example, by depositing metal atoms with vapors of some of these halides.

References

1. K. Utsumi, *Publ. Astron. Soc. Jpn.* **22**, 93 (1970).
2. W. A. Fowler, *Int. Astron. Union, Symp.* **26**, 335 (1966).
3. L. H. Allen, I. S. Bowen, and R. Minkowski, *Astrophys. J.* **122**, 62 (1955).

4. L. H. Aller, *NASA Doc.* **N62-14**, 813 (1962); from *NASA Tech. Publ., Announce.* **2**, 734 (1962).
5. A. M. Bonch-Bruevich and Y. A. Imas, *Exp. Tech. Phys.* **15**, 323 (1967).
6. S. I. Anisimov, A. M. Bonch-Bruevich, M. A. Elyashevich, Y. A. Imas, N. A. Partenko, and G. S. Romanov, *Zh. Tekh. Fiz.* **36**, 1273 (1966).
7. A. I. Korunchikov and A. A. Yankovskii, *Zh. Prikl. Spektrosk.* **5**, 586 (1966).
8. J. L. Dumas, *Rev. Phys. Appl.* **5**, 795 (1970).
9. M. Burden and P. A. Walley, *Vacuum* **19**, 397 (1969).
10. L. Bachmann, *Naturwissenschaften* **49**, 34 (1962).
11. E. B. Graper, *J. Vac. Sci. Technol.* **8**, 333 (1971).
12. S. Namba, *Proc. Symp. Electron Beam Technol., 4th, 1962* p. 304 (1962).
13. A. Schram, *Adv. Vac. Sci. Technol., Proc. Int. Congr., 1st, 1958* Vol. 1, p. 446 (1960).
14. M. Auberg, M. Brabers, M. Heuset, and M. Meulenans, *Mem. Sci. Rev. Metall.* **62**, 373 (1965).
15. M. Aubecq, M. Brabers, M. Heuset, and M. Meulemans, *Mem. Sci. Rev. Metall.* **62**, 373 (1965).
16. R. K. Koch and W. E. Arable, *U. S., Bur. Mines, Rep. Invest.* **7063**, (1968).
17. R. Thum and J. B. Ramsey, *Natl. Symp. Vac. Tech.* **6**, 192 (1959).
18. A. I. Chernenko, *Izv. Vyssh. Uchebn. Zaved., Fiz.* No. 1, p. 140 (1958).
19. E. I. Nikonva, V. K. Prokofiew, and V. P. Zakharov, *Fiz. Sb., L'vovGos. Univ.* No. 4, p. 64 (1958).
20. B. L. Vallee and R. W. Peattie, *Anal. Chem.* **24**, 434 (1952).
21. S. Nagata, T. Nasu, and Y. Tomoda, *Oyo Butsuri* **27**, 459 (1958).
22. B. Vodar, S. Minn, and S. Offret, *J. Phys. Radium* **16**, 811 (1955).
23. H. F. Sterling, *Vide* **21**, 121 (1966).
24. W. A. Fischer, D. Janke, and K. Stahlschmidt, *Arch. Eisenhuettenwes.* **45**, 757 (1974).
25. E. N. Grishin and V. P. Sinev, *Fiz. Khim. Obrab. Mater.* No. 6, p. 12 (1974).
26. Handbook of Chemistry and Physics, 56th ed. CRC Press, Cleveland, Ohio, 1975-1976.
27. B. Siegel, *Q. Rev., Chem. Soc.* **19**, 77 (1965).
28. E. B. Owens and A. M. Sherman, *U. S. Dep. Commer., Off. Tech. Serv., AD* **275**, 468 (1962).
29. P. N. Walsh, *AEC Accession* No. 4 6956, Rep. No. NP-15514 (Vol. 1) (1965).
30. R. J. Ackermann and E. G. Rauh, *J. Chem. Phys.* **36**, 448 (1962).
31. O. H. Krikorian, *J. Phys. Chem.* **67**, 1586 (1963).
32. R. K. Koch, W. E. Anable, and R. A. Beall, *U.S., Bur. Mines, Rep. Invest.* **7125** (1968).
33. M. B. Panish and L. Reif, *J. Chem. Phys.* **38**, 253 (1963).
34. W. Reichelt, *Angew. Chem., Int. Ed. Engl.* **14**, 218 (1975).
35. P. L. Timms, *in* "Cryochemistry", (M. Moskovits and G. A. Ozin, eds.), p. 61. Wiley (Interscience), New York, 1976.
36. K. J. Klabunde, *in* "Reactive Intermediates" (R. A. Abramovitch, ed.), Plenum, New York, 1979.
37. J. Feldman, M. Friz, and F. Stetten, *Res. Dev.*, p. 49 (1976).
38. P. L. Timms, *Angew. Chem., Int. Ed. Engl.* **14**, 273 (1975).
39. P. L. Timms, *Chem. Commun.* p. 258 (1968).
40. M. J. McGlinchey and P. S. Skell, *in* "Cryochemistry", (M. Moskovits and G. A. Ozin, eds.) p. 137. Wiley (Interscience) New York, 1976.
41. R. Kirk and P. L. Timms, *J. Am. Chem. Soc.* **91**, 6315 (1969).
42. P. S. Skell and P. W. Owen, *J. Am. Chem. Soc.* **94**, 5434 (1972).
43. F. W. S. Benfield, M. L. H. Green, J. S. Ogden, and D. Young, *J. Chem. Soc., Chem. Commun.* 866 (1973).

44. M. T. Anthony, M. L. H. Green, and D. Young, *J. Chem. Soc. Dalton Trans.* p. 1419 (1975).
45. F. G. Cloke, M. L. H. Green, and G. E. Morris, *J. Chem. Soc., Chem. Commun.* p. 72 (1978).
46. E. A. Koerner von Gustorf, O. Jaenicke, O. Wolfbeis, and C. R. Eady, *Angew. Chem., Int. Ed. Engl.* **14**, 278 (1975).
47. E. Koerner von Gustorf and O. Jaenicke, *Nachr. Chem. Tech.* **21**, 95 (1973).
48. E. Koerner von Gustorf, O. Jaenicke, and O. E. Polansky, *Angew. Chem., Int. Ed. Engl.* **11**, 532 (1972).
49. J. Gladysz, J. Fulcher, and S. Togashi, *J. Org. Chem.* **41**, 3647 (1976).
50. J. Gladysz, J. Fulcher, and S. Togashi, *Tetrahedron Lett.* p. 521 (1977).
51. J. Gladysz, private communications.
52. T. O. Murdock and L. Wescott, unpublished work from these laboratories.
53. T. Chivers and P. L. Timms, *J. Organomet. Chem.* **118**, C37 (1976).
54. J. S. Roberts, Ph. D. Thesis, University of North Dakota, Grand Forks (1975); also unpublished work of J. S. Roberts in this laboratory.
55. J. R. Blackboro, R. Grubbs, A. Miyashita, and A. Scrivanti, *J. Organomet. Chem.* **120**, (3), C49 (1976).
56. P. L. Timms, *Adv. Inorg. Chem. Radiochem.* **14**, 121 (1972).
57. P. L. Timms, *Chem. Commun.* p. 1033 (1969).
58. M. J. D'Aniello, Jr. and E. K. Barefield, *J. Organomet. Chem.* **76**, C 50 (1974).
59. E. M. Van Dam, W. N. Brent, M. P. Silvon, and P. S. Skell, *J. Am. Chem. Soc.* **97**, 465 (1975).
60. J. R. Blackborow, R. H. Grubbs, A. Miyashita, A. Scrivanti, and E. A. Koerner von Gustorf, *J. Organomet. Chem.* **122** (1), C6 (1977).
61. P. S. Skell and M. J. McGlinchey, *Angew. Chem., Int. Ed. Engl.* **14**, 195 (1975).
62. M. Heberhold, *Chem. Unserer Zeit* **10**, 120 (1976).
63. P. S. Skell, D. L. Williams-Smith, and M. J. McGlinchey, *J. Am. Chem. Soc.* **95**, 3337 (1973).
64. J. A. Gladysz, J. G. Fulcher, S. J. Lee, and A. B. Bocarsley, *Tetrahedron Lett.* No. 39, p. 3421 (1977).
65. P. S. Skell, E. M. Van Dam, and M. P. Silvon, *J. Am. Chem. Soc.* **96**, 626 (1974).
66. M. Yevitz and P. S. Skell, *Int. Union Crystallography Intercong. Symp. Intra-Intermol. Forces, 1974* P. State Univ., Ser. 2, Vol. 2, Pap. E9 (1974).
67. V. M. Akhmedov, M. T. Anthony, M. L. H. Green, and D. Young, *J. Chem. Soc., Dalton Trans.* p. 1412 (1975).
68. V. M. Akhmedov, M. T. Anthony, M. L. H. Green, and D. Young, *J. Chem. Soc., Chem. Commun.* p. 777 (1974).
69. E. M. Van Dam, W. N. Brent, M. P. Silvon, and P. S. Skell, *J. Am. Chem. Soc.* **97**, 465 (1975).
70. P. L. Timms and T. W. Turney, *J. Chem. Soc., Dalton Trans.* p. 2021 (1976).
71. P. L. Timms and T. W. Turney, *Adv. Organomet. Chem.* **15**, 53 (1977).
72. S. P. Kolesnikov, J. E. Dobson, and P. S. Skell, *J. Am. Chem. Soc.* **100**, 999 (1978).
73. H. F. Efner, R. R. Smardzewski, D. E. Tevault, and W. V. Fox, private communications.
74. P. L. Timms, *Chem. Commun.* p. 1033 (1969).
74. E. O. Fischer and W. Hafner, *Naturforsch.*, B **10**, 665 (1955).
75. P. L. Timms, *J. Chem. Educ.* **49**, 782 (1972).
76. K. J. Klabunde and H. F. Efner, *Inorg. Chem.* **14**, 789 (1975).
77. R. Middelton, J. R. Hull, S. R. Simpson, C. H. Tomlinson, and P. L. Timms, *J. Chem. Soc., Dalton Trans.* p. 120 (1973).

78. H. F. Efner, unpublished results from this laboratory.
79. A. N. Neameyanov, N. N. Zaitseva, G. A. Dormrachev, V. D. Zinovev, L. P. Yureva, and I. I. Tverdokhlebova, *J. Organomet. Chem.* **121**, (3), C52 (1977).
80. M. P. Silvon, E. M. Van Dam, and P. S. Skell, *J. Am. Chem. Soc.* **96**, 1945 (1974).
81. F. G. Cloke, M. L. H. Green, and D. H. Price, *J. Chem. Soc., Chem. Commun.*, p. 431 (1978).
82. L. Radonovich, C. Zuerner, H. F. Efner, and K. J. Klabunde, *Inorg. Chem.* **15**, 2976 (1976).
83. L. Radonovich and C. Zuerner, unpublished work.
84. V. Graves and J. J. Lagowski, *J. Organomet. Chem.* **120**, 397 (1976).
85. V. Graves and J. J. Lagowski, *Inorg. Chem.* **15**, 577 (1976).
86. Private communications with P. L. Timms.
87. M. L. H. Green, private communications.
88. T. Groshens, unpublished work from this laboratory.
89. K. J. Klabunde and H. F. Efner, *J. Fluorine Chem.* **4**, 115 (1974).
90. M. J. McGlinchey and T. S. Tan, *Can. J. Chem.* **52**, 2439 (1974).
91. M. J. McGlinchey and T. S. Tan, *J. Am. Chem. Soc.* **98**, 2271 (1976).
92. T. S. Tan and M. J. McGlinchey, *J. Chem. Soc., Chem. Commun.*, 155 (1976).
93. A. Agarwal, M. J. McGlinchey, and T. S. Tan, *J. Organomet. Chem.* **141**, 85 (1978).
94. N. Hao and M. J. McGlinchey, *J. Organomet. Chem.* **161**, 381 (1978).
95. M. J. McGlinchey and T. S. Tan, *J. Am. Chem. Soc.* **98**, 2271 (1976).
96. L. H. Simons, P. E. Riley, R. E. Davis, and J. J. Lagowski, *J. Am. Chem. Soc.* **98**, 1044 (1976).
97. P. E. Riley and R. E. Davis, *Inorg. Chem.* **15**, 2735 (1976).
98. G. Essenmacher (with P. Treichel), Ph. D. Thesis, University of Wisconsin, Madison (1976); collaborative work with H. Efner and K. J. Klabunde.
99. P. Treichel, G. Essenmacher, H. Efner, and K. J. Klabunde, unpublished results.
100. K. J. Klabunde, *Trans. N. Y. Acad. Sci.* [2] **295**, 83 (1977).
101. V. Graves and J. J. Lagowski, *Inorg. Chem.* **15**, 577 (1976).
102. J. Boyd, J. Lavoie, and D. M. Gruen, *J. Chem. Phys.* **60**, 4088 (1974).
103. H. F. Efner, D. E. Tevault, W. B. Fox, and R. R. Smardzewski, *J. Organomet. Chem.* **146**, 45 (1978).
104. H. F. Efner, W. B. Fox, R. R. Smardzewski, and D. E. Tevault, *Inorg. Chem. Acta* **24**, L93 (1977).
105. P. L. Timms, *J. Chem. Soc. A* p. 2526 (1970).
106. R. B. King and M. Chang, *Inorg. Chem.* **18**, 364 (1979).
107. D. Gladkowski and F. R. Scholar, *171st (Centen.) Meet., Natl. Am. Chem. Soc.* Paper INOR 133 (1976).
108. G. C. Pimentel, *Angew. Chem., Int. Ed. Engl.* **14**, 199 (1975).
109. M. Moskovits and G. A. Ozin, in "Cryochemistry", (M. Moskovits and G. A. Ozin, eds.), p. 9. Wiley (Interscience), New York, 1976.
110. M. Moskovits and G. A. Ozin, in "Cryochemistry" (M. Moskovits and G. A. Ozin, eds.), p. 261. Wiley (Interscience), New York, 1976.
111. B. Mayer, "Low Temperature Spectroscopy." Am. Elsevier. New York, 1971.
112. H. Hallam, "Vibrational Spectroscopy of Trapped Species." Wiley, New York, 1972.
113. A. M. Bass and H. P. Broida, "Formation and Trapping of Free Radicals." Academic Press, New York, 1960.
114. G. C. Pimentel, *Spectrochim. Acta* **12**, 94 (1958); *Pure App. Chem.* **4**, 61 (1962).
115. E. Whittle, D. A. Dows, and G. C. Pimentel, *J. Chem. Phys.* **22**, 1943 (1954).
116. A. B. P. Lever and G. A. Ozin, *Inorg. Chem.* **16**, 2012 (1977).

117. R. Busby, W. Klotzbucher, and G. A. Ozin, *Inorg. Chem.* **16**, 822 (1977).
118. L. Hanlan, H. Huber, and G. A. Ozin, *Inorg. Chem.* **15**, 2592 (1976).
119. T. A. Ford, H. Huber, W. Klotzbucher, M. Moskovits, and G. A. Ozin, *Inorg. Chem.* **15**, 1666 (1976).
120. R. L. DeKock, *Inorg. Chem.* **10**, 1205 (1971).
121. R. N. Perutz and J. J. Turner, *J. Am. Chem. Soc.* **97**, 4791 and 4800 (1975).
122. M. A. Graham, M. Poliakoff, and J. J. Turner, *J. Chem. Soc. A*, p. 2939 (1971).
123. E. P. Kundig and G. A. Ozin, *J. Am. Chem. Soc.* **96**, 3820 (1974).
124. H. Huber, E. P. Kundig, G. A. Ozin, and A. J. Poe, *J. Am. Chem. Soc.* **97**, 308 (1975).
125. H. Huber, E. P. Kundig, and G. A. Ozin, *J. Am. Chem. Soc.* **96**, 5585 (1974).
126. J. H. Darling, M. B. Garton-Sprenger, and J. S. Ogden, *J. Chem. Soc., Faraday Trans. 2 Symp.*, p. 75 (1973).
127. D. McIntosh and G. A. Ozin, *Inorg. Chem.* **15**, 2869 (1976).
128. G. A. Ozin, *Acc. Chem. Res.* **10**, 21 (1977).
129. D. W. Green, R. V. Hodges, and D. M. Gruen, *Inorg. Chem.* **15**, 970 (1976).
130. C. G. Francis and P. L. Timms, *J. Chem. Soc., Chem. Commun.*, p. 466 (1977).
131. R. Niedermayer, *Angew. Chem., Int. Ed. Engl.* **14**, 212 (1975).
132. E. P. Kundig, M. Moskovits, and G. A. Ozin, *Angew. Chem., Int. Ed. Engl.* **14**, 292 (1975).
133. T. A. Ford, H. Huber, W. Klotzbucher, M. Moskovits, and G. A. Ozin, *Inorg. Chem.* **15**, 1666 (1976).
134. M. Moskovits and G. A. Ozin, *in* "Cryochemistry", (M. Moskovits and G. A. Ozin, eds.), p. 261. Wiley (Interscience), New York, 1976.
135. W. E. Klotzbucher, S. A. Mitchell, and G. A. Ozin, *Inorg. Chem.* **16**, 3063 (1977).
135a. W. E. Klotzbucher and G. A. Ozin, *J. Am. Chem. Soc.* **100**, 2262 (1978).
135b. W. E. Klotzbucher and G. A. Ozin, *Inorg. Chem.* **16**, 984 (1977).
135c. W. E. Klotzbucher, G. A. Ozin, J. G. Norman, and H. J. Kolari, *Inorg. Chem.* **16**, 2871 (1977).
135d. T. A. Ford, H. Huber, W. Klotzbucher, E. P. Kundig, M. Moskovits, and G. A. Ozin, *J. Chem. Phys.* **66**, 524 (1977).
135e. R. Busby, W. Klotzbucher, and G. A. Ozin, *J. Am. Chem. Soc.* **98**, 4013 (1976).
135f. H. Huber, T. A. Ford, W. Klotzbucher, and G. A. Ozin, *J. Am. Chem. Soc.* **98**, 3176 (1976).
136. J. Katzer, private communications.
137. J. Gauzit, *Ann. Astrophys.* **18**, 354 (1955).
138. A. D. Thackeray, *Nature (London)* **160**, 370 (1947).
139. Y. B. Patrikeev, V. A. Morozova, G. P. Dudchik, O. G. Polychenok, and G. I. Novikov, *Zh. Fiz. Khim.* **47**, 266 (1973).
140. I. A. Ratkovskii, L. N. Novikova, and T. A. Pribytkova, *Khim. Khim. Tekhnol. (Lvov)* **9**, 19 (1975).
141. K. Krasnov and N. I. Giricheva, *Teplofiz. Vys. Temp.* **10**, 1321 (1972).
142. P. W. Gilles, H. F. Franzen, G. D. Stone, and P. G. Wahlbeck, *J. Chem. Phys.* **48**, 1938 (1968).
143. B. R. Conrad, J. E. Bennett, and P. W. Gilles, *J. Chem. Phys.* **63**, 5502 (1975).
144. P. W. Gilles, K. D. Carlson, H. F. Franzen, and P. G. Wahlbeck, *J. Chem. Phys.* **46**, 2461 (1967).
145. P. W. Gilles, G. H. Rinehart, and R. I. Sheldon, *J. Chem. Phys.* **66**, 2229 (1977).
146. R. I. Sheldon and P. W. Gilles, *J. Chem. Phys.* **66**, 3705 (1977).
147. J. G. Edwards, H. F. Franzan, and P. W. Gilles, *J. Chem. Phys.* **54**, 545 (1971).
148. S. A. Sachukanev, M. A. Oranskaya, T. A. Tolmacheva, and A. K. Yakhkind, *Zhr. Neorg. Khim.* **1**, 30 (1956).

149. M. Vasile, G. R. Jones, and W. E. Falconer, *Adv. Mass. Spectrom.* **6**, 557 (1974).
149a. J. Drowart, A. Pattoret, and S. Smoes., *Proc. Brit Ceram. Soc.* No. 8, p. 67 (1967).
150. G. A. Semenov, K. E. Frantseva, and E. K. Shalkova, *Vestn. Leningr. Univ., Fiz., Khim.* No. 3, p. 82 (1970).
151. H. Schafer and L. Bayer, *Z Anorg. Allg. Chem.* **277**, 140 (1954).
152. O. Glemser and H. Weizenkorn, *Naturwissenshaften* **48**, 715 (1961).
153. G. P. Pudchik and O. G. Polyacherok, *Zh. Neorg. Khim.* **14**, 3165 (1969).
154. R. D. Wesley and C. W. DeKock, *J. Phys. Chem.* **77**, 466 (1973).
155. P. N. Walsh, H. W. Goldstein, and D. White, *J. Am. Ceram. Soc.* **43**, 229 (1960).
156. N. Nimmagaddo and R. F. Bunshah, *J. Vac. Sci. Technol.* **8**, VM85 (1971).
157. V. I. Tsirelnikov, *J. Less-Common Met.* **19**, 287 (1969).
158. N. D. Denisova, E. K. Safronov, and O. N. Bystrova, *Zh. Neorg. Khim.* **11**, 2185 (1966).
159. K. A. Sense, M. J. Synder, and R. B. Filbert, *J. Phys. Chem.* **58**, 995 (1954).
160. S. S. Berdonosov, A. V. Lapitskii, and E. L. Bakov, *Zh. Neorg. Khim.* **10**, 322 (1965).
161. K. Rinke, M. Klein, and H. Schaefer, *J. Less-Common Met.* **12**, 497 (1967).
162. F. Henning, *Naturwissenchaften* **13**, 661 (1925).
163. E. H. P. Cordfunke, *Thermodyn. Nucl. Mater., Proc. Symp., 1962* p. 465 (1963).
164. G. N. Dubrovskaya, *UKr. Khim. Zh.* **33**, 997 (1967)
165. K. M. Alexander and F. Fairbrother, *J. Chem. Soc.* p. S223 (1949).
166. F. Fairbrother and W. C. Frith, *J. Chem. Soc.* p. 3051 (1951).
167. I. S. Gotkis, A. V. Gusarov, and L. N. Gorkhov, *Zh. Neorg. Khim.* **20**, 1250 (1975).
168. J. Fawcett, A. J. Hewitt, J. Holloway, and M. A. Stephen, *J. Chem. Soc., Dalton Trans.* No. 23, p. 2422 (1976).
169. S. A. Shcukanev, G. A. Seminov, and K. E. Frantseva, *Izv. Vyssh. Uchebn. Zaved., Khim. Tekhnol.* **5**, 691 (1962).
170. P. E. Blackburn, M. Hoch, and H. L. Johnston, *J. Phys. Chem.* **62**, 769 (1958).
171. E. K. Kazenas and Y. V. Tsvelkov, *Zh. Neorg. Khim.* **14**, 11 (1969).
172. E. K. Kazenas, D. Chizhika, and Y. V. Tsvetkov, *Termodin. Kinet. Protsessov Vosstanov. Met., Mater. Konf., 1969* p. (1972).
173. T. V. Charlu and O. J. Kleppa, *J. Chem. Thermodyn.* **3**, 697 (1971).
174. S. C. Schaefer, A. H. Larson, and A. W. Schlechten, *Trans. AIME* **230** (3), 594 (1964).
175. F. D. Stevenson, C. E. Wicks, and F. E. Block, *U.S., Bur. Mines, Rept. Invest.* **6367** (1964).
176. S. A. Shehudarev, G. I. Novikov, and G. A. Kokovin, *Zh. Neorg. Khim.* **4**, 2184 (1959)
177. G. I. Novikov and S. A. Shehukanev, *Ser. Khim. Nauk.* No. 12, p. 37 (1953).
178. A. Buechler, P. E. Blackburn, and J. L. Stauffer, *J. Phys. Chem.* **70**, 685 (1966).
179. R. W. Lins and R. J. Sime, *High Temp. Sci.* **5**, 56 (1973).
180. V. I Deer and V. I. Smirnov, *Dokl. Akad. Nauk SSSR* **140**, 822 (1961).
181. E. K. Kazenas, D. M. Chizhikov, and Y. V. Tsvetkov, *Issled. Protsessov. Metall. Tsvetn. Redk. Met.* p. 30 (1969).
182. G. A. Semenov and K. V. Ovchinnikov, *Zh. Obshch. Khim.* **35**, 1517 (1965).
183. H. B. Skinner and A. W. Searcy, *J. Phys. Chem.* **77**, 1578 (1973).
184. V. S. Vinogradov, V. V. Ugarov, and N. G. Rambidi, *Zh. Strukt. Khim.* **13**, 715 (1972).
185. W. T. Smith, Jr., L. E. Line, Jr., and W. A. Bell, *J. Am. Chem. Soc.* **74**, 4964 (1953).
186. P. J. Ackermann and E. G. Rauh, *J. Phys. Chem.* **67**, 2596 (1963).
187. E. G. Wolff and C. B. Alcock, *Trans. Br. Ceram. Soc.*, **61**, 667 (1962).
188. C. B. Alcock and M. Peleg, *Trans. Br. Ceram. Soc.* **66**, 217 (1967).
189. J. Drowart, A. Pattoret, and S. Smoes, *Proc. Br. Ceram. Soc.* No. 8, p. 67 (1967).
190. O. P. Charkin and N. E. Dyatkina, *Zh. Strukt. Khim.* **6**, 579 (1965).
191. R. F. Packard, *Trans., Vac. Metall. Conf., 6th 1963* p. 175 (1964).

CHAPTER 5

Late Transition Metals, Metal Halides, Metal Oxides, and Metal Sulfides (Group VIII)

I. Late Transition Metal Atoms (Fe, Co, Ni, Ru, Rh, Pd, Os, Ir, and Pt)

A. Occurrence, Properties, and Techniques

As with the early transition metals, the vaporization of these elements is energetically costly, especially for the second- and third-row transition elements. Therefore, the natural occurrence of these metal vapors is only possible in the high-temperature atmosphere of stars and the sun, and in these cases it is often unclear whether the vapor species are atoms or ions.[1]

The d-orbitals of the late transition metals are approaching the filled configuration, and so the formation of stable low-valent complexes is greatly favored by strong π-acid ligands. Also, these transition elements, in particular Ni, Pd, and Pt, are capable of forming strong σ-bonds with electronegative groups, and so oxidative addition is a preferred type of reaction and has been studied a great deal.

For purposes other than the study of the chemistry of these atoms, these metals have been vaporized by a wide variety of methods including laser,[2-5] e-beam,[6-9] induction heating or levitation,[10-12] and arcs.[13-17] (cf. Table 5-1).[2,3,5-8,10-37] Novel variations of resistive heating vaporization apparatus have also been reported,[22,23] including an important finding that coating normal ZrO_2, MgO, or Al_2O_3 crucibles with BN gives the crucibles more resistance to decomposition, cracking, etc.[24] Platinum, a particularly difficult metal to vaporize because it does not sublime and one which when molten attacks crucible materials, can be vaporized from Pt-electroplated W wires.[37] Field evaporation methods[25] and various reports of evaporation for shadow casting scanning electron microscope samples have also been published.[38] And in 1963, Liebl and Herzog reported an important discovery regarding

TABLE 5-1

Preferred Vaporization Techniques and Vapor Compositions of the Late Transition Metals (Group VIII)

Element	mp (°C)[b]	bp (°C)[b]	Heat of Vap (kcal/mole)[a]	Vaporization methods employed	Vapor comp	References
Fe	1535	2750		Laser, electron beam, induction, arc, resistive, field evap.	Fe	2, 3, 5, 8, 10–14, 16, 17, 21–27
Ru	2310	3900	154.9	Resistive	Ru	21, 28–30
Os	3045	5020	187.4		Os	21, 31
Co	1495	2870		Electron beam, induction, arc, resistive, field evap.	Co	6, 10, 16, 21, 22, 25, 26
Rh	1966	3727	132.5	Arc, resistive	Rh	17, 21, 28, 29, 32, 33
Ir	2410	4130	157.9	Electron beam, arc	Ir	7, 15, 21, 28, 32
Ni	1453	2732	107.7	Laser, electron beam, induction, arc, resistive, field evap.	Ni	2, 3, 5, 8, 11, 16, 17, 21–27, 29, 31, 34, 34a
Pd	1552	3140	91.2	Arc, resistive	Pd	17, 29, 34, 35
Pt	1772	3827	134.9	Electron beam, arc, resistive	Pt	7, 14, 15, 17, 21, 23, 32, 36, 37

[a] See also Figure 1-1 in Chapter 1. Given here are some of the original references.
[b] See references 18–20.

the now extremely useful argon ion sputtering evaporation of a variety of metals, including some Group VIII metals as well as insulators.[39]

For atom chemistry studies, only resistive heating techniques have been reported for the Group VIII metals. For Fe, Co, Ni, and Pd, $W-Al_2O_3$ crucibles are perfectly suited for either small- or large-scale vaporizations. However, for Ru, Rh, Os, Ir, and Pt, resistive heating techniques are very limited in scale because of a lack of satisfactory insulator crucible materials that can survive the temperatures needed to vaporize these metals. Therefore, W wires are generally electrochemically coated, or wires of these metals are directly wrapped around the W filaments. An example is a stranded W coil electrodeposited with Rh and heated directly to vaporize about 0.1 g Rh.[40] Another example is an approximately 3-mm W rod wrapped with about 15 turns of 0.3-mm Pt wire heated directly to vaporize about 0.1 g Pt.[41] The problems with these resistive heating methods are serious in that only small amounts of metal can be vaporized, which is a limitation for macro-scale work. Also, alloying with the hot W occurs and much of the precious Pt or Rh is lost. And lastly, some Pt–W and Rh–W species and small amounts of W are also vaporized at the temperatures employed.

Recently, a simple but useful method for putting powdered metals on W wires was published.[30] This entailed mixing the metal, Ru, with epoxy cement and allowing it to harden on the W wire. Slow heating drove off the carbonaceous residue leaving Ru coated on the W. The Ru could then be

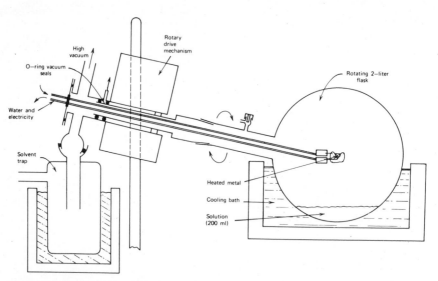

Figure 5-1.   Design of a solution metal atom reactor (after Timms).[42]

vaporized with more vigorous resistive heating. This is useful since only powders of some metals are commercially available.

In this chapter the solution metal atom reactor is introduced. Figure 5-1[42] illustrates the methodology introduced by Timms and co-workers.[43] Note that metal is vaporized in the normal way from resistance-heated $W-Al_2O_3$ crucibles (Fe, Co, Ni, Pd) and that the vapor moves upward into a cold film of solution on the top of the rotating flask. This solution contains the reactant and solvent and is continually renewed because of the rotating action of the flask. This method is useful for a variety of reasons, perhaps the most important being that nonvolatile reactants can be studied. This allows studies of salts and large organic molecules, and could allow facile preparation of heterogeneous catalysts. However, one serious problem with the method is in choosing proper solvent which dissolves an adequate amount of the desired reactant, has a low vapor pressure at a usable temperature (often $-100°C$), and does not react with metal atoms. Nevertheless, this method constitutes a very important addition to the metal atom chemists arsenal of techniques.[44]

B. Chemistry

1. ELECTRON-TRANSFER PROCESSES

In metal atom studies related to "molecular metals,"[45] we have cocondensed nickel atoms with very electron-demanding ligands, such as hexafluorobenzene, C_6F_6, and tetracyanoquinodimethane (TCNQ). In the case of C_6F_6, a green nonvolatile explosive substance was formed that behaves somewhat like a 1:1 Ni–C_6F_6 π-complex.[48] However, its thermal stability indicated that the strong π-acceptor character of C_6F_6 was very important and that this material may be a borderline case between complete electron transfer $(Ni^+C_6F_6^-)$ and simple π-complexation with strong $d \rightarrow \pi^*$ backdonation.

In the case of a still more electron-demanding ligand, TCNQ was codeposited with Ni atoms,[49] and complete electron-transfer yielded the anhydrous version of the known $Ni^{2+}(TCNQ^-)_2$,[50] which was isolated as the hydrated complex by addition of water, followed by filtration and crystallization. The anhydrous $Ni^{2+}(TCNQ^-)_2$ produced an emerald green film on the walls of the reactor as it formed. (A cocondensation apparatus with four electrodes

TCNQ

was employed so that TCNQ could be sublimed as Ni was vaporized.[44]) The presence of $Ni^{2+}(TCNQ^{2-})$ was not detected by IR under anaerobic conditions in CH_3CN.[51]

There remains a great deal of interesting work with the later transition metal atoms with electron-demanding ligands. Borderline cases between complete electron transfer and simple orbital mixing should be quite interesting.

2. ABSTRACTION PROCESSES

It seems that the only low-temperature abstraction process reported for the late transition metal atoms is the abstraction of oxygen from cyclohexene oxide by Ni and Co atoms.[52] Cobalt was found to be about two times as efficient as Ni in this process (equivalents O removed/M atom, Co = 1.2, Ni = 0.6).

Under a strict definition, the reactions of some alkyl halides with Group VIII transition metal atoms could be considered abstraction processes, since the first step often appears to be $RX + M \rightarrow \overline{R^{\cdot} \times M^{\cdot}}$. However, RMX is eventually formed, and so these reactions will be covered under the Oxidation Addition Processes Section of this chapter.

High temperature fast flow reactors (HTFFR) have been employed for the study of Fe atom–O_2 reactions at elevated temperature. (This technique is described in more detail in Chapter 7.) The $Fe—O_2 \rightarrow FeO + O$ reaction at $1330°C$ proceeds with a rate coefficient of 3.6×10^{-3} ml/molecule sec.[46, 47] Obviously, a wide variety of similar studies of the late transition metals with a variety of substrates need to be carried out.

3. OXIDATIVE ADDITION PROCESSES

a.   Saturated Alkyl Halides

$$RX + M \longrightarrow RMX$$

$R = CH_3, C_2H_5, CH_3CH_2CH_2, (CH_3)_3C, (CH_3)_3CCH_2, CF_3, C_2F_5, CF_3CF_2CF_2, (CF_3)_2CF$

There should be obvious interest in these nonsolvated, coordinatively unsaturated RMX species. These and related arylmetal halides have been proposed on numerous occasions as possible reactive intermediates in catalysis or organometallic synthetic schemes.[53-55] The species proposed had never been isolated, trapped, or spectroscopically detected in anyway, however, prior to our metal atom work.

We believed these RMX species could be prepared utilizing metal atom methods at low temperature in the absence of solvents or other complicating

TABLE 5-2
Oxidative Addition of Saturated Alkyl Halides to Ni and Pd Atoms

RMX	Comments	References
CH_3NiI	CH_4 formation during $-196°C$ deposition	57
C_2H_5NiI	CH_4 and/or H_2 formation during $-196°C$ deposition	57
$CF_3NiBr(I)$	$< -80°C$ stability	57
$CH_3PdBr(I)$	$< -100°C \rightarrow PdBr_2, CH_4, C_2H_4, C_2H_6$	56
C_2H_5PdI	$< -100°C \rightarrow PdI_2, C_2H_4, C_2H_6$	56
CF_3PdI	$85°C$ stability, reactive	56, 59
C_2F_5PdI	Similar to CF_3PdI	56, 59
$CF_3CF_2CF_2PdI$	Similar to CF_3PdI	56, 59
$(CF_3)_2CFPdI$	Unstable $< -80°C$	56

ligands. We carried out a survey of RX–M atom reactions, where M usually was Pd.[56,57] Indeed, RPdX and RNiX species were produced. The efficiency of their formation was in the order RI > RBr > RCl ≫ RF. A tremendous range of thermal stabilities for RPdX and RNiX was found. For normal alkyl groups $[R = CH_3, C_2H_5, (CH_3)_3C,$ and $(CH_3)_3CCH_2]$ extremely unstable RPdX and RNiX species formed which decomposed either during the macroscale codeposition or, more often, on matrix warm-up. Evidence for the existence of RMX was sometimes obtained by trapping experiments where triethylphosphine was added at different temperatures as the matrix warmed. Table 5-2 summarizes the RMX species studied.

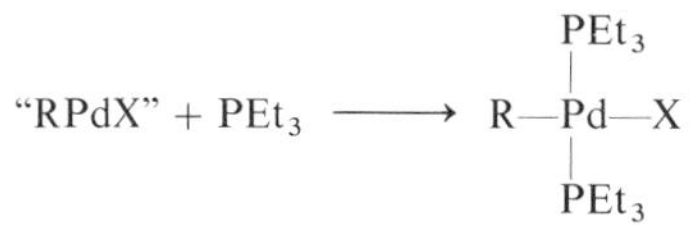

$$\text{"RPdX"} + PEt_3 \longrightarrow \begin{array}{c} PEt_3 \\ | \\ R{-}Pd{-}X \\ | \\ PEt_3 \end{array}$$

stable

To our considerable surprise, when perfluoroalkyl halides were codeposited with Ni or Pd, *much more* thermal stability of the RMX species was encountered. In fact, in the case of Pd these formally coordinatively unsaturated species, CF_3PdI, C_2F_5PdI, and $CF_3CF_2CF_2PdI$,[49,58,59] were isolable as reactive, sensitive red solids. Molecular-weight studies indicate, as expected, that these materials are trimers and tetramers in solution, and that they are probably polymeric in the solid state. However, in solution they exhibit chemistry as if they were truly coordinatively unsaturated. Thus, almost any added ligand is scavenged by R_fPdI to yield $R_fPdI(L)_2$, where L can be PEt_3, $(CH_3)_2S$, $(CH_3)_2NH$, C_5H_5N, $(CH_3)_2C{=}O$, or others.

We have attempted to learn what mechanism these oxidative addition reactions follow. Macroscale cocondensation techniques were employed, and a summary of this mechanistic study is enumerated as follows:[60,61]

1. Pd atoms condensed with CF_3I or C_2F_5I yielded stable, isolable CF_3PdI and C_2F_5PdI, respectively. Some PdI_2 was also formed, but no gases such as CF_4, C_2F_6, or C_4F_{10} were formed from decomposition processes. Since the hydrido-analogs did yield decomposition gases, it may be concluded that the gases came from decomposition of RPdX, not from a prior process *leading to* RPdX.

2. Detailed analyses of gaseous products in combination with HX doping experiments (HX, RX, Pd all condensed together, or HX deposited on top of RX–Pd matrix) showed that the probable decomposition pathway for C_2H_5PdX was HPdX elimination, with the HPdX capable of reducing RX to RH.

$$C_2H_5X + Pd\ atom \longrightarrow [C_2H_5PdX] \longrightarrow C_2H_4 + HPdX$$
$$C_2H_6 + PdX_2 \xleftarrow{\quad\quad}_{C_2H_5X}$$

3. Neopentyl bromide $[(CH_3)_3CCH_2Br]$ with Pd yielded as major products methane and isobutylene, apparently involving a vibrationally excited radical species $[(CH_3)_3C\overset{\cdot}{C}H_2{}^*]$ which split out $\cdot CH_3$, which in turn picked up H$\cdot$ from the matrix.

4. Doping experiments with radical scavengers did not affect product yields or distributions. Therefore, radical chain processes were not operating.

5. The use of bare tungsten crucibles rather than the normal Al_2O_3-coated crucibles had no effect on product yield or distribution with C_2H_5I–Pd. So photolytic energy from the crucibles apparently had no effect.

6. *Tertiary* halides reacted more efficiently than primary halides as shown by competition experiments, indicating an S_N2 process is probably *not involved*.

7. From comparisons with Ag–RX reactions discussed in Chapter 6, (Ag being very similar to Pd in properties), it is likely that upon cocondensations a R–X—Pd complex forms which converts to RPdX on warming *via* a close-radical pair mechanism, with the R$\cdot$ instantaneously possessing excess vibrational energy. A proposed general mechanistic scheme for *normal alkyl halide*–Pd atom reactions involving caged radicals was proposed, and RPdX was the crucial species formed in each case.[60,61]

Matrix-isolation spectroscopic studies would be very helpful in further elucidation of these oxidative addition processes. Nothing has been done yet, although such studies are being initiated in our laboratory.

b. Aryl and Vinyl Halides. Carrying mechanistic considerations further, it is currently believed that prior π-complexation with the metal takes place with aryl and vinyl halides before oxidative addition.[58,60,62] At $-196°C$, the π-complex readily forms with very low Ea, and not until warming to $> -100°C$ does the higher Ea oxidative addition take place. In a qualitative sense, visual observation of the matrices on warming supports this concept, as drastic color changes occur on warming. The matrices look very similar

$$\text{C}_6\text{H}_5\text{—Br} + \text{M} \xrightarrow{-196°C} \underset{\text{Br}}{\text{C}_6\text{H}_5}\text{—M} \xrightarrow{> -100°C} \text{C}_6\text{H}_5\text{—M—Br}$$

at $-196°$ whether a C—X bond is present or not in the arene. Upon warming, if a C—X bond is present dark red ArMX is usually formed, whereas if no C—X bond is present (e.g., toluene), black M—M particles, due to cluster formation (cf. Section I,B.6), form on warming.

In a more quantitative sense, matrix-isolation spectroscopy has shown that Fe, Co, and Ni–arene π-complexes do readily form at $10°–77°K$.[63]

Vinyl halide–metal atom complexes have also been examined spectroscopically, and π-complexation was found,[64] without evidence of oxidative addition having occurred in the temperature range of about $8°–50°K$. Using macroscale methods, we do know that oxidative addition does occur, but with $CF_2=CFBr$, for example, not until about $-100°C$ or higher.

We have studied macroscale oxidative additions of C_6H_5Br, C_6F_5Br, and $CF_2=CFBr$ with Ni and Pd. In each case initially an unstable π-complex forms, and upon warming ArMBr is efficiently produced. Table 5-3 summarizes the compounds prepared.[56–58,65–68] With C_6H_5Br and Pd, a thermally unstable C_6H_5PdBr species was formed. This compound was trapped at $-100°C$ with Et_3P to form isolable $C_6H_5PdBr(PEt_3)_2$. Above about $-100°C$ only $PdBr_2$ was trappable with PEt_3, and biphenyl was found. The following sequence is possible:

$$\text{Pd} + \text{C}_6\text{H}_5\text{Br} \xrightarrow{-196°C} \underset{\text{Br}}{\text{C}_6\text{H}_5}\text{—Pd} \xrightarrow[\text{warming}]{\text{slight}} \text{C}_6\text{H}_5\text{PdBr}$$

with $-100°C$ PEt_3 branch to $\underset{PEt_3}{\overset{PEt_3}{C_6H_5PdBr}}$ and $> -100°C$ PEt_3 branch to $C_6H_5\text{—}C_6H_5 + PdBr_2 + Pd$

In contrast to C_6H_5Br, C_6F_5Br and Pd yielded a stable, albeit reactive, organometallic compound C_6F_5PdBr. This material is similar to CF_3PdI

TABLE 5-3

Oxidative Addition of Aryl and Vinyl Halides to Co, Ni, Pd, and Pt Atoms

Metal	Halide	Species isolated or trapped	Comments	References
Pd	$C_6H_5Br(Cl)$	C_6H_5PdBr	Trapped at $-100°C$ with PEt_3	56
Pd	$C_6F_5Br(Cl, I)$	C_6F_5PdBr	Stable to $130°C$, reaction with many ligands to give $C_6F_5PdBr(L)_2$	56–58, 65
Pd	CF_2=$CFBr(Cl, I)$	CF_2=$CFPdBr$	Stable at $<25°C$	66
Pt	C_6F_5Br	C_6F_5PtBr	Reacts with PEt_3 to give *cis*- and *trans*- $C_6F_5PtBr(L)_2$; stable to $72°C$	58
Ni	$C_6F_5Cl(Br)$	C_6F_5NiCl	$-80°C$ stability, can be trapped	57
	C_6F_5Br	$(C_6F_5)_2Ni$–toluene (mesitylene)	Isolated from toluene solution, stable to $140°C$	67
Co	C_6F_5Br	$(C_6F_5)_2Co$–toluene (mesitylene)	Isolated from toluene solution, stable to $135°C$	68

in that it is telomerized in solution, and probably polymerized in the solid state. This coordination probably takes place through Br bridging and possibly π-arene bridging[65]:

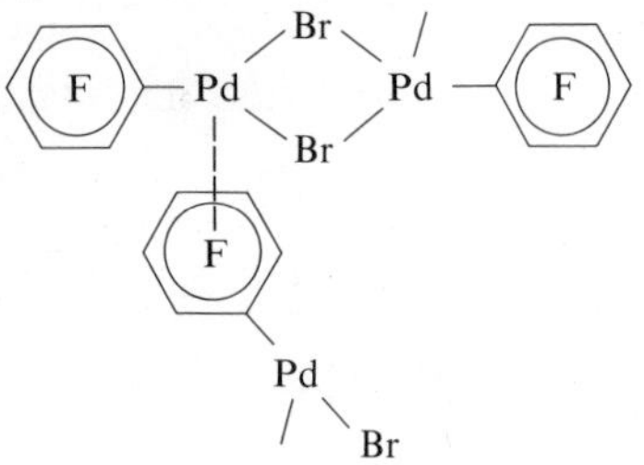

In acetone solution C_6F_5PdBr is quite stable, as acetone acts as a weak stabilizing ligand. In benzene or toluene, however, C_6F_5PdBr is slowly decomposed.[56] However, addition of almost any stabilizing ligand causes the breakup of the telomer and quantitative formation of $C_6F_5PdBr(L)_2$. If bidendate ligands are employed cis-complexes are obtained, whereas monodentate ligands always yield trans complexes.

Similarly, CF_2=$CFPdBr$ was found to be a stable species, which is probably also telomeric (no MW studies have been done in this case).

Macroscale codeposition of C_6F_5Br with Pt also yields a stable species C_6F_5PtBr. Not much work has been done with this compound other than

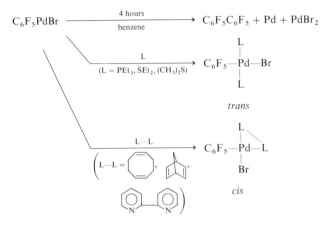

addition of PEt_3, which yielded both the *cis*- and *trans*-$C_6F_5PtBr(PEt_3)_2$ in contrast to the Pd case.

Nickel with C_6F_5Br is an intriguing system. In the approximate range of $-150°C$ to $-100°C$, oxidative addition to yield $C_6F_5NiBr(Cl)$ takes place. However, in the absence of solvent or other stabilizing ligand, this molecule decomposes to $C_6F_5C_6F_5$, Ni, and $NiBr_2$ at $\sim -80°C$. Trapping experiments at $-80°C$ or lower with PEt_3 were successful.[57] However, in the presence of toluene or other arene, C_6F_5NiBr disproportionates and binds an arene molecule as shown below.[67]

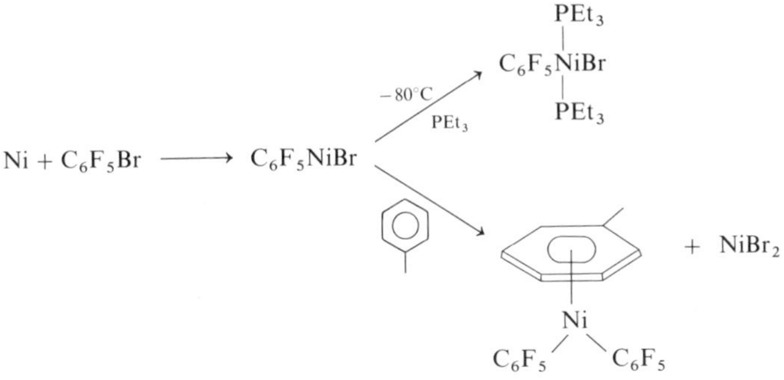

$$Ni + C_6F_5Br \longrightarrow C_6F_5NiBr$$

The η^6-$C_6H_5CH_3Ni(C_6F_5)_2$ molecule can be obtained in very high yield and 3–4 g can be easily prepared in half a day. It is an unusual compound in that it is the first example of a π-arene complex with Ni(II). The π-arene ligand is extremely labile, able to exchange with other arenes at room temperature, and readily displaced by stronger ligands such as cyclooctadiene, norbornadiene, butadiene, PEt_3, and even THF. The toluene complex is an active catalyst for the polymerization of norbornadiene, trimerization of butadiene, and hydrogenation of toluene.[67,69] Some of this fascinating chemistry is outlined below:

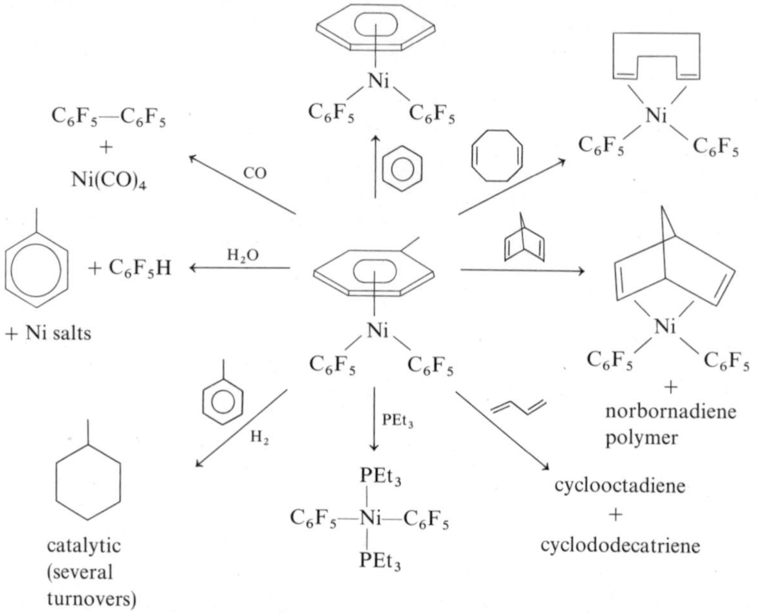

C_6F_5—C_6F_5

+

$Ni(CO)_4$ $\xleftarrow{\text{CO}}$

(central complex) $Ni(C_6F_5)(C_6F_5)$ with η^6-arene

(benzene) + C_6F_5H $\xleftarrow{\text{H}_2O}$

+ Ni salts

(cyclooctadiene) $\longrightarrow$ $Ni(C_6F_5)(C_6F_5)$

(toluene) Ni C_6F_5 C_6F_5

(norbornadiene) $\longrightarrow$ $Ni(C_6F_5)(C_6F_5)$

+

norbornadiene polymer

(toluene) $\xrightarrow{\text{H}_2}$

catalytic (several turnovers)

$\downarrow$ PEt_3

C_6F_5—Ni—C_6F_5 with PEt_3 above and below

$\xrightarrow{}$ (propene) cyclooctadiene

+

cyclododecatriene

Similar results were obtained in the analogous Co–C_6F_5Br system. Again, in the presence of toluene, efficient disproportionation took place to yield $(\eta^6$-$C_6H_5CH_3)Co(C_6F_5)_2$ + $CoBr_2$.[68] In this case, however, we have not yet been successful at trapping the presumed intermediate C_6F_5CoBr. Apparently, it is very short-lived, even in a pure C_6F_5Br matrix.

Single crystal X-ray studies show that the π-arene Co and Ni compounds are isostructural. According to bond lengths, the Co atom binds the π-arene more strongly than Ni, but Ni binds the σ-C_6F_5 groups more strongly than Co. The chemistry of the Co system seems to reflect this stronger π-arene bond in that it is slightly more difficult to displace the arene.[70] The Co compound is a 17-electron paramagnetic system showing a broad featureless EPR spectrum at room temperature, but Co 7/2 fine structure at low temperature, implying a low spin complex,[71] which is also supported by a magnetic susceptibility measurement of BM = 1.7.[65]

These are unique π-arene systems in that they are the first examples of such complexes for Co(II) and Ni(II). The scope of the π-arene ligands that can be tolerated is quite large (mesitylene, toluene, benzene, anisole, fluorobenzene, etc.). However, to date we have found that the σ-bonded ligand must be C_6F_5. No other σ-bonding group we have tested has yielded a stable complex.[67] Thus, there is something unique in the bonding of C_6F_5. We are currently involved in X-ray[72] and spectroscopic studies, and theoretical calculations[73] so that we might gain a better understanding of these and related C_6F_5M and CF_3M complexes.

c. Allyl and Benzyl Halides.

A useful high-yield synthesis of $(\eta^3\text{-}CH_2CHCH_2NiCl)_2$ is achieved through the codeposition of Ni atoms with allyl chloride.[74] This compound can be prepared by other means, but the metal atom method is probably the most convenient, if a metal atom reactor is available. Palladium[58] and platinum[75] behave similarly, and high yields of the analogous compounds are obtained. However, under some conditions even bulk Pd will react to yield the complex,[76] and thus Pd vaporization is not really necessary. In the case of Pt,[75] only small amounts of the complex were prepared since the amount of Pt vaporized was necessarily small (resistive heating of Pt coated W wire). All of these complexes were previously known before these studies.

Somewhat related to the allyl dimer synthesis is one of the hitherto unknown η^3-benzylpalladium chloride derivative.[77,78] It is the only new non-florinated RPdX that we found isolable and readily characterized. This compound exhibits η^3-bonding to the benzyl group and is dimeric. Spectroscopic studies show that this compound is very fluxional at room temperature and even at $-90°C$ (facile movement so that bonding between two allylic positions, 2-1-7 and 6-1-7 carbons, are equivalently populated.[78] However, a 3,4-dimethyl substituted analog did show a temperature-dependent

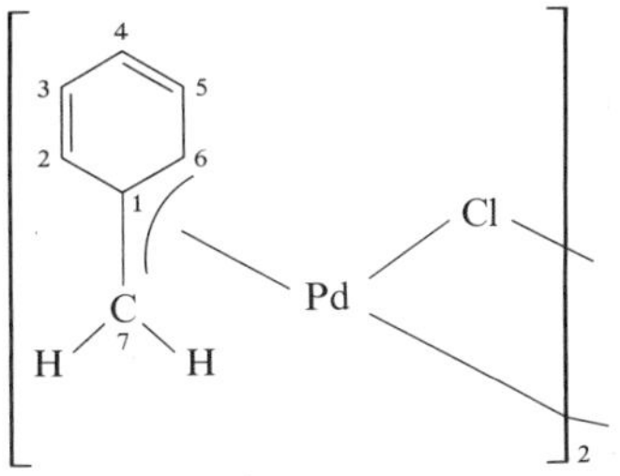

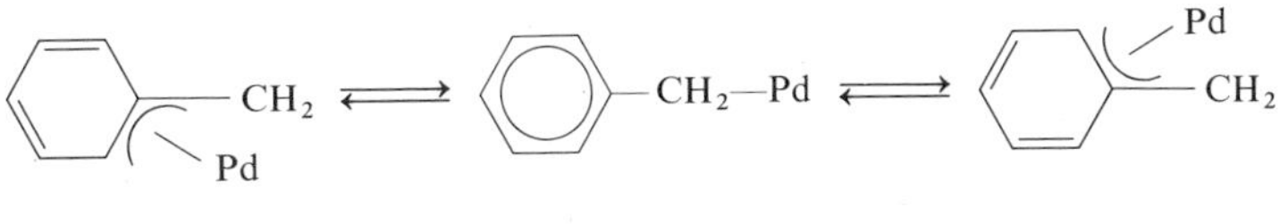

NMR spectrum between $+40°C$ and $-61°C$. A preferred conformation was exhibited at about $-60°C$ where the 6-1-7 allylic position was favored. Careful analysis of the spectrum indicated that the mechanism for the fluxionality (equilibrium between 2-1-7 and 6-1-7 bonding positions) was due to a rapid $\pi \to \sigma \to \pi$ rearrangement.[78] Addition of ligands such as PEt_3 or C_5H_5N caused the destruction of the π-benzyl bond followed by cleavage of the Pd$\diagdown$ $\overset{Cl}{\underset{Cl}{\diagup \diagdown}}$ Pd bridge, and after four equivalents had been added, *trans*

σ-$C_6H_5CH_2PdCl(L)_2$ was formed quantitatively.

In terms of metal atom chemistry, it is very interesting that this type of η^3-benzyl bonding is possible. The RPdX species formed initially is co-ordinatively unsaturated, and in this case in order to partially satisfy open coordination sites, drastic distortion of the aromatic systems occurs. This allows greatly enhanced stability over similar RPdX species such as C_6H_5PdBr and CH_3PdI and demonstrates the unique possibilities in metal atom chemistry for generating unusually reactive organometallics.

Again, it is believed the π-complexation precedes oxidative addition for unsaturated organic halides and metal atoms.[60] In the case of benzyl chloride and Pd, a red matrix initially forms. The presence of this red matrix does not necessarily indicate product formation, as it has been shown that the formation of $(C_6H_5CH_2PdCl)_2$ is very sensitive to the presence of pyrolysis products formed when certain types of metal vaporization sources are used.[79] The red complex can transform to metal particles or product, and it is believed the initial red material is a π-arene–Pd complex.

Studies of 5-bromo-1-hexene with Pd also showed that a π-complex formed initially (clear matrix) which, on warming, yielded only Pd particles and isomerized bromohexene.[60]

Recent matrix-isolation spectroscopy studies support the π-complex mechanism.[64] Allyl chloride and Ni yielded a π-complex upon cocondensation at 12°K. Warming of the matrix showed no sign of oxidative addition occurring until the onset of metal cluster formation. Are metal clusters needed in order for oxidative addition to occur?[64] Further work is certainly needed in this area.

d. Acyl Halides.   Acyl chlorides oxidatively add to palladium atoms and nickel atoms.[56,57] In all examples studied the resultant RCOMCl species very readily liberated CO. By carrying out low-temperature trapping experi-

$$\overset{O}{\overset{\|}{RC}}-Cl + M \text{ atom} \longrightarrow \overset{O}{\overset{\|}{RC}}-M-Cl$$

$$M = Ni, Pd$$
$$R = CH_3, CF_3, C_6H_5, C_6F_5, n\text{-}C_3F_7$$

ments with Et_3P we were able to draw some conclusions about thermal stabilities and decomposition pathways. Table 5-4 summarizes our findings. The thermal stabilities of the RCOMCl species are remarkably dependent on R, with the most stable being R = n-C_3F_7 > CF_3 > C_6F_5 > C_6H_5 >

CH_3. The decarbonylation process leads directly to RMCl, and the reaction probably takes places as follows:

$$RCMCl \longrightarrow R-M \begin{matrix} CO \\ \\ Cl \end{matrix} \xrightarrow{-CO} RMCl \begin{matrix} \xrightarrow{PEt_3} & RMCl & PEt_3 \\ & & | \\ & & PEt_3 \\ \xrightarrow{warm} & RCl + R-R + MCl_2 + M \end{matrix}$$

$R = C_6H_5$

In cases where PEt_3 was added at a temperature above the decomposition temperature of RCOMCl, RMCl was trapped. However, if RMCl was also thermally unstable, the expected decomposition products were formed.

TABLE 5-4

Oxidative Addition of Acyl Halides to Ni and Pd Atoms

Reactants	Comments
Pd, CF_3COCl	$CF_3COPdCl$ eliminates CO at $-80°$ at which point Et_3P addition yielded almost equal quantities of $(Et_3P)_2PdCl(CF_3)$, $(Et_3P)_2PdCl(COCF_3)$, and $(Et_3P)_2PdCl_2$
Pd, $CF_3CF_2CF_2COCl$	$CF_3CF_2CF_2COPdCl$ stable above $-80°C$ (but $<0°C$) where it was efficiently trapped to yield $(Et_3P)_2PdCl(CO-n-C_3F_7)$
Pd, C_6F_5COCl	$C_6F_5COPdCl$ lost CO at $\sim -50°C$ to efficiently yield C_6F_5PdCl which could be trapped with PEt_3 at $40°C$
Pd, C_6H_5COCl	$C_6H_5C_6H_5$, C_6H_5Cl and $(Et_3P)_2PdCl_2$ found after trapping at $40°C$
Pd, CH_3COCl	CO loss at $< -100°C$, only $(Et_3P)_2PdCl_2$ found after trapping at $0°C$ or $-78°C$
Ni, CF_3COCl	Et_3P trapping at $-80°C$ yielded small amount of $(Et_3P)_2NiCl(CF_3)$ and mostly $(Et_3P)_2NiCl_2$ but no $(Et_3P)_2NiCl(COCF_3)$
Ni, CH_3COCl	CO loss at $< -100°C$, and Et_3P trapping at $-80°C$ yielded only $(Et_3P)_2NiCl_2$

e. Acid Anhydrides. Hexafluoroacetic anhydride codeposited with Pd atoms yielded a complex that slowly deposited a Pd metal mirror while standing at room temperature in acetone.[80] We believe that this complex is zero valent in Pd, but its structure has not yet been determined. Addition of Et_3P to a fresh acetone solution of the complex yielded cis-bis(triethylphosphine)perfluorodiacetatopalladium(II)$[(Et_3P)_2Pd(OCOCF_3)_2]$. Thus, two molecules of the anhydride must be bonded in the complex. The preference for the cis-compound is quite interesting. Formally, this is an example of C–O oxidative addition to a palladium atom.[58, 80]

Efner and Fox have extended these studies and have found a number of stable perfluoroanhydride–metal complexes, the structures of which are not yet elucidated.[81]

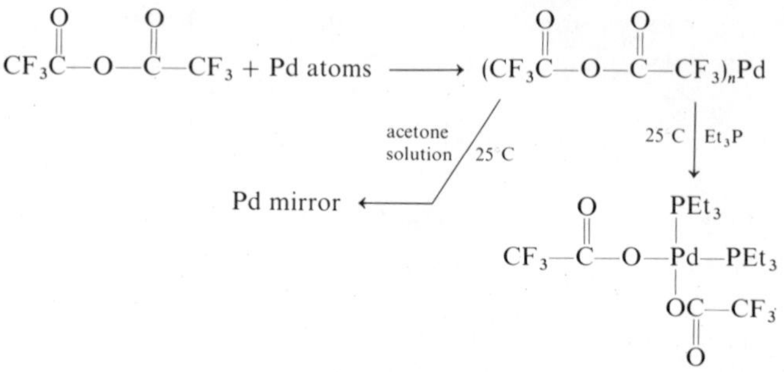

$$CF_3\overset{\overset{\displaystyle O}{\|}}{C}-O-\overset{\overset{\displaystyle O}{\|}}{C}-CF_3 + \text{Pd atoms} \longrightarrow (CF_3\overset{\overset{\displaystyle O}{\|}}{C}-O-\overset{\overset{\displaystyle O}{\|}}{C}-CF_3)_n Pd$$

acetone solution / 25 C

25 C | Et$_3$P

Pd mirror ←

$$CF_3-\overset{\overset{\displaystyle O}{\|}}{C}-O-\underset{\underset{\displaystyle O}{OC-CF_3}}{\overset{\overset{\displaystyle PEt_3}{|}}{Pd}}-PEt_3$$

cis

Acetic anhydride codeposited with Pd atoms followed by Et$_3$P addition yielded only PdCl$_2$(PEt$_3$)$_2$. No RCOPdCl was trappable even at $-78°$C.[82] Similar results were found with Ni.

Oxidative additions with the Group VIII transition metals still need a great deal of study. Functional groups other than C–X need attention. Also, matrix isolation spectroscopic studies are badly needed to gain mechanistic information.

4. LIGAND-TRANSFER PROCESSES

Very closely related to oxidative addition are ligand-transfer processes. The metal atom is oxidized in the process, while another metal is reduced. One interesting example has been reported by Timms,[83] in which tetra-allyltin codeposited with Ni atoms yield bis(π-allyl)nickel.

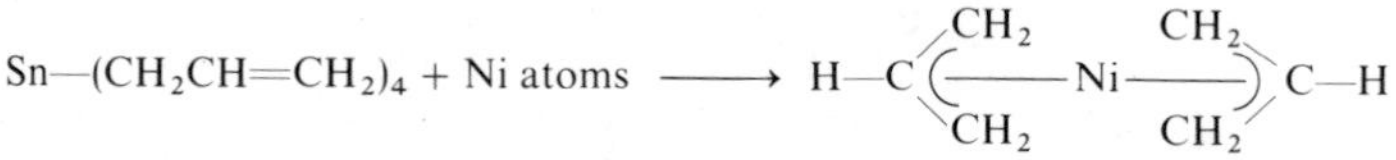

$$Sn-(CH_2CH=CH_2)_4 + \text{Ni atoms} \longrightarrow H-C\underset{\diagdown CH_2}{\overset{\diagup CH_2}{\Big(}}\!-Ni-\underset{CH_2}{\overset{CH_2}{\Big)}}\!C-H$$

This type of reaction appears to have great potential for production of many M(R)$_n$ systems, and warrants further investigation.

5. SIMPLE ORBITAL MIXING PROCESSES

Simple complexation to donor–acceptor ligands has been studied exten-sively with the Group VIII transition metal atoms. Since these metals are somewhat rich in d-electrons, which can be donated to the ligand, π-acid ligands generally yield the most stable complexes. However, essentially all molecules possessing π- or nonbonding electrons form discrete complexes at low temperature with transition metal atoms.

a. Alkene Reactions.   Ozin and his co-workers[84–86] have carried out matrix isolation investigations of metal atom–ethylene complexes (cf. Table V-5).[84,87–89] Based on spectroscopic changes during annealing, isotope labeling experiments, and on dilution studies, these workers concluded that they have observed mono-, bis-, and tris-ethylene complexes for several metals. In the $Ni-C_2H_4$ system, it was particularly interesting that the bis-complex was favored in the region $40°-60°K$ (tris-complex is most stable normally, decomposing at $0°C$). This peculiar behavior was rationalized in terms of matrix packing requirements.

$$C_2H_4/Ar + M \text{ atom} \longrightarrow C_2H_4-M \longrightarrow (C_2H_4)_2M \longrightarrow (C_2H_4)_3M$$

Comparison of ν_{C-O} stretching frequencies (Table 5-5) shows the order tris > mono > bis for nickel, while tris > bis > mono for copper.[84,85] This unexpected order for the nickel complexes was rationalized in terms of $C_2H_4-C_2H_4$ interaction in the bis-complex. In comparing tris-complexes for Co and Ni, $\nu_{C=C}$ was in the order Ni > Co (16- and 15-electron systems respectively), which is consistent with the $M(d_\pi \rightarrow C_2H_4-\pi^*)$ back bonding picture (less back bonding for Ni < Co). Also, the close similarities in the IR spectra of these tris-ethylene complexes indicate that the compounds are isostructural, probably with planar D_{3h} geometry, which was recently predicted, theoretically, by Rosch and Hoffman.[90]

Further studies by Ozin and Power with a wide variety of alkenes indicated that essentially all alkenes, with the exception of C_2F_4, form simple Ni–(alkene)$_n$ (where $n = 1, 2, 3$) π-complexes.[64] However, the insensitivity of the UV absorption to the type of substituent on the alkene suggested that backbonding to π^*-orbitals of the alkene may in fact not be very important.

Timms[89,91] and Ozin[84] and their co-workers have reported that tris-(ethylene)nickel(0) can be synthesized by cocondensing Ni atoms with C_2H_4 on a macroscale. This interesting compound was first isolated and characterized by Fischer, Jonas, and Wilke[92] from bis(cyclooctadiene)nickel(0) and ethylene. Atkins, MacKenzie, Timms, and Turney[89] have extended the work to $Pd-C_2H_4$ matrices and other M–olefin complexes, and were able to isolate tris(bicyclo[2.2.1]heptene)palladium on a macroscale. The norbornene ligand is unique in allowing the formation of relatively stable

tris(bicyclo[2.2.1]heptenepalladium(0)

(olefin)$_3$M complexes. This apparently results from the slight strain energy of the $C=C$ bond which in turn allows stronger $\pi-M$ interaction.

Styrene and Ni depositions have yielded tris(styrene)nickel(0).[93] This complex is only stable to about $-20°C$.

TABLE 5-5

Ethylene Metal Atom Complexes of the Group VII Metals as Studied by Matrix Isolation Spectroscopic Methods

Complex	Prep method	$v_{C=C}$ (cm^{-1})	Comments	Thermal stability	References
$(C_2H_4)_3Ni$	Matrix isolation in argon/15°K	1512	Least π–Ni interaction, (D_{3h} planar)	0°C	84
$(C_2H_4)_2Ni$	Matrix isolation in argon/15°K	1465	Most π–Ni interaction, favored complex at 20°–60°K	20°–60°K	84
$(C_2H_4)Ni$	Matrix isolation in argon/15°K	1496	Intermediate π–Ni interaction (out of order)	25°K	84
$(C_2H_4)_3Co$	Matrix isolation in argon/10°K	1499	π–Co interaction $> \pi$–Ni $>$ π–Cu (D_{3h} planar)		84
$(C_2H_4)_3Pd$	Codeposit at 77°K	1513, 1522	Less stable than Ni complex	−70°C	87–89

Polystyrene incorporating small Ni particles can be prepared by depositing styrene and Ni atoms in the presence of Kaowool crucible insulation, which apparently encourages pyrolysis processes that in turn cause the presence of free radicals in the matrices that catalyze polymerization[79] (larger reaction flasks can minimize this process). This stryene–Ni polymer is soluble in organic solvents and usable as a homogeneous catalyst.

A series of macroscale cocondensation experiments were carried out where propene was cocodensed at $-196°C$ with metals Co, Ni, Pd, Pt, Al, Dy, Er, and Zr.[94] These were initial experiments intended to demonstrate differences and similarities between the metals. Organometallic products were not isolated and characterized, but rather were decomposed with D_2O to mark C—M bonds. It was believed that π-type M–propene bonds would be destroyed by D_2O to simply release unlabeled propene while σ-type M–propene bonds would yield deutereopropane. For Co, Ni, Pd, and Pt, very little deuterium incorporation was found, indicating mainly the formation of π-type bonds with these metals rather than σ-type. 1-Butene was catalytically isomerized to *cis*- and *trans*-2-butene.[95,96] It was proposed that Bonneman-type[97] π-allylnickel hydrides were short-lived intermediates. Deuterium scrambling occurred when C_3H_6–C_3D_6 mixtures were codeposited with Ni. Also, when 2-deuteropropene and 3-deuteropropene were studied, scrambling occurred only in the latter case. These experiments support the idea that –Ni–H species are involved, and the inter- as well as intramolecular hydride exchange processes take place.

$$CH_3CH_2CH{=}CH_2 \underset{\underset{Ni}{|}}{} \rightleftharpoons CH_3CH\underset{\underset{Ni}{|}}{\overset{CH}{\diagup\backslash}}CH_2 \rightleftharpoons CH_3CH{=}CHCH_3$$

The only direct evidence for propene forming an allyl bond with a Group VIII metal atom was Timms' experiment wherein propene, PF_3, and Co atoms were codeposited.[83] The missing hydrogen was apparently taken up in biproduct $HCo(PF_3)_4$.

$$Co + CH_3CH{=}CH_2 + PF_3 \longrightarrow H{-}C\underset{CH_2}{\overset{CH_2}{\diagup\backslash}}Co(PF_3)_3$$

Perfluoroalkenes with Ni, Pd, and Pt atoms form metalocyclopropanes instead of π-bonded complexes.[58,80] Palladium with octafluoro-2-butene yielded an (alkene)$_3$Pd complex stable to $-30°C$. At this temperature the complex decomposed cleanly back to free alkene and bulk palladium, However, at $-78°C$ after vacuum removal of excess alkene, ligands could be added to efficiently produce ligand-stabilized metallocyclopropanes.[58]

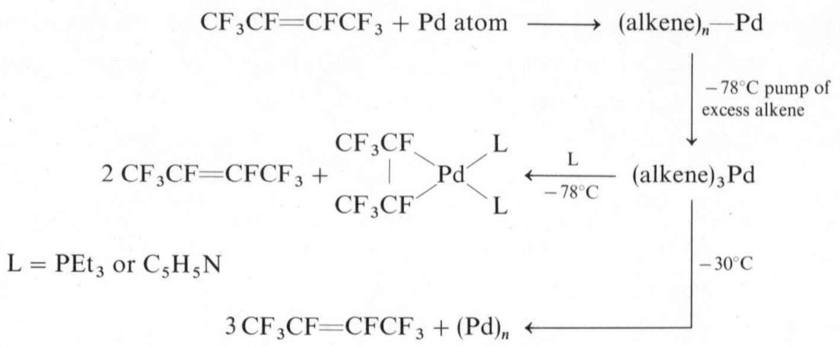

$$CF_3CF{=}CFCF_3 + Pd \text{ atom} \longrightarrow (\text{alkene})_n{-}Pd$$

$$2\,CF_3CF{=}CFCF_3 + \underset{CF_3CF}{\overset{CF_3CF}{\underset{|}{\|}}}\!\!\Big\rangle Pd \overset{L}{\underset{L}{\Big\langle}} \xleftarrow[-78°C]{L} (\text{alkene})_3Pd$$

$-78°C$ pump of excess alkene

$L = PEt_3 \text{ or } C_5H_5N$

$-30°C$

$$3\,CF_3CF{=}CFCF_3 + (Pd)_n \longleftarrow$$

Matrix-isolation studies of tetrafluoroethylene–Ni atom matrices have been carried out by Ozin and co-workers.[98] In this case, a combination of π-complexation and metallocyclopropane formation seemed most likely from the observed spectra. A possible structure for $(C_2F_4)_3Ni$ is shown below, and if this structure is correct it represents one of the rare cases of like ligands bonding differently to a metal atom.[98]

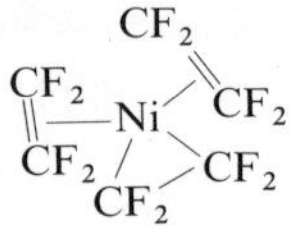

$$\begin{array}{c} CF_2 \\ \overset{CF_2}{\underset{CF_2}{\|}}{-}Ni\overset{CF_2}{\underset{CF_2}{\diagdown}} \\ CF_2 \end{array}$$

b. Diene Reactions.   Cyclopentadiene reacts with metal atoms in a combination of oxidative addition and simple orbital mixing processes. In 1969, Timms showed that Fe atoms with cyclopentadiene yielded ferrocene plus hydrogen gas. In the cases of Co and Ni, hydrogen transfers took place yielding *tetrahapto-* and *trihapto*-complexes.[99,100] It is evident that intermediates in such reactions are very reactive M—H species, and that rearrangements and hydrogen transfers occur readily eventually to form the most stable systems.

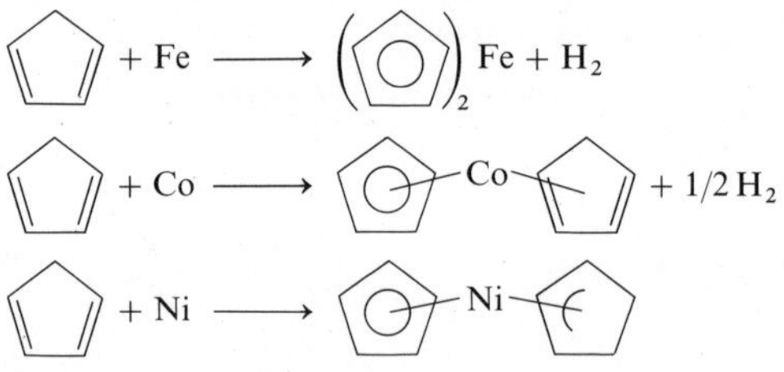

$$\text{[cyclopentadiene]} + Fe \longrightarrow \text{[ferrocene]}_2 Fe + H_2$$

$$\text{[cyclopentadiene]} + Co \longrightarrow \text{[Co complex]} + 1/2\,H_2$$

$$\text{[cyclopentadiene]} + Ni \longrightarrow \text{[Ni complex]}$$

1,5-Cyclooctadiene is a unique organometallic bidentate ligand for Group VIII metals. Classical synthetic methods have yielded the bis-(cyclooctadiene)Ni[101] and more recently the analogous Pd and Pt compounds.[102] Bogdanovic, Wilke, and co-workers[103] have demonstrated the unusual reactivity of $(COD)_2Ni$, and because of this reactivity have termed this complex a source of "naked nickel" (to mean a very weakly ligand-stabilized system of high reactivity). An example of the high reactivity is shown by reaction with C_2H_4 to yield $(C_2H_4)_3Ni$ and two COD molecules.[92,104]

Of course, naked nickel is a term better applied to Ni atoms; and, as would be expected, codeposition of Ni with COD yields $(COD)_2Ni$. Likewise, $(COD)_2Pd$ and $(COD)_2Pt$ can be prepared in this manner.[89,105–107]

The chemistry of Ni atoms and $(COD)_2Ni$ is often very similar.[44] For example, C_2H_4 with either Ni atoms or $(COD)_2Ni$ yields $(C_2H_4)_3Ni$, as discussed previously.[84,88,89,91,92,104,108] Also, CO or PR_3 yield $Ni(CO)_4$ and $Ni(PR_3)_4$, respectively, with either Ni atoms or $(COD)_2Ni$.[83,92,104,109] However, differences are found when more selective ligands such as $CF_3C\equiv CCF_3$,[92,104,110] or C_5H_6,[92,99,100,104] are employed.

$$Ni + CF_3C\equiv CCF_3 \longrightarrow (CF_3C\equiv CCF_3)_n\!-\!Ni$$

$$(COD)_2Ni + CF_3C\equiv CCF_3 \longrightarrow [(CF_3)_6C_6]Ni(COD)$$

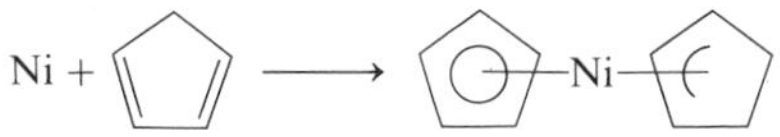

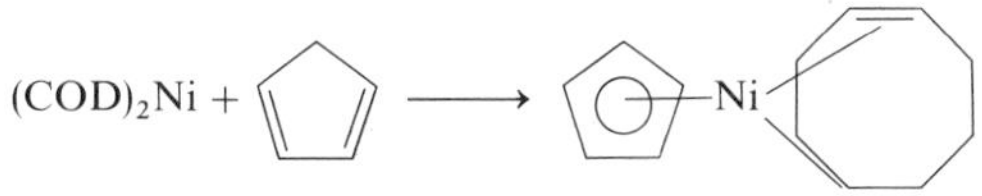

$$\sigma + \pi$$

Further comparisons of $(COD)_2Ni$ and Ni atoms, as well as $(COD)_2Pd$ and Pt, with Pd and Pt atoms would be very useful.

In the first published solution metal atom reaction, (cf. p. 84), Mackenzie and Timms[43] synthesized the previously unknown unique complex $(COD)_2Fe$. The preparation was carried out by evaporating Fe into a cold 10% solution of COD in methylcyclohexane in a rotating reactor. This method gave a reasonably high yield of the complex, although normal codeposition methods will also yield $(COD)_2Fe$. The compound $(COD)_2Fe$ is stable to ca-30°C in hexane, decomposing to yield an Fe mirror above that temperature. It is stable to 0°C in COD solution. The compound appears to be paramagnetic, and is extremely reactive with phosphines, cyclooctatetraene, cycloheptatriene, and butadiene, all of which displace COD partially or completely. For example, treatment with PF_3 yielded $(COD)Fe(PF_3)_3$, and with cyclooctatetraene (COT), $(COT)_2Fe$ was formed.[43] Diphos-$[(C_6H_5)_2PCH_2CH_2P(C_6H_5)_2]$ with $(COD)_2Fe$ in the presence of N_2

yielded (diphos)$_2$FeN$_2$. Further chemistry is outlined below (note the isomerization of 1,5-COD to 1,3-COD in some cases)[111,112]:

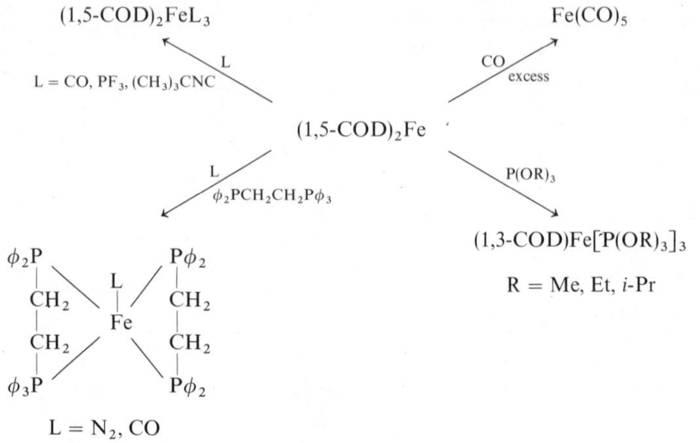

The complex (COD)$_2$Fe is formally coordinatively unsaturated (16-electron system) which explains its high reactivity with many ligands and which is similar in that sense to RPdX compounds previously discussed.

Smaller cyclic dienes, when reacted with metal atoms, do not yield stable complexes. Norbornadiene and Ni atoms yielded an exo-dimer of norbornadiene.[96] 1,3-Cyclohexadiene catalytically disproportionated to benzene and cyclohexene in the presence of Fe, Co, or Ni atoms.[95] It is not clear whether atoms or particles of these metals are the effective catalysts for these hydrogen-transfer disproportionations.

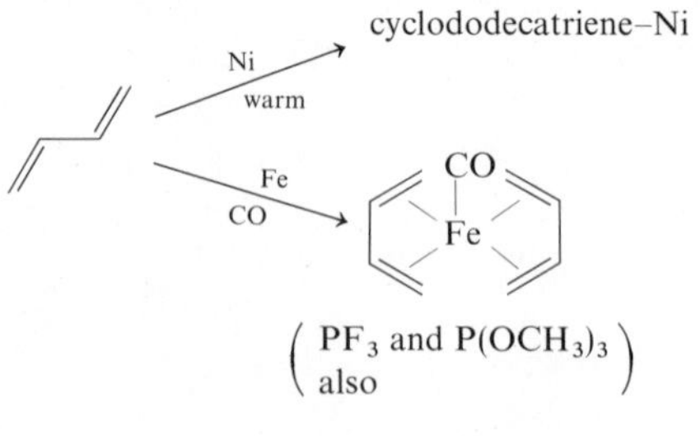

1,3-Butadiene, codeposited with Fe or Ni, yielded unstable diene–M complexes. Upon matrix warm-up of the 1,3-butadiene–Ni system, a diene–Ni polymer formed that, in the presence of excess diene, finally yielded 2,6,10-dodecatrienylnickel,[95] again showing that Ni atom reactions can mimic those of (COD)$_2$Ni[103] A small amount of a volatile organonickel compound was also formed, originally believed to be bis(1,3-butadiene)-Ni(O),[96] but which is actually a bis(crotyl)Ni derivative.[98] In the Fe case, bis(1,3-butadiene)Fe–L derivatives can be formed if L (PF_3 or CO) is added at low temperature to the diene–Fe matrix.[113]

M. L. H. Green and co-workers have briefly investigated the capacity of Fe, Co, and Ni atoms to oligomerize 1,3-butadiene.[114,115] Benzene and diluent were also simultaneously deposited, and presumably arene—M complexes are intermediates. Fe formed mainly diene trimers, whereas Co and Ni yielded dimers. Addition of Et_2AlCl to the matrix mixture caused the formation of diene polymer as the main product with Fe and Ni. However, for Co, Et_2AlCl appeared to have no effect. Several other catalytic systems were investigated as well: M atom–Al atom, M atom–Al atom–EtCl, M atom–EtCl, and M atom–$P(C_6H_5)_3$.[114,115]

1,3-Butadiene in the presence of isobutane (a source of readily abstractable hydrogen) and Co atoms yielded a $(diene)_2$–CoH complex.[95] Nonconjugated acyclic dienes are rapidly isomerized to conjugated systems, and similar $(diene)_2$Co–H systems are produced:

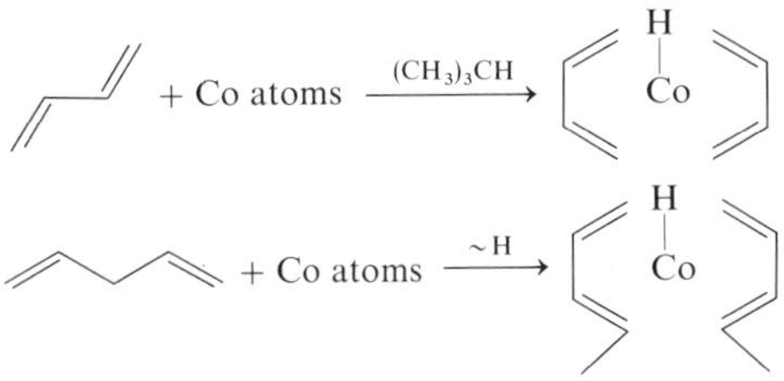

c. Triene Reactions. Cycloheptatriene codeposited with Fe atoms yielded the η^7-η^3-Fe sandwich complex shown below.[116] Similarly, Co depositions with cycloheptatriene and PF_3 yielded an η^3-bonding system plus $HCo(PF_3)_4$.[83]

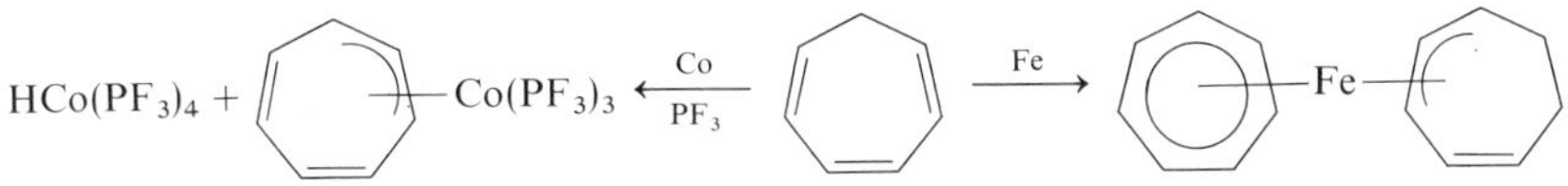

Tan, Fletcher, and McGlinchey[117] have studied a unique triene, 7,7-dimethylfulvene, that was condensed with Fe atoms to yield a coupled ferrocene as well as a reduced ferrocene. The hydrogen required was obtained from excess substrate.

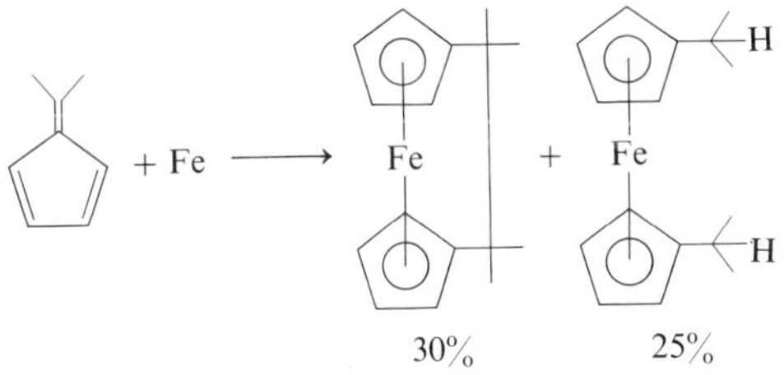

These triene reactions again demonstrate how readily hydrogen-transfer processes occur during metal atom reactions. These transfers must proceed through very reactive metal hydride intermediates that would be of great interest to study by matrix-isolation spectroscopy methods.

d. Alkyne Reactions.   Very little work with alkynes and metal atoms has been published. Generally, alkynes are efficiently trimerized by Group VIII metal atoms, but few stable metal complexes have been found.[118] Lagowski has reported that Fe—RC≡CR codepositions yield an $FeC_{10}R_{10}$ formulation, possibly of the ferrocene type.[119] A further report by Simons and Lagowski demonstrates that a complex, black Ni–alkyne organometallic species (from Ni atom–alkyne codepositions) is an effective homogeneous catalyst for oligomerization of terminal alkynes to novel organic materials.[120] The structure of this catalyst is not known but it is probably a mixture of compounds.

Hexafluoro-2-butyne (HFB) has, among alkynes, been investigated the most extensively.[110] Low valent, stable HFB complexes were prepared for Co, Ni, Pd, Pt, Cu, and Ag, although only in the case of Ni and Pd were the compounds amenable to characterization. All of these complexes were extremely susceptible to decomposition to metal particles and $(CF_3)_6C_6$. For Ni, an HFB–Ni–solvent complex was formed according to MW, decomposition, and spectral studies. For Pd, a $(HFB)_2Pd_2$–solvent complex was isolated. Before solvent (acetone or acetophenone) was added, these complexes had 1:1 HFB:M ratios, as shown by HCl decomposition studies. Currently, it is believed that in the solid state the HFB–Ni and HFB–Pd species are telomeric, while in solutions of acetone or acetophenone, one molecule of solvent is complexed.[110]

Addition of CO to the HFB–Ni or HFB–Pd complexes before warm-up of the matrices caused the formation of volatile (HFB)–$M(CO)_2$ complexes that spontaneously converted to $M_4(HFB)_3(CO)_4$ clusters, which were soluble in hexane and readily characterized. In the Ni case, the cluster was previously known, but it is a new cluster in the Pd case.[110]

These studies clearly indicated the superior complexing ability of HFB for the later transition metal atoms (cf. Cluster Formation Processes of this chapter).

$$CF_3C{\equiv}CCF_3$$

$$(CF_3C{\equiv}CCF_3)Ni\cdot solv \qquad\qquad (CF_3C{\equiv}CCF_3)_2Pd_2\cdot solv \qquad\qquad (CF_3C{\equiv}CCF_3)_x(M)$$

Ni solvent ⟋   Pd solvent ↓   Co, Pt, Cu, Au ⟍

e. Arene Reactions.   The codeposition of benzene, toluene, or mesitylene, with Fe, Co, or Ni yields π-arene complexes that may be 1:1 in stoichiometry, as examined by matrix isolation spectroscopic methods.[63] These are unstable reactive complexes, as first demonstrated by Timms,[99] where an Fe–benzene complex sometimes decomposed explosively at ca $-50°C$. Addition of H_2 to this unstable complex produced some cyclohexane. This material was believed to be bis(benzene)iron(0). A more stable analog was obtained when Ru–C_6H_6 codepositions were carried out, yielding $(C_6H_6)_2Ru$, stable to approximately $0°C$.[30]

The use of toluene has, because of its low mp of $-96°C$, some definite advantages. The arene complexes of Fe, Co, and Ni are stable generally to about $-70°C$ to $-20°C$, and so are stable in cold toluene solution. Thus, although not readily isolable, these complexes are very useful organometallic intermediates, and have been called "solvated metal atoms."[121] Ligands can be added to the Ni–toluene complex which will displace the toluene and yield NiL_4 complexes.[121–124] In general, the toluene complex of Ni is an excellent source of Ni(0) and should find many uses in the future. It has advantages over Ni atoms in that a high excess of the sometimes precious ligand (L) is not necessary, and toluene is even more readily displaced than COD when compared with $(COD)_2Ni$. In fact, addition of COD to the Ni–toluene complex allows the formation of some $(COD)_2Ni$.[125]

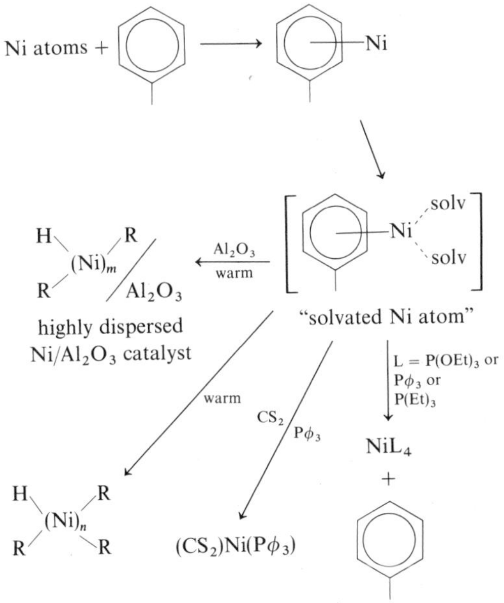

In some cases, the arene ligand remains bound to the metal atom, and the added ligand serves to stabilize the arene η^6-metal bond. These experiments have been particularly useful in the Fe system.[83,113,126]

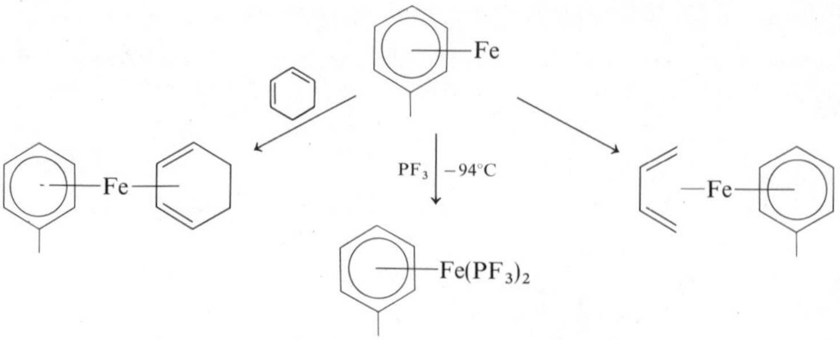

In the case of mesitylene, hydrogen transfers from excess mesitylene allowed the formation of a $\eta^6-\eta^4$ sandwich complex (18-electron rule satisfied).[83]

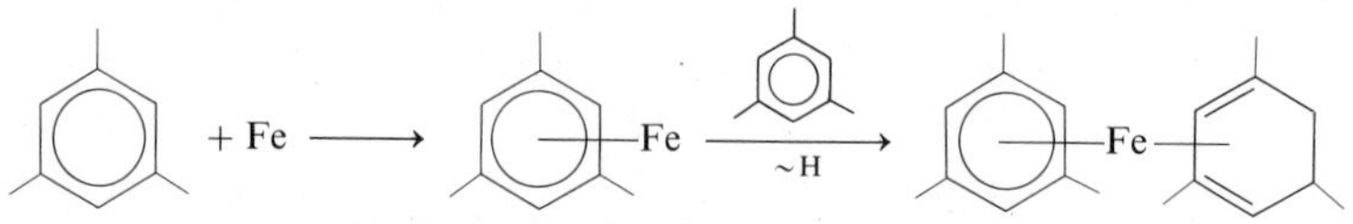

More electron-demanding arenes allow the Fe, Co, and Ni–arene complexes to be more thermally stable. Trifluoromethylbenzene and bis-1,3-ditrifluoromethylbenzene form complexes with Ni that are stable at least 20°C higher than the Ni–toluene complex.[127] Hexafluorobenzene with Ni atoms yielded a thermally stable, possibly polymeric, 1:1 C_6F_6:Ni complex.[128] Addition of CO and $P(OEt)_3$ caused the release of one mole of C_6F_6 and one mole of NiL_4. Addition of H_2O also released C_6F_6. However, H_2 addition caused partial reduction of C_6F_6. The C_6F_6–Ni complex was extremely reactive, and addition of reagents had to be carried out at very low temperatures or explosions invariably resulted.[128]

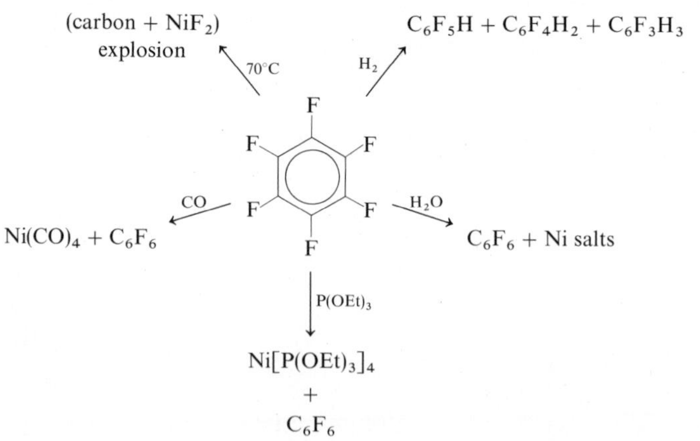

Apparently more electron-demanding arenes allow stronger M–arene bonding for the later transition metals. This is reasonable since these arenes would have better π-acid characteristics (stronger d $\rightarrow$ π^* backbonding).[128]

f. Phosphine and Phosphite Reactions.   The macroscale codeposition of PF_3, which supposedly imitates CO in its ligand properties, has yielded a series of Group VIII M–PF_3 complexes. And, although some of these complexes can be prepared by conventional high-pressure methods, the metal atom method serves as an attractive alternative procedure. Furthermore, some of these M–PF_3 complexes have only been prepared by the metal atom method.[129] New, mixed PF_3–PH_3 complexes have also been prepared in this way, showing that mixtures of weak and strong π-acid ligands behave satisfactorily in metal atom reactions.

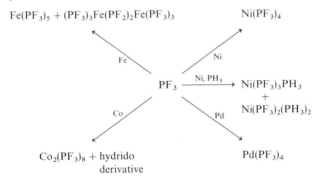

Some other interesting examples include a PF_2H–Ni reaction. The $Ni(PF_2H)_4$ complex, prepared by Parry and Stampin,[130] is a perfectly stable species, although the ligand itself is only stable at temperatures well below $0°C$. This again illustrates the unusual synthetic capabilities of the metal atom method.

Very electron-rich M–phosphine and phosphite complexes have been prepared by M–$P(CH_3)_3$ and M–$P(OCH_3)_3$ depositions, as shown below.[42,83,131] Note that the Co atom took up only four phosphine molecules. The resultant complex is apparently too sterically hindered to dimerize to the $Co_2(PR_3)_8$ configuration. Note also that in the Fe system with $P(CH_3)_3$ some C—H oxidative addition took place. This was not the case in Fe–$P(OCH_3)_3$ reactions.[131]

Two unusual aminophosphines have been codeposited with Fe atoms to yield homoleptic FeL_5 and FeL_4 complexes. In the latter case, one diphosphine ligand served as a bidentate ligand while the other three served as monodentate ligands.[132] Extension of these studies to Co, Ni, and Cr have been reported.[133] A series of new homoleptic complexes were prepared by

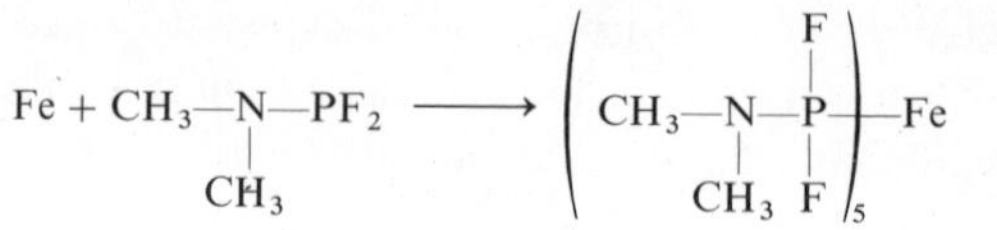

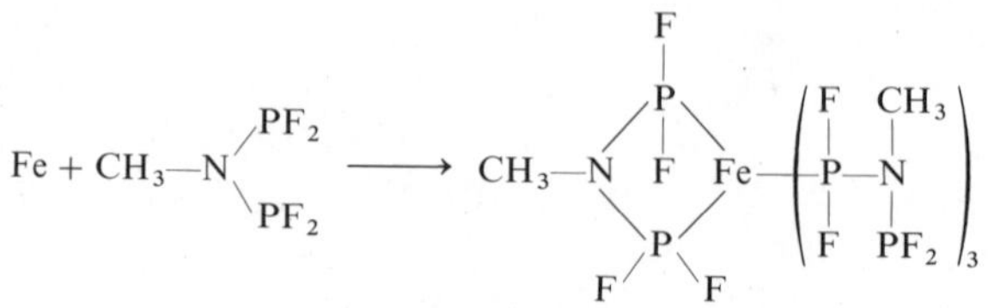

$$Fe + CH_3-\underset{\underset{CH_3}{|}}{N}-PF_2 \longrightarrow \left(CH_3-\underset{\underset{CH_3}{|}}{N}-\underset{\underset{F}{|}}{\overset{\overset{F}{|}}{P}}-Fe \right)_5$$

$$Fe + CH_3-N \overset{PF_2}{\underset{PF_2}{\diagdown}} \longrightarrow CH_3-N \overset{\overset{F}{\overset{P}{\diagup}}\diagdown}{\underset{\underset{F}{\overset{P}{\diagdown}}\diagup F}{F}} Fe-\left(\underset{\underset{F}{\overset{F}{\overset{|}{P}}}}{\overset{F}{\overset{|}{\underset{F}{|}}}}-\underset{\underset{PF_2}{\overset{|}{N}}}{\overset{CH_3}{\overset{|}{N}}} \right)_3$$

$Co_2[(PF_2)_2NCH_3]_5$, $Ni[(PF_2)_2NCH_3]_n$, $Fe[PF_2N(CH_3)_2]_3[(PF_2)_2NCH_3]$,

$Co_2[PF_2N(CH_3)_2]_2[(PF_2)_2NCH_3]_3$, $Ni[PF_2N(CH_3)_2]_3[(PF_2)_2NCH_3]$,

$Ni_2[PF_2N(CH_3)_2][(PF_2)_2NCH_3]_3$

codeposition of each aminophosphine separately with metal atoms, or by codeposting a mixture of the two aminophosphines. The complexes obtained exhibited very good thermal and oxidative stabilities (cf. Table V-6).[132,133]

Oxidative addition to P–C and/or C–H bonds also occurs in metal atom–phosphine reactions. This is primarily why phosphines such as $P(Et_3)_3$ have not been studied extensively, since decomposition to H_2, C_2H_4, and other products is an important pathway.[134,135]

Phosphines have been employed very profitably for trapping unstable intermediates in metal atom reactions. Some of these have been discussed previously under the Oxidative Addition Section of this chapter. However, the phosphine adducts themselves are important and valuable products, many of which are new compounds. Table 5-6 summarizes the $M(PR_3)_n$, $RMX(PR_3)_2$, and (arene)$M(PR_3)_n$ complexes prepared by macroscale metal atom methods.[42,56,57,68,77,78,80,83,113,129–133,136–139]

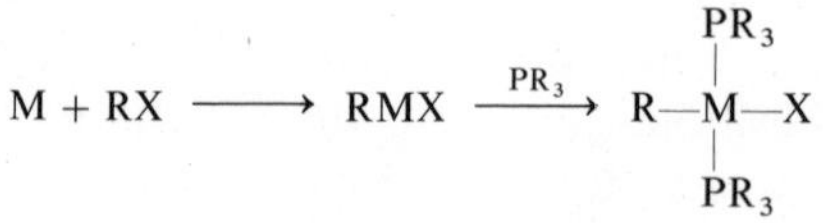

$$M + RX \longrightarrow RMX \xrightarrow{PR_3} R-\underset{\underset{PR_3}{|}}{\overset{\overset{PR_3}{|}}{M}}-X$$

Another important example is the previously discussed $(COD)_2Fe$ complex, which when allowed to react with diphos under N_2, formed an Fe–dinitrogen complex.[136]

$$(COD)_2Fe + \underset{\text{diphos}}{(C_6H_5)_2PCH_2CH_2P(C_6H_5)_2} \xrightarrow{N_2} (diphos)_2FeN_2$$

TABLE 5-6

M–(PR₃)ₙ, RMX(PR₃)₂, and Cp, Arene, N₂–M(PR₃)ₙ Complexes Prepared by Macroscale Metal Atom Methods (Group VIII)

Ligands	Metal	Pertinent compounds formed	References
PF_3	Co	$Co_2(PF_3)_8$, $HCo(PF_3)_4$	129
PF_3	Ni	$Ni(PF_3)_4$	129
PF_3	Pd	$Pd(PF_3)_4$	129
PF_3	Fe	$Fe(PF_3)_5 + (PF_3)_3Fe(PF_2)_2Fe(PF_3)_3$	129
PF_2Cl	Ni	$Ni(PF_2Cl)_4$	129
PF_3, PH_3	Ni	$Ni(PF_3)_2(PH_3)_2 + Ni(PF_3)_3(PH_3)$	129
PF_2H	Ni	$Ni(PF_2H)_4$	130
$P(CH_3)_3$	Ni	$Ni[P(CH_3)_3]_4$	83
$P(CH_3)_3$	Co	$Co[P(CH_3)_3]_4$	83
$P(CH_3)_3$	Fe	$Fe[P(CH_3)_3]_5$ or $[(CH_3)_3P]_3Fe(H)CH_2(PCH_3)_2$	42, 83
$P(C_6H_5)_3$	Ni	$Ni[P(C_6H_5)_3]_4$[a]	83
$P(OCH_3)_3$	Fe	$Fe[P(OCH_3)_3]_5$	131
$(CH_3)_2NPF_2$	Fe	$Fe[PF_2N(CH_3)_2]_5$	132, 133
$CH_3N(PF_2)_2$	Fe	$Fe[(PF_2)_2NCH_3]_4$	132, 133
$CH_3N(PF_2)_2$	Co	$Co_2[(PF_2)_2NCH_3]_5$	133
$CH_3N(PF_2)_2$	Ni	$Ni[(PF_2)_2NCH_3]_n$	133
$CH_3N(PF_2)_2$, $(CH_3)_2NPF_2$ 1:4	Fe	$Fe(PF_2N(CH_3)_2]_3[(PF_2)_2NCH_3]$	133
$CH_3N(PF_2)_2$, $(CH_3)_2NPF_2$	Co	$Co_2[PF_2N(CH_3)_2][(PF_2)_2NCH_3]_3$	133
$CH_3N(PF_2)_2$, $(CH_3)_2NPF_2$	Ni	$Ni[PF_2N(CH_3)_2]_3[(PF_2)_2NCH_3] + Ni_2(PF_2N(CH_3)_2][(PF_2)_2NCH_3]_3$	133
PF_3, C_6H_6	Fe	$C_6H_6Fe(PF_3)_2$	113
CF_3Br, PEt_3	Ni	$CF_3NiBr(PEt_3)_2$	56, 57
CF_3I, PEt_3	Pd	$CF_3PdI(PEt_3)_2$	56, 57
CF_3Br, PEt_3	Pd	$CF_3PdBr(PEt_3)_2$	56, 57
C_6H_5Br, PEt_3	Pd	$C_6H_5PdBr(PEt_3)_2$	56, 57
C_6F_5Cl, PEt_3	Pd	$C_6F_5PdCl(PEt_3)_2$	56, 57
C_6F_5Cl, PEt_3	Ni	$C_6F_5NiCl(PEt_3)_2$	56, 57
C_6F_5I, PEt_3	Pd	$C_6F_5PdI(PEt_3)_2$	56, 57
$n\text{-}C_3F_7I$, PEt_3	Pd	$n\text{-}C_3F_7PdI(PEt_3)_2$	56, 57
C_2F_5I, PEt_3	Pd	$C_2F_5PdI(PEt_3)_2$	56, 57
CF_3COCl, PEt_3	Pd	$CF_3COPdCl(PEt_3)_2$, $CF_3PdI(PEt_3)_2$	56, 57
$n\text{-}C_3F_7COCl$, PEt_3	Pd	$n\text{-}C_3F_7COPdCl(PEt_3)_2$	56, 57, 80
CF_3COCl, PEt_3	Ni	$CF_3NiCl(PEt_3)_2$	56, 57
$(CF_3CO)_2O$, PEt_3	Pd	$cis\text{-}(CF_3CO)_2)Pd(PEt_3)_2$	80
$C_6H_5CH_2Cl$, then PEt_3	Pd	$C_6H_5CH_2PdCl(PEt_3)_2$	77, 78
$(C_6H_5)_2PCH_2CH_2P(C_6H_5)_2$	Fe	$(diphos)_2FeN_2$	136
$N_2(cyclooctadiene)$ $C_6H_5CH_3$, then PF_3	Fe	$C_6H_5CH_3Fe(PF_3)_2$	113
$CH_2{=}CHCH{=}CH_2$	Fe	$(CH_2{=}CHCH{=}CH_2)_2Fe(PF_3)$	113, 137
C_5H_6, then PF_3	Co	$C_5H_5Co(PF_3)_2$	138, 139
C_3H_6, then PF_3	Ni	$C_3H_6Ni(PF_3)_2$	139
$CH_3CH_2CH{=}CH_2$, then PF_3	Co	$(CH_3CHCHCH_2)Co(PF_3)_3$	138
$CF_3CF{=}CFCF_3$, PEt_3	Pd	$(CF_3CF{=}CFCF_3)Pd(PEt_3)_2$	80
C_6F_5Br, PEt_3	Ni	$C_6F_5NiBr(PEt_3)_2$	57
C_6F_5Br, PEt_3	Pd	$C_6F_5PdBr(PEt_3)_2$	56, 57

[a] Prepared in methylcyclohexane solution.

g. Nitric Oxide Reactions.   Nitric oxide (NO), a three-electron ligand, is too volatile to be cocondensed at $-196°C$ for macroscale preparations. The BF_3 adduct of NO, which is less volatile, has been used. Depositions of BF_3–NO, PF_3, and Fe and Co yielded $Fe(NO)_2(PF_3)_2$ and $Co(NO)(PF_3)_3$, respectively.[126] Attempted preparations of $Co(NO)_3$, $(C_6H_6)Co(NO)$, and $(C_6H_6)Cr(NO)_2$, however, failed.

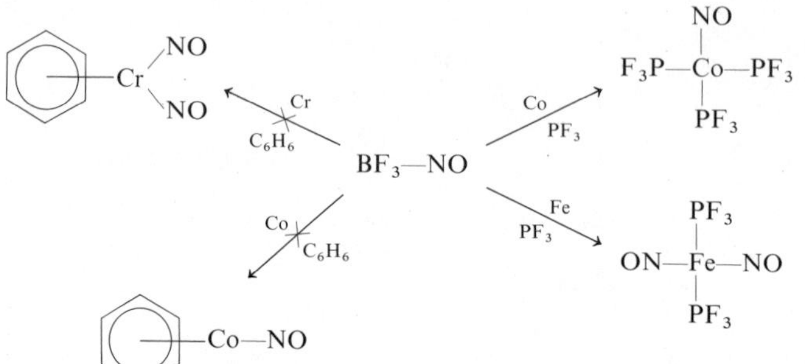

Recent microscale studies of the Fe–NO reaction have been carried out.[140] Utilizing this system as a model of chemisorbed NO on an Fe surface, Bandon, Onishi, and Tamaru observed two broad bands in the IR at $1800 \, cm^{-1}$ and $1720 \, cm^{-1}$. These bands were attributable to NO on oxidized Fe and on metallic iron, respectively.

h. Isonitrile Reactions.   A very brief disclosure by Gladkowski and Scholar[141] describes the preparation of NiL_4 and FeL_5 complexes on a macroscale where L = tert-butyl isocyanide, methyl isocyanide, cyclohexyl isocyanide, and vinyl isocyanide. The complexes did undergo exchange reactions with phosphines and phosphites. Codepositions of isocyanides and PF_3 yielded mixed complexes.

i. Carbon Monoxide Reactions.   As already briefly discussed in Chapter 3, carbon monoxide is a "super" ligand, having excellent π-acid characteristics. It is an obvious choice for study in metal atom chemistry and has been studied extensively by matrix-isolation techniques. However, on a macroscale, where $-196°C$ is the lowest temperature attainable for practical reasons, CO is too volatile to study.

By matrix isolation studies, employing high dilution, unsaturated M–$(CO)_n$ species have been prepared for the later transition metals. A great deal has been learned about bonding from these studies since the M–C and C–O bonds in $M(CO)_n$ species are amenable to bonding analysis through study of $\nu_{C=O}$ in the IR.

Often, in order to know the value of n in an $M(CO)_n$ complex, ^{12}CO–^{13}CO mixtures must be deposited. These and related techniques have been described by Moskovits and Ozin,[142] Pimentel,[143] and references cited therein.[144]

Table V-7 summarizes the M–CO matrix isolation work for the later transition metal (Group VIII) atoms.[102,145–155]

Some fascinating comparisons can be made from the data in Table 5-7. Considering the Ni, Pd, Pt triad, only $Ni(CO)_4$ is a stable complex. Comparison of the spectra (v_{C-O} and v_{M-C}) indicate that the effectiveness of CO as a σ-donor and π-acceptor toward M is in the order Ni > Pt > Pd. It is interesting that this is also the stability order for most π-complexes of these metals.

TABLE 5-7

Metal Carbonyl Complexes of the Late Transition Metals (Group VIII) Prepared by Matrix Isolation Metal Atom Techniques

Complex	v_{C-O}	Comments	References
Fe(CO)	1898		
$Fe(CO)_3$	2042, 1936 (Ar)	C_{3v}, low spin	145
$Fe(CO_4$	1999, 1994, 1974	C_{3v}, Interacts with rare gases and CH_4	146
$Fe(CO)_5$	2014, 2034	D_{3h}, stable	147
Co(CO)	1952, 1944	$C_{\infty v}$	148, 149
$Co(CO)_2$	1914	$D_{\infty h}$	148
$Co(CO)_3$	1977	C_{3v}	148
$Co(CO)_4$	2021, 2011	C_{3v}	146, 148, 150
Ni(CO)	1996	$C_{\infty v}$	149, 151
$Ni(CO)_2$	1967	$D_{\infty h}$	151
$Ni(CO)_3$	2017	D_{3h}	151
$Ni(CO)_4$	2052	T_d, stable	151
Rh(CO)			102
$Rh(CO)_2$			102
$Rh(CO)_3$			102
$Rh(CO)_5$			102
Ir(CO)			102
$Ir(CO)_2$			102
$Ir(CO)_3$			102
$Ir(CO)_4$			102
Pd(CO)	2050		152, 153
$Pd(CO)_2$	2044	$D_{\infty h}$	152, 153
$Pd(CO)_3$	2057	D_{3h}	152, 153
$Pd(CO)_4$	2071	T_d, stable to 80°K	152, 153, 154
Pt(CO)	2052	$C_{\infty v}$	155
$Pt(CO)_2$	2058	$D_{\infty h}$	155
$Pt(CO)_3$	2049	D_{3h}	155
$Pt(CO)_4$	2053	T_d	154, 155

Note in the M(CO) comparison for Ni, Pd, Pt, that the Ni–CO interaction is much stronger than the Pd or Pt–CO interaction, which also holds true throughout the M(CO), $M(CO)_2$, $M(CO)_3$ series. Geometries of all of these species appear to be the same (when n is the same). Calculated IR frequencies agree very well with those found.[102]

Similar application of spectroscopic and isotope labeling experiments showed that $Co(CO)_4$ probably has C_{3v} symmetry, likewise $Fe(CO)_4$ has C_{3v} symmetry.[146,148,150]

Moskovits and Ozin believe that by the study of M–CO matrix species, a controversy concerning chemically absorbed CO has been resolved.[102] That is, for CO chemisorbed on certain metal surfaces, two major bands have been found. Eischens, Francis, and Plisken[156] assigned these to terminal and bridged CO, while Blyholder and Allen have proposed CO adsorbed on flat or central metal sites vs edge sites.[157] In the former case (terminal and bridged) a higher frequency $v_{C=O}$ should be observed for the terminal adsorbed CO. In the latter case (edge vs flat surface adsorption), a higher frequency should be observed for the central or flat sites than for edge sites, because the higher the n for $(M)_n$–CO, the less back bonding to CO would be necessary. In a single metal atom M–CO system, the strongest π^*-back bonding overlap would be expected, thus making $v_{C=O}$ lower than for terminal CO on any metal surface sites. Since in almost every case, the $v_{C=O}$ for M atom–CO was found to be between the two chemisorbed bands, it was proposed that Eischens original ideal was correct. That is, only bridged CO could fall at a lower $v_{C=O}$ than the $v_{C=O}$ for M atom–CO.[102]

j. Dinitrogen Reactions.   Again, because of volatility problems, dinitrogen has been mainly studied by microscale techniques. Table V-8 summarizes the complexes observed in the matrix spectroscopically. Both "end-on" and "side-on" bonding have been observed. For example, $Co(N_2)$ is believed to be triangular,[158] whereas Ni, Pd, and $Pt(N_2)$ are believed to be linear.[159] Note also in Table 5-8[158–162] that the Ni–N interaction is stronger than Pt–N which is greater than Pd–N,[159–161] the same trend found for M–C in the M–CO series. This trend holds for the $M(N_2)$, $M(N_2)_2$, $M(N_2)_3$ series. However, it is believed that in the case of Pd, symmetrical end-on bonding is present ($D_{\infty h}$), while with Pt both end-on and side-on bonding exist.[159–161,163] However, addition of one more N_2 causes the formation of $Pt(N_2)_3$ with complete end-on bonding, which is also the case for $Pd(N_2)_3$.[159]

Mixed N_2–CO depositions allowed the formation of mixed complexes where π-bonding as well as σ-bonding was favored for CO over N_2. However, it was believed that the force constants for bonded CO involved a $(\sigma - \pi)$ term, but those of bonded N_2 a $(\sigma + \pi)$-term, which explains why N_2 shows a greater decrease in force constant on complexation than CO.[162,164]

TABLE 5-8

Metal–Dinitrogen Complexes of the Later Transition Metals (Group VIII) Prepared by Matrix Isolation Metal Atom Techniques

Complex	$v_{N=N}(cm^{-1})$	Comments	References
$Co(N_2)$	2101	Side-on bonding	158
$Ni(N_2)$	2090	$v_{M-N} = 466$, linear	159
$Ni(N_2)_2$	2106		159
$Ni(N_2)_3$	2134		159
$Ni(N_2)_4$	2175	T_d, end-on bonded, but distorted in pure N_2 matrix	159
$Pd(N_2)$	2213	$v_{M-N} = 378$, linear	159
$Pd(N_2)_2$	2234	$D_{\infty h}$	159
$Pd(N_2)_3$	2242	D_{3h}, C_2 in solid N_2	159
$Pt(N_2)$	2170	$v_{M-N} = 394$, linear	159–161
$Pt(N_2)_2$	2198	End-on and side-on bonding	159–161
$Pt(N_2)_3$	2212	D_{3h}, end-on bonding	159–161
$Ni(N_2)_m(CO)_{4-m}$			162

k. Dioxygen Reactions.   Matrix-isolation methods have shown that both end-on and side-on bonding can occur with O_2 and the Group VIII transition metal atoms. Generally side-on bonding is preferred for O_2, whereas, generally, end-on bonding is preferred for N_2. For Ni, Pd, and Pt atoms, $M(O_2)_2$ species of D_{2d} symmetry were proposed, and this bonding would be predicted by Dewar-Chatt-Duncanson reasoning.[102,165–168] In these

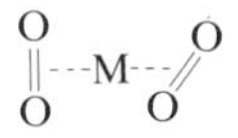

$$\begin{matrix} O & & O \\ \| \cdots M \cdots \| \\ O & & O \end{matrix}$$

complexes $v_{O=O}$ shows that the strength of intereaction is Pt > Ni > Pd. In the $M(O_2)$ complexes side-on bonding was also found, and the bonding strength is in the same order Pt > Ni > Pd. (cf. Table 5-9).

Mixed O_2–N_2 depositions with Ni were carried out. Mixed side-on (O_2) and end-on (N_2) species were formed.[102] Table 5-9 tabulates the spectroscopic data for these complexes. It appears that the presence of an N_2 group

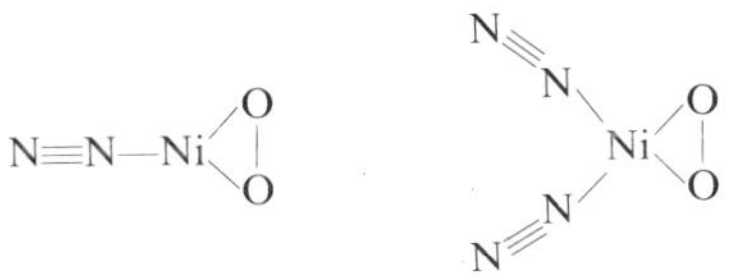

causes the O_2 group to be bonded even more strongly [compare $Ni(O_2)$, $Ni(O_2)_2$, and $Ni(N_2)(O_2)$].

TABLE 5-9

Metal–Dioxygen and Dioxygen–Dinitrogen Complexes of the Late Transition Metals (Group VIII) Prepared by Matrix Isolation Metal Atom Techniques

Complex	Spectral features and comments	References
$Ni(O_2)$	$\nu_{O=O} = 996$ cm^{-1}, side-on, C_{2v}	102, 165, 166
$Ni(O_2)_2$	$\nu_{O=O} = 1062$ cm^{-1}, side-on, D_{2d} or D_{2h}	165, 166
$Pd(O_2)$	$\nu_{O=O} = 1024$ cm^{-1}, side-on, C_{2v}	102, 165, 167
$Pd(O_2)_2$	1111 cm^{-1}, side-on D_{2d}	102, 165, 167
$Pt(O_2)$	$\nu_{O=O} = 927$ cm^{-1}, side-on, C_{2v}	102, 165, 167
$Pt(O_2)_2$	$\nu_{O=O} = 1050$ cm^{-1}, side-on, D_{2d}	165, 166
$Ni(N_2)(O_2)$	$\nu_{O=O} = 977$ cm^{-1}, $\nu_{N=N} = 2243$ cm^{-1}	102
$Ni(N_2)_2(O_2)$	$\nu_{O=O} = 972$, $\nu_{N=N} = 2260$ cm^{-1}	102
$Rh(O_2)$	$\nu_{O=O} = 900$ cm^{-1}, side-on, C_{2v}	168
$Rh(O_2)_2$	$\nu_{O=O} = 1045$ cm^{-1}, side-on, D_{2d}	168

l. Carbon Disulfide Reactions. Macroscale codepositions of Ni with excess CS_2 yielded a large amount of NiS and $(CS)_n$ polymer.[163]

$$Ni + CS_2 \xrightarrow{\;-196°\;} NiS + CS$$

Microscale Ni–CS_2 depositions at 10–12°K have yielded complexes of unknown structure with the probable stoichiometry of $Ni(CS_2)$, $Ni(CS_2)_2$, and $Ni(CS_2)_3$.[164]

6. CLUSTER FORMATION PROCESSES

Metal cluster growth under equilibrium conditions has been discussed by Niedermayer.[169] Kohlschutter[170] pointed out the relationship between metal atom chemistry and cluster studies (cf. Klabunde,[44] for some discussion of these reports).

The major difference between previous studies of metal clusters under high vacuum and in equilibrium is that at very low temperatures in a matrix, equilibrium conditions certainly do not exist. Once clusters form in a low-temperature matrix, they do not degrade to smaller particles.

In a low-temperature matrix metal atoms can diffuse at remarkably low temperatures.[169] The smaller the metal atom and the lighter the diluent gas, the more readily the diffusion can occur (diffusion order Ne > Ar > Kr > Xe).[171–175]

Depending on dilution conditions and rate of deposition, Group VIII transition metal atoms can all diffuse readily in the quasi-liquid layer formed as metal atoms and matrix material are being codeposited. If a reactive ligand is present, metal dimers and trimers can be often trapped, as discussed below.

a. **Metal Carbonyl Dimers and Trimers.** Studies of Fe, Co, Ni, Rh, and Ir dimerization and trimerization in matrices of inert gases mixed with varying amounts of reactive gases have been carried out. Table 5-10[148,172,176–178] tabulates the complexes reported. Usually, limited structural information can be learned from these experiments because of the combined presence of the many interferring bands of mononuclear species, dimers, trimers, and higher oligomers. However, to a degree, kinetic analysis of the dimerization process has been possible. Both a statistical frozen-matrix approach (calculable probability that M and M are neighbors and react to give M_2) and a high mobile metal atom approach (diffussion is rapid in quasi liquid layer) have been used.[171] It was found, not unexpectedly, that the diffusion mechanism approach appears to be corroborated best by experimental results, and that the eventual M_2 concentration is proportional to the square of the M/substrate ratio. Concentration of higher metal aggregates vary as some higher power of the M/substrate ratio (in these analyses "reactive matrices" are assumed). Analysis of the nonreactive matrices kinetically is somewhat more involved (e.g., M + argon + CO). A simple model based on both statistical and kinetic approaches has been outlined by Moskovits and Hulse.[179] Cluster growth in CO, N_2, and Ar was considered. In particular, these treatments have aided Moskovits and Hulse[178] in their attempts to establish the cluster size in matrix experiments where $(Ni)_n CO$ species were produced. For $Ni_2 CO$, two IR ν_{CO} absorptions were observed (1973 and 1938 cm^{-1}), which could be due to linear Ni_2–CO and bridged Ni_2–CO. However, it is disconcerting to compare the high frequency Ni–CO ν_{CO} band at 1999 cm^{-1} (highest for series) with ν_{CO} for CO adsorbed on silica-supported Ni particles, which show several bands

TABLE 5-10

Metal Carbonyl Dimers and Trimers Formed with CO and the Later Transition Metals (Group VIII)

Complex	$\nu_{C=O}$(cm^{-1})	Comments	References
$Fe_2(CO)_8$		CO bridged	176
$Fe_2(CO)_9$		D_{3h}, CO bridged	176
$Co_2(CO)_8$		Both CO bridged and	148
		Co–Co bonded forms	148
$Rh_2(CO)_8$		CO bridged; $Rh_4(CO)_{12}$ at $-48°C$	177
$Ir_2(CO)_8$		CO bridged; forms $Ir_2(CO)_{12}$ at $-58°C$	177
$Ni_2(CO)$	1973	Linear	178
$Ni_2(CO)$	1938	Bridged	178
$Ni_3(CO)$	1970	Terminal	178
$Ni_3(CO)$	1963	Terminal	178
$Ni(CO)$	1999	For comparison	172, 178

above 2000 cm^{-1}. Moskovits, Ozin, and Hulse[172,178] question whether silica-supported $(Ni)_n$ is zero valent, or whether just one CO/Ni is adsorbed under such conditions. It appears that a great deal of work is still needed to unravel these puzzling questions.

b. Metal–C_2H_4 Dimers and Trimers. Ethylene–Co and ethylene–Ni co-depositions with high concentrations of metal atoms have produced $(Co)_2$–C_2H_4, $(Co)_2$–$(C_2H_4)_2$, $Ni_2(C_2H_4)$, and $Ni_2(C_2H_4)_2$. The presumed structures of these complexes are shown below, and the $v_{C=C}$ values compared with

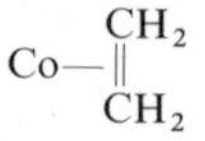

$$
\begin{array}{c}
\text{CH}_2 \\
\text{Co}—\| \\
\text{CH}_2
\end{array}
\qquad\qquad
\begin{array}{c}
\text{CH}_2 \\
\text{Ni}—\| \\
\text{CH}_2
\end{array}
$$

1504 cm^{-1} 1499 cm^{-1}

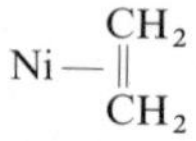

C_{2v} C_{2v}

$$
\begin{array}{c}
\text{CH}_2 \\
\text{Co}—\text{Co}—\| \\
\text{CH}_2
\end{array}
\qquad\qquad
\begin{array}{c}
\text{CH}_2 \\
\text{Ni}—\text{Ni}—\| \\
\text{CH}_2
\end{array}
$$

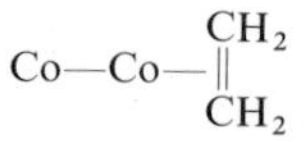

1484 cm^{-1} 1488 cm^{-1}

C_{2v}

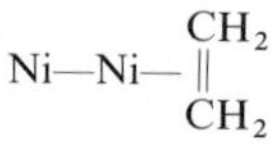

$$
\begin{array}{ccc}
\text{CH}_2 & & \text{CH}_2 \\
\|—\text{Co}—&\text{Co}—\| \\
\text{CH}_2 & & \text{CH}_2
\end{array}
\qquad
\begin{array}{ccc}
\text{CH}_2 & & \text{CH}_2 \\
\|—\text{Ni}—&\text{Ni}—\| \\
\text{CH}_2 & & \text{CH}_2
\end{array}
$$

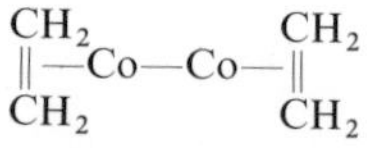

1508 cm^{-1} 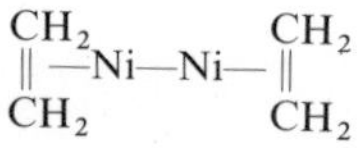   1504 cm^{-1}

mononuclear systems.[180] If the assumption that $v_{C=C}$ reflects binding strength is correct (see reference 180 for discussion), then the strongest M–C_2H_4 binding is observed in the M_2–C_2H_4 systems, which would *not* have readily been predicted since M–M bonding would presumably lessen one metal atoms "need" for orbital mixing with the π-ligand. Further studies of the Ni–C_2H_4 and Ni_2–C_2H_4 systems have been carried out,[181] and theoretical calculations suggest 14-kcal and 27-kcal interactions, respectively. Furthermore, these calculations suggest that there is very little geometrical perturbation of the C_2H_4 molecule upon complexation to Ni or Ni_2.[181] Furthermore, it seems most likely that the C_2H_4 molecule prefers side-on bonding to *one* Ni atom of Ni_2, which differs with a variety of previous reports regarding C_2H_4 chemisorbed on Ni surfaces.

c. Matrix Isolated (Inert Gas) Bare Metal Dimers and Trimers. Spectroscopic identification of matrix isolated dimers and trimers of Group VIII metals has been carried out by Ozin and Moskovits and their co-workers. For example, the molecules Rh_2 and Rh_3 have been characterized by UV–visible absorption spectroscopy.[168,182] These small molecules were then allowed to react with O_2 to yield $Rh_2(O_2)_n$ (where $n = 1, 2, 3, 4$) and $Rh_3(O_2)_m$ (where m is probably 2–6). These molecules are probably metal–metal bound[182]:

Rh—Rh

$$\begin{array}{c} Rh\text{—}Rh \\ \diagdown \diagup \\ Rh \end{array}$$

And, as mentioned previously, further statistical and kinetic treatments of metal atom dimerization in rare gases and in reactive matrices have been carried out by Moskovits and Hulse,[179] and simple models have been proposed. Furthermore, the molecular orbital picture (HOMO–LUMO) for Ni_2, Ni_3, and other higher clusters has been examined by UV–visible spectroscopy studies of these species in argon matrices.[183]

d. Small Discrete Organometallic Clusters (Macroscale). The potential of utilizing macroscale metal atom methods for the synthesis of small discrete organometallic clusters is quite high. There are three possibilities: (1) trapping clusters as they are formed by warming or photolyzing the

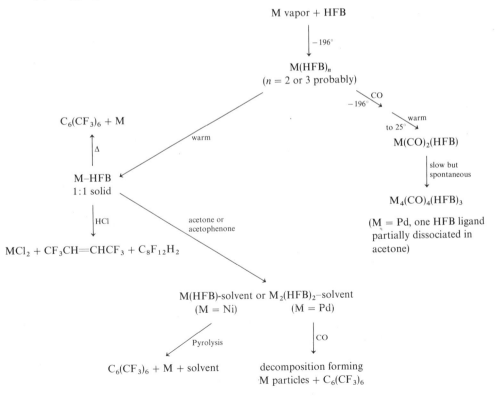

matrix. This can be done by rapid addition of a trapping ligand to the warming matrix in an exact temperature range. Some experiments we have done indicate that Co_2, Co_4, Co_6, and higher clusters are trappable in small amounts as their carbonyls. The clusters form at $-95°C$ in a toluene matrix. This process is hindered as a synthetic method because of the large number of species formed; (2) formation of new metal clusters by reaction of metal-containing compounds with metal atoms; (3) formation of $M(L)_n$ or $M(L)_m(L)_n$ dispersions that kinetically or thermodynamically favor the formation of small, discrete clusters on warming. This reaction sequence is followed when Ni and Pd–hexafluoro-2-butyne (HFB) matrices are treated with CO.[110] The formation of $M(CO)_2(HFB)$ is followed by spontaneous production of $M_4(CO)_4(HFB)_3$. In the case of Pd–HFB matrices without CO addition, a stable dimeric product $(Pd)_2(HFB)_2$ spontaneously formed. These processes are summarized above and represent the first examples of the synthesis of discrete organometallic clusters by macro metal vapor techniques.[110]

e. Large Metals Clusters (Metal Slurries, Small Particles or Crystallites).
(i). Active Metal Powders and Slurries.   The process of activating a metal by cleaning its surface and/or making it high in surface area is a very common practice. Recall techniques for activating Mg chips for Grignard reactions, or the preparation of Raney metals by extraction of Al from M–Al alloys. Overall, there are now six general ways of producing active metal powders or slurries for use in organometallic or organic synthesis schemes or as catalysts: (1) mechanical reduction of particle size (grinding, etc.); (2) use of alloys or couples; (3) addition of catalysts or activators (e.g., I_2 or metal salts); (4) chemical cleansing (e.g., HCl or $BrCH_2CH_2Br$), or the Raney procedure where Al is chemically cleansed away from the Al–M alloy of interest; (5) metal salt reduction techniques (e.g., $K + MgCl \rightarrow KCl + Mg^*$); and (6) metal vapor–solvent codeposition techniques.[184] For the Group VIII metals the most important method employed has been the Raney procedure.

These metals can also be activated by metal vaporization procedures in organic media. Dispersing metal atoms in an excess of weakly complexing organic media followed by warming allows, first, the formation of "solvated metal atoms", which is followed by partial crystallization to form tiny crystallites. However, *chemisorption* of and *reaction with* the solvent prevents the metal crystallites from growing to larger sizes. Depending on the solvent and the metal employed, the final crystallites can be > 100 Å or so small that the metal is essentially amorphous. Large amounts of organics can be retained in the final stable metal powder, sometimes approaching C:M ratios of $1:2$.

Of the Group VIII metals, nickel has been most studied. Nickel has been crystallized from hexane, toluene, or THF. Each sample has a vastly different appearance, chemical reactivity,[44,49,67] and catalytic reactivity.[185]

$$M \text{ atoms} + \text{solvent vapor} \xrightarrow[\text{cocondense}]{-196^\circ C} \underset{\text{(colored)}}{M\text{–solvent complex}}$$

$$\Big\downarrow \text{melt}$$

$$\underset{\text{(slurry-black)}}{(M)_n\text{–solvent}} \xleftarrow[\text{warming}]{\text{further}} \underset{\text{(solvated)}}{M\text{–solution}}$$

$$\Big\downarrow \begin{array}{l}\text{vaporize} \\ \text{excess solvent, } 25^\circ C\end{array}$$

$$\underset{\substack{\text{small crystallites} \\ (R = \text{fragments of solvent})}}{\begin{array}{c}R \diagdown \quad \diagup R \\ (M)_n\text{–solvent} \\ R \diagup \quad \diagdown R\end{array}} \xrightarrow{\text{pyrolyze}} \underset{\text{larger crystallites}}{(M)_n + \text{organics}}$$

Alkanes (pentane, hexane, etc.) react with nickel to yield stable "pseudo organometallic" powders that are extremely active hydrogenation catalysts, much more active than Raney nickel for alkene hydrogenation.[185] They are not effective alkene isomerization catalysts, however. These pseudo-organometallic powders are stable to over 200°C and are apparently stabilized toward sintering by the presence of strongly bound organic fragments of the starting solvent.[185] These fragments are present in remarkably large amounts; with Ni:C:H ratios in the powder generally in the

$$Ni + C_5H_{12}$$

$$\Big\downarrow -196^\circ C$$

$$\begin{array}{c}R \diagdown \\ (Ni)_1(C_5H_{12})_n \\ R \diagup\end{array} \xleftarrow{K'} (Ni)_1(C_5H_{12})_n$$

$$\Big\downarrow K_1$$

$$\begin{array}{c}R \diagdown \quad \diagup R \\ (Ni)_2(C_5H_{12})_n \\ R \diagup\end{array} \qquad (Ni)_2(C_5H_{12})_n$$

$$\Big\downarrow K_2$$

$$\begin{array}{c}R \diagdown \quad \diagup R \\ (Ni)_3(C_5H_{12})_n \\ R \diagup \quad | \\ \quad R\end{array} \xleftarrow{K_2'} (Ni)_3(C_5H_{12})_n$$

$$\Big\downarrow K_n$$

$$\begin{array}{c}R \quad R \\ R \diagdown | \\ (Ni)_x(C_5H_{12})_n \\ R \diagup \quad \diagdown R\end{array} \xleftarrow{K_n'} (Ni)_x(C_5H_{12})_n$$

2–5:1:2 composition range.[186] Hydrolysis, hydrogenation, and ESCA studies suggest that these organic fragments are alkyl-, alkenyl-, and carbenoid-like, not carbide-like. Also, the nickel–alkane reaction to form the final $(Ni)_x(R)_y(H)_z$ framework takes place most effectively about $-130°C$, and apparently involves bare Ni clusters rather than Ni atoms.[186] Apparently, a competition between Ni cluster growth and reaction of the clusters with pentane takes place, with the pentane reaction at $-130°C$ being quite fast.[186]

These alkane reactions are the first examples of C—C cleavage in alkanes by small metal clusters at low temperature. It is believed that only clusters undergo this reaction at $-130°C$ and not nickel atoms.[186] This idea was supported in later matrix isolation work that implied that Fe_2 and not Fe atoms reacted with methane by C—H oxidative addition[187]:

$$\text{Fe—Fe} + CH_4 \longrightarrow \underset{\underset{H}{|}}{\text{Fe—Fe}}\text{—CH}_3 \text{ or } H\text{–Fe—Fe—CH}_3$$

These $(Ni)_n$ and Fe_2 studies strongly suggest that more than one atom is needed in order for C—H and C—C cleavages in alkanes to take place, especially under these very low temperature conditions.

Nickel crystallized from toluene also yielded a pseudo-organometallic powder that showed good stability and interesting catalytic activity. Its selectivity was significantly different than the Ni–pentane catalyst in that alkenes were rapidly isomerized by Ni–toluene, and 1,3-butadiene was converted to butenes before final hydrogenation to butane (with Ni–pentane, butane was formed immediately).[185]

Nickel crystallized from THF yielded a powder very high in organic content. Matrix isolation spectroscopy studies have suggested that Ni atoms react with THF even at $-196°C$.[188] A yellow complex formed at $-196°$, but on warming, amorphous black Ni particles were produced. This black powder possessed very poor catalytic activity, although it was extremely reactive with alkyl and allyl halides.[44,49,189] Pyrolysis of the Ni–THF powder yielded THF, butanol, furan, butanal, ethylene, and other small molecular fragments.

(ii). *Highly Dispersed Nickel Catalysts on Alumina.* Employing "solvated metal atoms" (Ni–toluene solutions at $-20°$ to $-90°C$), we have permeated catalyst supports with Ni metal. The Ni–toluene (or other arene) complex is stable on melt down, and so the solution can be allowed to stir and warm with Al_2O_3 or other catalyst support. Finally, the Ni–toluene solution deposits Ni atoms and/or Ni clusters in the pores of the support.[121,190] This serves as a new method for preparation of zero-valent highly dispersed catalysts at low temperatures in the absence of H_2O or O_2 and without an H_2-reduction step (normally the case in conventional preparations).[190]

Studies of the activity of these Ni–toluene–Al_2O_3 catalysts in hydrogenation reactions have been made and high activity has been found. Variations in Ni loading showed that for toluene hydrogenation to methylcyclohexane, a 0.4% Ni/Al_2O_3 loading was optimal. This suggested that an optimal cluster size was being produced at this loading level. However, direct crystallite size measurements have so far been fruitless because of the complications arising from the presence of toluene in the Al_2O_3.

Activities of the catalysts increased upon heating under vacuum. Simultaneous measurements of organics, H_2O, and CO_2 released upon the heat treatment suggest that the release of these materials was opening effective catalytic sites. At 500°C preheat treatment, optimum activity was found.

Above this temperature, severe sintering took place and catalytic activity dropped precipitously.[190]

II. Later Transition Metal (Group VIII) Halide, Oxide, and Sulfide Vapors

A. Occurrence, Properties, and Techniques

Only under extreme conditions are the vapors of these species found naturally, such as in the atmosphere of stars or the sun.[191] Parson has stated that the vapor pressures of Fe, CaO, MgO, and SiO_2, the most important substances in planetary beginnings, are sufficiently high to have been important in the nucleation processes for the genesis of a planet. [192]

Table 5-11 summarizes available vaporization data in the literature and includes the vapor composition when available.[114,193–212] Many of the metal halides of this series are quite volatile, but vaporize as telomers. Many of the oxides and sulfides are not thermally stable.

Dewing[213] has reported that the volatilities of MCl_2 (M = Mg, Mn, Co, Ni, Cd) are greatly enhanced in the presence of $AlCl_3$ or $FeCl_3$ vapor. The existence of gaseous MAl_2Cl_8, MAl_3Cl_{11}, and MFe_2Cl_8 was proposed.

B. Chemistry

Van Leirsburg and Dekock[214,215] studied by matrix-isolation spectroscopy the interaction of MX_2 (MX_2 = CaF_2, NiF_2, $NiCl_2$, CrF_2, MnF_2, CuF_2, and ZnF_2) with CO, N_2, NO, and O_2. Their work was preceded by a number of matrix isolation studies of MX_2 molecules themselves, without added reactive ligands.[216–224]

TABLE 5-11

Vaporization Data for the Later Transition Metal (Group VIII) Halides, Oxides, and Sulfides

Compound	mp (°C)	bp (°C)	Heat of vap. (kcal/mole)	Vapor composition[a]	References
$FeBr_2$	d 684			$FeBr_2$, some dimer	193
$FeBr_3$	d subl			Fe_2Br_6, Br_2, $FeBr_2$	194, 195
$FeCl_2$	670 subl	1023	32.1		196
$FeCl_3$	306	d 315	35.2		196
FeF_2	>1000				
FeF_3	>1000				
FeI_2	Red heat			FeI_2, $(FeI_2)_2$	197
FeO	1420				
Fe_2O_3	1565			O_2, lower oxides	198
Fe_3O_4	d 1538			O_2, lower oxides	198
FeS	1197				199
FeS_2	1171				
Fe_2S_3	d				
$CoBr_2$	678			$CoBr_2$, $(CoBr_2)_2$	196, 197
$CoCl_2$	724	1049	30.6	$CoCl_2$, $(CoCl_2)_2$	197, 200
$CoCl_3$	subl				
CoF_2	1200	1400			
CoF_3					
CoI_2	515	570			
CoO	1935		130	Co, O_2, CoO	201
Co_2O_3	d 895				
CoS	1100				199
$NiBr_2$	963			$NiBr_2$, linear	197, 202, 208
$NiCl_2$	1001	subl 973	53.4	$NiCl_2$	196, 197
NiF_2	subl 1000		60		203
NiI_2	797				
NiO	1990		117	Ni, O_2, O, NiO	114, 204, 205
NiS	797				
Ni_2S_3	790				
$RuCl_3$	d 500		46		206
RuF_5	101	250			
RuO_2	d				207
RuO_4	25	d 108			
RuS_2	d 1000				
$RhCl_3$	d 475	subl 800			
RhF_3	subl 600				
RhO_2				Rh, RhO, RhO_2	208
Rh_2O_3	d 1100				207
RhS	d				
Rh_2S_3	d				

[a] Charkin and Dyatkina[212] predicts that the MX_2 species in the vapor state are angular in ground state for Sc, Ti, V, Cr, Y, Zr, Nb, Mo, La, Hf, Ta, W, Ru, Rh, Pd, Re, Os, and Ir, but linear for Mn, Fe, Co, and Ni.

TABLE 5-11 (*continued*)

Compound	mp (°C)	bp (°C)	Heat of vap. (kcal/mole)	Vapor composition[a]	References
$PdBr_2$	d			$(PdBr_2)_4$, $(PdBr_2)_6$	209
$PdCl_2$	d 500			$(PdCl_2)_6$	210
PdF_2	vol red heat				
PdF_3	d				
PdI_2	d 350				
PdO	870			Pd, PdO	208, 207
PdS	d 950				
PdS_2	d				
Pd_2S	d 800				
$OsCl_3$	d 500–600				
OsF_6	32	46			
Os_2O_3	d				
OsO_4	40	130			
OsS_4	d				
$IrBr_4$	d				
$IrCl_2$	d 773				
$IrCl_4$	d				
$IrCl_3$	d 703				
IrF_6	44	53			
IrI_4	d 100				
IrI_3	d				
IrO_2	d 1100				207
IrS_2	d 300				
Ir_2S_3	d				
$PtBr_2$	d 250				
$PtBr_4$	d 180			decomp	211
$PtCl_2$	d 581			Pt_5Cl_{10}, Pt_6Cl_{12}, Pt_4Cl_8	210, 211
$PtCl_4$	d 370				
$PtCl_3$	435				
PtF_6	58				
PtF_4	d red heat				
PtI_2	d 360				
PtI_4	d 130				
PtI_3	d 270				
PtO	d 550				
PtO_2	450				207
Pt_3O_4	d				
PtS_2	d 225				
PtS	d				
Pt_2S_3	d				

With NiF_2 all the small molecules—CO, N_2, NO, and O_2—were studied in argon matrices. The ratio of small molecule/NiF_2 had to be kept reasonably high ($\sim 50/1$) to avoid the facile dimerization of NiF_2.

Upon complexation of CO to the MX_2 molecules, $\nu_{C=O}$ shifted to higher frequencies. This type of pertubation had been studied for CO–solid interactions, and it has been determined that the extent of the pertubation was dependent on the strength of the electric field near the metal cation.[225] It was also concluded that CO attaches to the metal ion, not the halide, and that carbon, not oxygen, was attached to the metal (the MX_2 species were assumed to be ionic in nature based on ESR evidence that CuF_2 is ionic in an argon matrix.)[226]

Comparing NiF_2 and $NiCl_2$, $\nu_{C=O}$ was determined as 2200 cm^{-1} and 2189 cm^{-1}, respectively. $NiCl_2$ is expected to be more covalent than NiF_2, leading to a smaller electric field strength for Ni^{2+} in $NiCl_2$, consequently yielding a lower $\nu_{C=O}$ frequently. It is believed that σ-donation by CO to Ni^{2+} strengthens the C=O bond, and since for Ni^{2+} π-back bonding to π^*-orbitals would not be important, the C=O bond would not be weakened as it is in Ni–CO complexes for Ni(0).

Perturbation of N_2, O_2, and NO was not observed upon complexation to NiX_2 molecules. Perhaps O_2 is a poorer σ-donor and better π-acceptor, and these effects cancel each other; the same may be true for NO.

The Van Leirsburg-Dekock study—along with the knowledge that Group VIII metal halides, and expecially oxides and sulfides are extremely important for catalytic processes with organic molecules—makes it obvious that there is a great deal of interesting work still to be done in this area.

We have carried out some preliminary macroscale studies where $NiCl_2$ was vaporized and allowed to condense with organic dienes. However, stable products were not produced, either by π-complexation or by Cl transfer processes.

$$NiCl_2 + \text{(norbornadiene)} \longrightarrow \text{N.R.}$$

$$NiCl_2 + \text{(butadiene)} \xrightarrow{\quad\times\quad} \text{(Cl–Ni–Cl complex)}$$

Attempts to use the vapors of NiO, FeO, and TiO_2 as butadiene polymerization catalysts have been reported.[114] Some success was encountered, but no real chemistry was elucidated. E-beam vaporization and concondensation ($-196°C$) procedures were employed.

References

1. F. B. Hearnshaw, *Mem. R. Astron. Soc.* **77**, 55 (1972).
2. S. I. Anisimov, A. M. Bonch-Bruevich, M. A. Elyashevich, Y. A. Imas, M. A. Pavlenko, and G. S. Romanov, *Zh. Tekh. Fiz.* **36**, 1273 (1966).
3. A. M. Bonch-Bruevich and Y. A. Imas, *Exp. Tech. Phys.* **15**, 323 (1967).
4. N. N. Rykalin and A. A. Uglov, *Teplofiz, Vys. Temp.* **9**, 575 (1971).
5. A. Korunchikov and A. A. Yan Kovskii, *Zh. Prikl. Spektrosk.* **5**, 586 (1966).
6. M. J. Burden and P. A. Walley, *Vacuum* **19**, 397 (1969).
7. L. Bachmann, *Naturwissenschaften* **49**, 34 (1962).
8. E. Graper, *J. Vac. Sci. Technol.* **8**, 333 (1971).
9. M. Burden and P. A. Walley, *Vacuum* **19**, 397 (1969).
10. J. van Audenhove, *Rev. Sci. Instrum.* **36**, 383 (1965).
11. N. D. Obradovic and G. Bennett, *J. Inst. Met.* **97**, 186 (1969).
12. W. A. Fisher, D. Janke, and K. Stahlschmidt, *Arch. Eisenhuettenwes.* **45**, 757 (1974).
13. M. N. Turko and I. I. Korshakevich, *Spektr. Anal. Geol. Geokhim., Mater. Sib. Soveshch. Spektrosk., 2nd, 1963* p. 53 (1963), (1967).
14. S. Nagata, T. Nasu, and Y. Tomoda, *Oyo Butsuri* **27**, 459 (1958).
15. B. Vodar, S. Minn, and S. Offret, *J. Phys. Radium.* **16**, 811 (1955).
16. B. L. Vallee and R. W. Peattie, *Anal. Chem.* **24**, 434 (1952).
17. M. N. Turko and I. I. Korshakevich, *Izv. Sib. Otd. Akad. Nauk SSSR, Ser. Tekh. Nauk* No. 3, p. 63 (1965).
18. "Handbook of Chemistry and Physics," 56th ed. CRC Press, Cleveland, Ohio, 1975–1976.
19. K. F. Purcell and J. C. Kotz, "Inorganic Chemistry." Saunders, Philadelphia, Pennsylvania, 1977.
20. M. N. Saha, *Nature (London)* **107**, 682 (1921).
21. B. Siegel, *Q. Rev., Chem. Soc.* **19**, 77 (1965).
22. W. L. Batten and G. A. Roberts, U.S. Patent 2,956,304 (1960).
23. G. Cano, *Mem. Sci. Rev. Metall.* **59**, 878 (1962).
24. R. W. Steeves, U.S. Patent 3,063,858 (1962).
25. D. F. Banofsky and E. W. Mueller, *Int. J. Mass Spectrom. Ion Phys.* **2**, 125 (1969).
26. R. J. Ackermann and R. J. Thorn, *Pap. Sect. Inorg. Chem., Congr. Pure Appl. Chem., 16th, 1957* p. 609 (1958).
27. A. Hejduk, Z. Marchwicki, T. Ohly, and M. Szreter, *Pr. Nauk. Inst. Technol. Elektron. Politech. Wroclaw.* **10**, 133 (1973).
28. A. A. Hasapis, A. J. Melveger, M. B. Panish, L. Reif, and C. L. Rosen, *U.S.A. E. C.* **WADD-TR-60-463** (Pt II) (1961).
29. T. Babeliowsky, *Physica (Utrecht)* **28**, 1160 (1962).
30. P. L. Timms and R. B. King, *J. Chem. Soc., Chem. Commun.* p. 898 (1978).
31. M. B. Panish and L. Reif, *J. Chem. Phys.* **37**, 128 (1962).
32. R. F. Hampson, Jr. and R. F. Walker, *J. Res. Natl. Bur. Stand. Sect. A* **65**, 289 (1961).
33. L. H. Dreger and J. L. Margrave, *J. Phys. Chem.* **65**, 2106 (1961).
34. O. C. Truison and P. O. Schissel, *J. Less-Common. Met.* **8**, 262 (1965).
34a. R. Baldock, *Adv. Mass Spectrom.* **3**, 749 (1966).
35. E. B. Owens and A. M. Sherman, *U.S., Dep. Commer., Off. Tech. Serv., AD* **275**, 468 (1962).
36. R. Vanselow and W. A. Schmidt, *Z. Naturforsch., Teil A* **21**, 1690 (1966).
37. J. Strong, *Phys. Rev.* **39**, 1012 (1932).
38. E. A. Rasauer and C. B. Wagner, *J. Appl. Phys.* **37**, 4103 (1966).
39. H. J. Liebl and R. F. K. Herzog, *J. Appl. Phys.* **34**, 2893 (1963).

40. P. S. Skell, private communications.
41. J. S. Roberts, unpublished results from this laboratory.
42. P. L. Timms, *in* "Cryochemistry" (M. Moskovits and G. A. Ozin, eds.), p. 61. Wiley (Interscience), New York, 1976.
43. R. MacKenzie and P. L. Timms, *J. Chem. Commun.* p. 650 (1974).
44. K. J. Klabunde, *in* "Reactive Intermediates" (R. A. Abramovitch, ed.). Plenum, New York, 1979 (in press).
45. K. J. Klabunde, "Metal Vapor Chemistry Related to Molecular Metals," NATO Conf. LesArcs, France, (1978).
46. A. Fontijn, S. C. Kurzius, J. J. Houghton, and J. A. Emerson, *Rev. Sci. Instrum.* **43**, 726 (1972).
47. A. Fontijn, S. C. Kurzius, and J. J. Houghton, *Symp. (Int.) Combust. [Proc.]* **14**, 167 (1973).
48. K. J. Klabunde and H. F. Efner, *J. Fluorine Chem.* **4**, 115 (1974).
49. K. J. Klabunde, *Ann. N.Y. Acad. Sci.* **295**, 83 (1977).
50. L. R. Melby, R. J. Harder, W. R. Hertler, W. Mahler, R. E. Benson, and W. E. Mochel, *J. Am. Chem. Soc.* **84**, 3374 (1962).
51. A. R. Siedle, *J. Am. Chem. Soc.* **97**, 5931 (1975).
52. J. Gladysz, J. Fulcher, and S. Tagashi, *J. Org. Chem.* **41**, 3647 (1976).
53. R. F. Heck, *J. Am. Chem. Soc.* **90**, 5518, 5526, 5531, 5535, 5538, and 5546 (1968).
54. P. M. Henry, *Tetrahedron Let.* p. 2285 (1968).
55. T. Hosokowa, C. Calvo, H. B. Lee, and P. M. Maitlis, *J. Am. Chem. Soc.* **95**, 4914 (1973). T. Hosokowa and P. M. Maitlis, *ibid.* p. 4924.
56. K. J. Klabunde and J. Y. F. Low, *J. Am. Chem. Soc.* **96**, 7674 (1974).
57. K. J. Klabunde, *Angew. Chem., Int. Ed. Eng.* **14**, 287 (1975).
58. K. J. Klabunde, *Acc. Chem. Res.* **8**, 393 (1975).
59. K. Neuenschwander and B. B. Anderson, unpublished results from this laboratory.
60. K. J. Klabunde and J. S. Roberts, *J. Organomet. Chem.* **137**, 113 (1977).
61. J. S. Roberts, Ph. D. Thesis, University of North Dakota; Grand Forks (1975); also unpublished work of J. S. Roberts in this laboratory.
62. K. J. Klabunde, *Chem. Tech.* **6**, 624 (1975).
63. H. F. Efner, D. E. Tevault, W. B. Fox, and R. R. Smardzewski, *J. Organomet. Chem.* **146**, 45 (1978).
64. G. A. Ozin and W. J. Power, *Inorg. Chem.* **17**, 2836 (1978).
65. B. B. Anderson, unpublished results from this laboratory.
66. M. Bader, unpublished results from this laboratory.
67. K. J. Klabunde, B. B. Anderson, M. Bader, and L. J. Radonovich, *J. Am. Chem. Soc.* **100**, 1313 (1978).
68. B. B. Anderson, C. Behrens, L. Radonovich, and K. J. Klabunde, *J. Am. Chem. Soc.* **98**, 5390 (1976).
69. R. Gastinger, unpublished results from this laboratory.
70. W. Martin, unpublished results from this laboratory.
71. B. M. Hoffman, D. L. Diemente, and F. Basolo, *J. Am. Chem. Soc.* **92**, 61 (1970).
72. Collaborative studies with L. J. Radonovich of this Department.
73. Collaborative with L. J. Radonovich of this department and T. Albright of the University of Houston.
74. M. J. Piper and P. L. Timms, *J. Chem. Soc., Chem. Commun.* p. 50 (1972).
75. P. S. Skell and J. J. Havel, *J. Am. Chem. Soc.* **93**, 6687 (1971).
76. B. L. Shaw and N. I. Tucker, "Comprehensive Inorganic Chemistry" Vol. 4, p. 763. Permagon, Oxford, 1973.
77. J. S. Roberts and K. J. Klabunde, *J. Organomet. Chem.* **85**, C-13 (1975).
78. J. S. Roberts and K. J. Klabunde, *J. Am. Chem. Soc.* **99**, 2509 (1977).

79. K. J. Klabunde, T. Groshens, H. F. Efner, and M. Kramer, *J. Organomet. Chem.* **157**, 91 (1978).
80. K. J. Klabunde, J. Y. F. Low, and H. F. Efner, *J. Am. Chem. Soc.* **96**, 1984 (1974).
81. H. F. Efner and W. B. Fox, private communications.
82. J. Y. F. Low, Ph.D. Thesis, University of North Dakota, Grand Forks (1975).
83. P. L. Timms, *Angew. Chem., Int. Ed. Engl.* **14**, 273 (1975).
84. H. Huber, G. A. Ozin, and W. J. Power, *J. Am. Chem. Soc.* **98**, 6508 (1976).
85. H. Huber, D. McIntosh, and G. A. Ozin, *J. Organomet. Chem.* **112**, C50 (1976).
86. D. McIntosh and G. A. Ozin, *J. Organomet. Chem.* **121**, 127 (1976).
87. H. Huber, G. A. Ozin, and W. J. Power, *Inorg. Chem.* **16**, 979 (1977).
88. G. A. Ozin and W. J. Power, *Inorg. Chem.* **16**, 212 (1977).
89. R. M. Atkins, R. MacKenzie, P. L. Timms, and T. W. Turney, *J. Chem. Soc., Chem. Commun.* p. 764 (1975).
90. N. Rosch and R. Hoffman, *Inorg. Chem.* **13**, 2656 (1974).
91. P. L. Timms, private communications.
92. K. Fischer, K. Jonas, and G. Wilke, *Angew. Chem.* **85**, 620 (1973); *Angew. Chem., Int. Ed. Eng.* **12**, 565 (1973).
93. J. R. Blackboro, R. Grubbs, A. Miyashita, and A. Scrivanti, *J. Organomet. Chem.* **120**, C49 (1976).
94. P. S. Skell, *Proc. Int. Congr. Pure Appl. Chem.* **23**(4), 215 (1971).
95. P. S. Skell and M. J. McGlinchey, *Angew. Chem., Int. Ed. Engl.* **14**, 195 (1975).
96. P. S. Skell, J. J. Havel, D. L. Williams-Smith, and M. J. McGlinchey, *J. Chem. Soc., Chem. Commun.* p. 1098 (1972).
97. H. Bonnemann, *Angew. Chem.* **82**, 699 (1970); *Angew. Chem., Int. Ed. Engl.* **9**, 736 (1970).
98. G. A. Ozin and W. J. Power, *Inorg. Chem.* **16**, 2864 (1977).
99. P. L. Timms, *J. Chem. Soc., Chem. Commun.* p. 1033 (1969)
100. P. L. Timms, *Adv. Inorg. Radiochem.* **14**, 121 (1972).
101. G. Wilke, *Angew. Chem., Int. Ed. Engl.*, **2**, 105 (1963).
102. M. Green, J. A. K. Howard, J. L. Spencer, and F. G. A. Stone, *J. Chem. Soc., Chem. Commun.* p. 449 (1975).
103. B. Bogdanovic, M. Kroner, and G. Wilke, *Justus Liebigs Ann. Chem.*, **699**, 1 (1966).
104. P. W. Jolly and G. Wilke, "The Organic Chemistry of Nickel," Vol. 1. Academic Press, New York, 1974.
105. P. S. Skell and J. J. Havel, *J. Am. Chem. Soc.* **93**, 6687 (1971).
106. Private communications with P. L. Timms.
107. Private communications with G. Wilke.
108. W. Klotzbucher and G. A. Ozin, *Inorg. Chem.* **15**, 292 (1976).
109. M. Moskovits and G. A. Ozin, *in* "Cryochemistry" (M. Moskovits and G. A. Ozin, eds.) p. 261 Wiley (Interscience), New York, 1976.
110. K. J. Klabunde, T. Groshens, M. Breżinski, and W. Kennelly, *J. Am. Chem. Soc.* **100**, 4437 (1978).
111. R. A. Cable, M. Green, R. E. Mackenzie, P. L. Timms, and T. W. Turney, *J. Chem. Soc., Chem. Commun.* p. 270 (1976).
112. A. D. English, J. P. Jesson, and C. A. Tolman, *Inorg. Chem.* **15**, 1730 (1976).
113. D. L. Williams-Smith, L. R. Wolf, and P. S. Skell, *J. Am. Chem. Soc.* **94**, 4042 (1972).
114. V. M. Akhmedov, M. T. Anthony, M. L. H. Green, and D. Young, *J. Chem. Soc., Dalton Trans.* p. 1412 (1975).
115. V. M. Akhmedov, M. T. Anthony, M. L. H. Green, and D. Young, *J. Chem. Soc., Chem. Commun.* 777 p. (1974).
116. E. A. Koerner von Gustorf, O. Jaenicke, O. Wolfbeis, and C. R. Eady, *Angew. Chem., Int. Ed. Engl.* **14**, 278 (1975).

117.　T. S. Tan, J. L. Fletcher, and J. McGlinchy, *J. Chem. Soc., Chem. Commun.* p. 771 (1975).
118.　T. Groshens and M. Brezinski, unpublished work from this laboratory.
119.　L. H. Simons and J. J. Lagowski, *Proc. Int. Conf. Organomet. Chem., 8th, 1977* Paper 5BOY. (1977).
120.　L. H. Simons and J. J. Lagowski, *J. Org. Chem.* **43**, 3247 (1978).
121.　K. J. Klabunde, H. F. Efner, T. O. Murdock, and R. Ropple, *J. Am. Chem. Soc.* **98**, 1021 (1976).
122.　P. S. Skell, *170th Natl. Meet. Am. Chem. Soc.* Paper INOR 57 (1975).
123.　P. S. Skell, private communications.
124.　T. Groshens, unpublished results from this laboratory.
125.　R. Gastinger, unpublished results from this laboratory.
126.　R. Middleton, J. R. Hull, S. R. Simpson, C. H. Tomlinson, and P. L. Timms, *J. Chem. Soc., Dalton Trans.* p. 120 (1973).
127.　H. F. Efner and D. Ralston, unpublished results from this laboratory.
128.　K. J. Klabunde and H. F. Efner, *J. Fluorine Chem.* **4**, 115 (1974).
129.　P. L. Timms, *J. Chem. Soc. A* p. 2526 (1970).
130.　D. Staplin and R. W. Parry, *170th Natl. Meet., Am. Chem. Soc.* Paper INOR 117 (1975).
131.　C. A. Tolman, L. W. Yarbrough, and J. G. Verkade, *Inorg. Chem.* **16**, 479 (1977).
132.　M. Chang, R. B. King, and M. G. Newton, *J. Am. Chem. Soc.* **100**, 998 (1978).
133.　R. B. King and M. Chang, *Inorg. Chem.* **18**, 364 (1979).
134.　P. L. Timms, private communications.
135.　W. J. Kennelly, unpublished results from this laboratory.
136.　R. Cable, M. Green, R. E. Mackenzie, P. L. Timms, and T. W. Turney, *J. Chem. Soc., Chem. Commun.* p. 270 (1976).
137.　E. Koerner von Gustorf, O. Jaenicke, and O. E. Polansky, *Angew. Chem.* **84**, 547 (1972); *Angew. Chem., Int. Ed. Engl.* **11**, 533 (1972).
138.　M. J. McGlinchey and P. S. Skell, *in* "Cryochemistry" (M. Moskovits and G. A. Ozin, eds.), p. 167. Wiley (Interscience), New York, 1976.
139.　J. J. Havel, Ph.D. Thesis, Pennsylvania State University, University Park (1972).
140.　H. Bandow, T. Onishi, and K. Tamaru *Chem. Lett.* p. 83 (1978).
141.　D. Gladkowski and F. R. Scholar, *171st Centen. Natl. Meet., Am. Chem. Soc.* Paper INOR 133 (1976).
142.　M. Moskovits and G. A. Ozin, *in* "Cryochemistry" (M. Moskovits and G. A. Ozin, eds.), p. 9. Wiley (Interscience), New York, 1976.
143.　G. C. Pimentel, *Angew. Chem., Int. Ed. Engl.* **14**, 199 (1975).
144.　B. Meyer, "Low Temperature Spectroscopy." Am. Elsevier, New York, 1971; H. Hallam, "Vibrational Spectroscopy of Trapped Species." Wiley, New York, 1972; A. M. Bass and H. P. Broida, "Formation and Trapping of Free Radicals." Academic Press, New York, 1960; G. C. Pimentel, *Spectrochim. Acta* **12**, 94 (1958); *Pure Appl. Chem.* **4**, 61 (1962); E. Whittle, D. A. Dows, and G. C. Pimentel, *J. Chem. Phys.* **22**, 1943 (1954).
145.　M. Poliakoff, *J. Chem. Soc., Dalton Trans.* p. 210 (1974).
146.　M. Poliakoff and J. J. Turner, *J. Chem. Soc., Dalton Trans.* p. 1351 (1973); p. 2276 (1974).
147.　G. Bor, *Acta Chim. Acad. Sci. Hung.* **34**, 315 (1962).
148.　L. Hanlan, H. Huber, E. P. Kundig, B. McGarvey, and G. A. Ozin, *J. Am. Chem. Soc.* **97**, 7054 (1975).
149.　L. Hanlan, H. Huber, and G. A. Ozin, *Inorg. Chem.* **15**, 2592 (1976).
150.　O. Crichton, M. Poliakoff, A. J. Rest, and J. J. Turner, *J. Chem. Soc., Dalton Trans.* p. 1321 (1973).
151.　R. L. DeKock, *Inorg. Chem.* **10**, 1205 (1971).
152.　E. P. Kundig, M. Moskovits, and G. A. Ozin, *Can. J. Chem.* **50**, 3587 (1972).

153. J. H. Darling and J. S. Ogden, *J. Chem. Soc., Dalton Trans.* p. 1079 (1973); *Inorg. Chem.* **11**, 666 (1972).
154. M. Moskovits and G. A. Ozin, *J. Mol. Struct.* **32**, 71 (1976).
155. E. P. Kundig, D. McIntosh, M. Moskovits, and G. A. Ozin, *J. Am. Chem. Soc.* **95**, 7234 (1973).
156. R. P. Eischens, S. A. Francis, and W. A. Pliskin, *J. Phys. Chem.* **60**, 194 (1956).
157. G. Blyholder and M. C. Allen, *J. Am. Chem. Soc.* **91**, 3158 (1969).
158. G. A. Ozin and A. Vander Voet, *Can. J. Chem.* **51**, 637 (1973).
159. W. Klotzbucher and G. A. Ozin, *J. Am. Chem. Soc.* **97**, 2672 (1975).
160. E. P. Kundig, M. Moskovits, and G. A. Ozin, *Can. J. Chem.* **51**, 2710 (1973).
161. D. W. Green, J. Thomas, and D. M. Gruen, *J. Chem. Phys.* **58**, 5453 (1973).
162. E. P. Kundig, M. Moskovits, and G. A. Ozin, *Can. J. Chem.* **51**, 2737 (1973).
163. K. J. Klabunde, unpublished results.
164. H. Huber, G. A. Ozin, and W. J. Power, *Inorg. Chem.* **16**, 2234 (1977).
165. H. Huber and G. A. Ozin, *Can. J. Chem.* **50**, 3746 (1972).
166. H. Huber, W. Klotzbucher, G. A. Ozin, and A. Vander Voet, *Can. J. Chem.* **51**, 2722 (1973).
167. D. McIntosh and G. A. Ozin, *Inorg. Chem.* **15**, 2869 (1976).
168. L. A. Hanlan and G. A. Ozin, *Inorg. Chem.* **16**, 2848 (1977).
169. R. Niedermayer, *Angew. Chem., Int. Ed. Engl.* **14**, 212 (1975).
170. H. W Kohlschutter, *Angew. Chem., Int. Ed. Engl.* **14**, 193 (1975).
171. E. P. Kundig, M. Moskovits, and G. A. Ozin, *Angew. Chem., Int. Ed. Engl.* **14**, 292 (1975).
172. M. Moskovits and G. A. Ozin, *in* "Cryochemistry" (M. Moskovits and G. A. Ozin, eds.), p. 395. Wiley (Interscience), New York, 1976.
173. L. Brewer and C. Chang, *J. Chem. Phys.* **56**, 1728 (1972).
174. J. M. Brom, W. D. Hewett, and W. Weltner, *J. Chem. Phys.* **62**, 3122 (1975).
175. D. W. Green and D. M. Gruen, *J. Chem. Phys.* **60**, 1797 (1974); **57**, 4462 (1972).
176. J. J. Turner and M. Poliakoff, *J. Chem. Soc. A* p. 2403 (1971).
177. L. Hanlan and G. A. Ozin, *J. Am. Chem. Soc.* **96**, 6324 (1974).
178. M. Moskovits and J. Hulse, *Surf. Sci.* **57**, 125 (1976).
179. M. Moskovits and J. E. Hulse, *J. Chem. Soc., Faraday Trans. 2* **73**, 471 (1977).
180. A. J. Hanlan, G. A. Ozin, and W. J. Power, *Inorg. Chem.* **17**, 3648 (1978).
181. G. A. Ozin, W. J. Power, T. H. Upton, and W. A. Goddard, *J. Am. Chem. Soc.* **100**, 4750 (1978).
182. A. J. Hanlan and G. A. Ozin, *Inorg. Chem.* **16**, 2857 (1977).
183. M. Moskovits and J. E. Hulse, *J. Chem. Phys.* **66**, 3988 (1977).
184. T. O. Murdock, Ph.D. Thesis, University of North Dakota, Grand Forks (1977).
185. K. J. Klabunde, S. Davis, H. Hattori, and Y. Tanaka, *J. Catal.* **54**, 254 (1978).
186. S. C. Davis and K. J. Klabunde, *J. Am. Chem. Soc.* **100**, 5973 (1978).
187. P. H. Barrett, M. Pasternak, and R. G. Pearson, *J. Chem. Soc.* **101**, 222 (1979).
188. S. C. Davis, K. J. Klabunde, R. Hauge, and J. L. Margrave, unpublished results.
189. T. O. Murdock, unpublished results from this laboratory.
190. K. J. Klabunde, D. Ralston, R. Zoellner, H. Hattori, and Y. Tanaka, *J. Catal.* **55**, 213 (1978).
191. A. Schadee, *Int. Astron. Union, Symp.* **26**, 92 (1966).
192. A. L. Parson, *Mon. Not. R. Astron. Soc.* **105**, 244 (1945).
193. R. F. Porter and R. C. Schoonmaker, *J. Phys. Chem.* **63**, 626 (1959).
194. N. W. Gregory and B. A. Thackrey, *J. Am. Chem. Soc.* **72**, 3176 (1950).
195. J. D. Christian and N. W. Gregory, *J. Phys. Chem.* **71**, 1579 (1967).
196. C. G. Maier, *U.S., Bur. Mines, Tech. Pap.* **360**, 1 (1925).
197. R. C. Schoonmaker, A. H. Friedmann, and R. F. Porter, *J. Chem. Phys.* **31**, 1586 (1959).

198. E. K. Kazenas, D. Chizhikov, and Y. U. Tsvetkov, *Termodin. Kinet. Protsessov. Vos-stavov. Metal., Mater., Konf., 1969* p. 14 (1972).
199. W. Biltz, *Z. Anorg. Chem.* **59**, 273 (1908).
200. Y. Saeki, R. Matsuzaki, and N. Aoyama, *J. Less-Common Met.* **55**, 289 (1977).
201. R. T. Grimley, R. P. Burns, and M. G. Inghram, *J. Chem. Phys.* **45**, 4158 (1966).
202. Z. Molnar, G. Schultz, J. Trammel, and I. Hargittai, *Acta Chim. Acad. Sci. Hung.* **86**, 223 (1975).
203. M. Farber, R. T. Meyer, and J. L. Margrave, *J. Phys. Chem.* **62**, 883 (1958).
204. R. T. Grimley, R. P. Burns, and M. G. Inghram, *J. Chem. Phys.* **35**, 551 (1961).
205. H. L. Johnston and A. L. Marshall, *J. Am. Chem. Soc.* **62**, 1382 (1940).
206. S. A. Shehukarev, N. I. Kolbin, and A. N. Ryabov, *Zh. Neorg. Khim.* **3**, 1721 (1958).
207. L. Brewer and G. M. Rosenblatt, *Chem. Rev.* **61**, 257 (1961).
208. J. H. Norman, H. G. Staley, and W. E. Bell, *J. Phys. Chem.* **68**, 662 (1964).
209. R. C. Williams and N. W. Gregory, *J. Phys. Chem.* **73**, 623 (1969).
210. H. Schafer, U. Wiese, K. Rinke, and K. Brendel, *Angew. Chem., Int. Ed. Engl.* **6**, 253 (1967).
211. S. A. Shchukarev, T. A. Tolmacheva, M. A. Oranskaya, and L. V. Komandrovskaya, *Zh. Neorg. Khim.* **1**, 8, 17 (1956).
212. O. P. Charkin and M. E. Dyatkina, *Zh. Strukt. Khim.* **6**, 579 (1965).
213. E. W. Dewing, *Metall. Trans.* **1**, 2169 (1970).
214. D. A. Van Leirsburg and C. W. Dekock, *J. Phys. Chem.* **78**, 134 (1974).
215. C. W. Dekock and D. A. Van Leirsburg, *J. Am. Chem. Soc.* **94**, 3235 (1972).
216. D. E. Milligan, M. E. Jacox, and J. D. McKinely, *J. Chem. Phys.* **42**, 902 (1965).
217. K. R. Thompson and K. D. Carlson, *J. Chem. Phys.* **49**, 4379 (1968).
218. J. W. Hastie, R. H. Hauge, and J. L. Margrave, *High Temp. Sci.* **1**, 76 (1969).
219. M. E. Jacox and D. E. Milligan, *J. Chem. Phys.* **51**, 4143 (1969).
220. J. W. Hastie, R. H. Hauge, and J. L. Margrave, *High Temp. Sci.* **3**, 257 (1971).
221. A. Loewenschuss, A. Ron, and O. Schnepp, *J. Chem. Phys.* **49**, 272 (1968).
222. V. Calder, D. E. Mann, K. S. Seshadri, M. Allavena, and D. White, *J. Chem. Phys.* **51**, 2093 (1969).
223. A. Snelson, *J. Phys. Chem.* **70**, 3208 (1966).
224. M. J. Linevsky, *U.S. Gov. Res. & Dev. Rep.* **AD-670-626** (1968).
225. C. L. Angell and P. C. Schaffer, *J. Phys. Chem.* **70**, 1413 (1966).
226. P. H. Kasai, E. B. Whipple, and W. Weltner, Jr., *J. Chem. Phys.* **44**, 2581 (1966).

CHAPTER **6**

Groups IB and IIB Metals (Cu, Ag, Au, Zn, Cd, Hg), Metal Halides, Oxides, and Sulfides

I. Groups IB and IIB Metal Atoms (Cu, Ag, Au, Zn, Cd, Hg)

A. Occurrence, Properties, and Techniques

Both these series of metals are fairly easy to vaporize, especially Group IIB. Mercury, of course, is so volatile that its vapors are ever present in laboratory vacuum systems, which is somewhat of a nuisance.[1] More importantly, since Hg is quite poisonous, Hg vapor is also found in the lower atmosphere, and its level of concentration usually coincides with smog levels.[2] Therefore, the source of Hg vapor in the atmosphere must be due at least partially to industrial processes. However, natural occurrences also pump Hg vapor into the air, and there is good evidence that an earthquake in the Soviet Union did so.[3]

As with most elements, the vapors of groups IB and IIB are detectable in stars, although it is often unclear as to whether metal atoms, ions, vapors of metal oxides, sulfides, etc. are involved.[4,5]

These metals are conveniently vaporized by resistive heating in $W-Al_2O_3$ crucibles, although many other vaporization techniques have been employed. These methods are summarized in Table 6-1 and[6-31] include electron-beam vaporizations,[6,12] induction heating or levitation,[11] and Nd-glass laser evaporations.[19] Studies of laser evaporations have shown that they can take place explosively. Heating of the metal, which can be as high as 10^{10} deg/sec, is followed by metal vaporization and then by the formation of a supersonic vapor jet.[20] Deep craters form very rapidly, accompanied by ejection of liquid as well as of gaseous metal.[21] Arc vaporizations,[7,10,22,25,27] glow discharge,[8] and resistive-heating vaporizations[24] have also been widely used.

For metal atom chemistry of these elements, simple resistive-heating vaporizations from $W-Al_2O_3$ or C crucibles are preferred, and in this way almost all the vapor is monoatomic.[17] Table VI-1 lists some of the original literature regarding the early studies of vapor compositions and the determination of heats of vaporizations.

TABLE 6-1

Vaporization Techniques and Vapor Compositions of the Groups IB and IIB Metals

Metal	mp (°C)	bp (°C)[a]	Heats of vap. (kcal/mole)[b]	Vaporization method[c]	Vapor comp (spectra)[d,e]	References
Cu	1083	2567	73.1	Elec. beam, pulsed or intermittant discharge, glow discharge, continuous discharge, induction heating, laser, resistive heating	Cu_1(1) Cu_2(0.0009)	7–23
Ag	961	2212	64.4	Elec. beam, induction heating, resistive heating, electric arc	Ag_1(1) Ag_2(0.0005)	6, 9–12, 14–18, 22, 24
Au	1064	2807	87.7	Elec. beam, intermittant discharge or arc, resistive heating	Au_1(1) Au_2(0.0007)	6, 10, 12, 14–18, 23, 25, 26
Zn	420	907	31.2	Elec. beam, intermittant discharge or arc, laser	Zn(1)	12, 17, 19–22, 25, 27–30
Cd	321	765		Electric arc, laser		17, 20–22, 27
Hg	−39	357		Electric arc, resistive heating		17, 22

[a] See Choong and other authors.[23,31]

[b] Also cf. p. 2 of Chapter 1. These are original literature references.

[c] Preferred vaporization method for metal atom chemistry applications is simple resistive heating from a W–Al_2O_3 crucible.

[d] Siegel[17] summarizes vapor composition data.

[e] Ag_1 3383A doublet absorption.[31] Ag_2 3281A doublet absorption. Nelson and Kuebler[23] record atomic absorption spectra for vaporized atoms of Cu, Au, B, Dy, Fe, Pb, and W.

The great volatility of the Group IIB metals allows them to be generated in vacuum systems under conditions in which they remain gaseous, and where they can then be used as sensitizers in photochemical processes.[32,33] Generally, the energy absorbed by the metal atom (*hv* from external source) is very efficiently transferred to some added reactant. These studies, best categorized as chemical studies of excited metal atoms, will be covered in the Chemistry section of this chapter (see especially Abstration Reactions).

Since the d-orbitals are filled (Group IIB) or nearly filled (Group IB) for these metals, there are no readily available empty orbitals. Back π-bonding with π-acid ligands is still a viable bonding mode, but synergism is much less important with these metals than with the Group VIII metals. Therefore, the simple orbital mixing process so important with the Group VIII metals is much less important with these metals. On the other hand, the Groups IB and IIB metals have a strong tendency to undergo abstraction and oxidative additions, reaction modes that reflect the ease of ionization and loss of electrons from the 4s, 5s, or 6s levels.

B. Chemistry

1. ELECTRON-TRANSFER PROCESSES

McIntosh and Ozin have codeposited, on microscale, Ag atoms and O_2. Two products were formed, both apparently involving close ion-pair formation (i.e., $Ag^+O_2^-$ and $Ag^+O_4^-$).[34]

When Ag–CO/O_2 depositions were carried out, OC–$Ag^+O_2^-$ was formed as an intermediate.[35] In this case, $v_{C=O}$ was observed at 2165 cm^{-1}, higher than free CO, indicating complexation to a cation site. Also, $v_{O=O}$ was in the same region as is found in the literature for O_2^- (1110 cm^{-1}). Thus, in this case Ag behaved as a strong σ-acceptor but a weak π-donor, with respect to CO.[35]

2. ABSTRACTION PROCESSES

In the earliest report of a synthetic application of the macroscale co-condensation technique (1968), Timms reported a very efficient abstraction

$$BCl_3 + 2\,Cu\ atoms \longrightarrow 2\,CuCl + B_2Cl_4$$

of halogen from BCl_3 by copper atoms.[36,37] The resultant ˙BCl_2 coupled to form B_2Cl_4 in high yields. This method was employed for synthesis of 10-g batches of B_2Cl_4, a considerable improvement over previous synthetic methods. Timms also used the method to prepare alkyl-substituted diboron chlorides in 40–70% yields.[38] With PCl_3, Cu atoms yielded smaller amounts of P_2Cl_4 (15%).[38]

Timms and co-workers have compared Cu atoms with Na atoms in de-halogenations of alkyl halides (using macro-cocondensation techniques at $-196°C$). Both metals reacted with alkyl iodides and bromides to yield a mixture of alkyl coupling products and disproportionation products.[38] Copper (and silver) atoms apparently coordinated to the incipient alkyl radical so that the radical was never in a free state; this was demonstrated by experiments with $R(-)$-sec-butyl chloride where in each case the coupling product $S,S(-)$-3,4-dimethylhexane was formed with about 70% optical purity.[38] Sodium atoms yielded an inactive product, which supports earlier discussion that Na atoms with alkyl halides indeed do yield "free" radicals.[39]

$$CH_3{-}CH_2{-}\overset{*}{C}H{-}CH_3 \xrightarrow[\text{or Ag atoms}]{\text{Cu atoms}} CH_3CH_2CH{-}CH{-}CH_2{-}CH_3$$

with Cl substituent on the starred carbon, and CH₃, CH₃ substituents on product

$R(-)$ Na atoms $S,S(-)$ [70% optical purity]

                              Same product but optically inactive

The implication of Timms' experiments with the optically active halide is that $R^{\cdot}$ in a Cu or Ag/RX matrix is never free, and possibly complexed as R–M at low temperatures. Since a number of meaningful papers have appeared demonstrating that some perfluoro R_fAg and Ar_fAg compounds exhibit good stability,[40-45] we attempted to synthesize and isolate R_fAg by the reaction of R_fI with Ag vapor.

$$R_fI + 2\,Ag \longrightarrow AgI + R_fAg$$

We were successful in this study, which served as the first example of a metal atom abstraction reaction being employed to produce an interesting *metal-containing* product (rather than just metal halide).[46] The yields of R_fAg ranged from 1–20%, and were favored by high R_fI/Ag ratios in the matrix, contrary to what would be expected from the observed reaction stoichiometry. These R_fI dilution experiments, along with observations of matrix appearance, convinced us that the abstraction reaction was taking place on warm-up rather than during deposition at $-196°C$, and that the first step was simple, R_fI—Ag complexation.[46] This complexation would discourage Ag wastage (due to repolymerization) and yields would thus go up with the R_fI/Ag ratio. On slow warm-up the rate-determining abstraction would take place with $R_f^{\cdot}$ quickly trapping a nearby complexed Ag atom.

$$R_f{-}I + Ag \xrightarrow{-196°C} R_f{-}I\ldots\ldots Ag$$

$$\downarrow \text{slow warmup}$$

$$R_f{-}I + R_tAg \xleftarrow{[R_fI\cdots Ag]} R_f^{\cdot} + IAg$$

Furthermore, it seemed most likely that the iodide group would serve as the Ag complexing site (at $-196°C$) since we had earlier observed that perfluoroalkanes did not complex at all with metal atoms at $-196°C$. Other findings in this work were that (1) new, more sensitive R_fAg, such as $CF_3CF_2CF_2Ag$ and CF_3Ag could be produced, (2) the R_fAg compounds were only isolable if complexed to a donor solvent CH_3CN, and (3) C_6F_5Cu and C_6F_5Ag would be produced and isolated in noncoordinating solvents such as toluene.[46]

Gold atoms also served to abstract halide from RX. We have produced $(C_6F_5Au)_n$ in this way, although careful characterization of this material is still needed. Timms[38] has made some interesting observations concerning the capacity of gold atoms to dehalogenate ethyl bromide, but not ethyl chloride. Thermodynamic calculations on the bulk $Au–C_2H_5Br$ reaction show it to be just slightly exothermic, whereas the bulk $Au–C_2H_5Cl$ reaction is endothermic. So it appears that at least in this case only a kinetic advantage is gained by using Au vapor, and not a thermodynamic one.[38]

No abstraction work with ground state Zn, Cd, or Hg has yet appeared in the literature. However, these metals, particularly Hg, can be promoted to excited states by photolysis in the gas phase (their volatility permits this experimentally). By far the overriding interest in this area has not been the chemistry of the excited metal atom, but simply the fact that energy transfer from the excited metal atom to small molecules such as N_2O, H_2, or alkanes is a very efficient process. In this way the production of H, O, and RH* have

$$Hg(6^1S_0) \xrightarrow[2537 \text{ Å}]{hv} Hg(6^3P_1) \xrightarrow{RH} Hg(6^1S_0) + RH^*$$

been throughly studied.[47] This is an extremely useful process, because reagents that normally do not absorb in the UV can be activated easily with these metal atoms. Table 6-2 summarizes the resonance levels for Zn, Cd, and Hg, and outlines some important energy considerations.

The excitation energy of the Hg atom (112 kcal; 2537 Å irradiation) is totally transferable to many added foreign molecules. This is sufficient energy to break a wide variety of chemical bonds, and the dissociations of these excited species is an interesting area of study. In fact, the dissociation may occur immediately upon energy transfer. Both C—C bonds and C—H bonds are readily cleaved.[47]

The quenching process of most interest in the present context is "quenching by compound formation."[47] This is formally an abstraction reaction by a metal atom, albeit an excited metal atom. No stable metal–H or metal–R compounds are produced, however, since the Hg—H bond strength is only

$$Hg(^3P_1) + RH \longrightarrow HgH + R^\cdot$$

TABLE 6-2

Energy Levels for Photochemical Excitation of Zn, Cd, and Hg Atoms

Metal	Resonance level	Wavelength of line (Å)	Excitation energy (kcal/mole)	Heat of formation of hydride (kcal/mole)	Max strength of bond that can be split (kcal/mole)	References
Zn	4^1P_1	2139	133.4	23.1	156.5	47
	4^3P_1	3076	92.5	23.1	156.5	
Cd	5^1P_1	2288	124.4	15.5	139.9	47
	5^3P_1	3261	87.3	15.5	102.8	
Hg	6^1P_1	1849	153.9	8.5	162.4	47
	6^3P_1	2537	112.2	8.5	120.7	
Na[a]	3^2P	5890/96	48.3	51.6	99.9	47

[a] For comparison.

about 8.5 kcal, and the compound rapidly dissociates even at room temperature. However, in principle the 8.5 kcal adds to the theoretical maximum strength of the C—H bond that can be broken (112.2 + 8.5 = 120.7 kcal). While the importance of this effect is not great for Hg, it is for Zn and Cd.

To employ Zn or Cd, the reaction zone must be maintained at 250°–300°C to generate enough vapor pressure ($\sim 10^{-3}$ mm) for effective photolysis studies to be carried out. For Cd, the excitation energy of the 5^3P_1 state is 87.3 kcal. This is too small to enable a photosensitized split of hydrogen by direct dissociation.

$$Cd(5^3P_1) + H_2 \longrightarrow Cd(5^1So) + 2 H^\cdot$$

If a Cd—H intermediate were formed, however, 102.8 kcal overall would be available, which is just sufficient to cleave the H—H bond. Indeed, Bender[48] and Olsen[49] have observed resonance excitation bands for CdH in Cd, H_2, $h\nu$ mixtures. And a similar situation occurs in Cd*–alkane systems; Cd* can only split the C—H bond if a Cd—H species is an intermediate,[47] and Cd* does split C—H bonds.

It must be concluded that excited Cd at least, and probably Zn* and Hg*, abstract H from H_2 and alkanes. The fate of the M—H species may be simple thermal dissociation to M atom + H$^\cdot$ or perhaps H$^\cdot$ + MH → H_2 + M.

It should be noted that Hg* can also be employed to polymerize acetylene or to hydrogenate acetylene through photosensitization processes. In both cases, H$^\cdot$ is believed to be a key intermediate.[47,50,51]

3. OXIDATIVE ADDITION PROCESSES

Only one report in the area of oxidative addition has appeared.[52] Zinc atoms codeposited with R_fI [$R_f = CF_3$, CF_3CF_2, $CF_3CF_2CF_2$, and

$(CF_3)_2CF]$ yielded R_fZnI at low temperature in a nonsolvated state. The chemistry of these species was quite different from the analogous ether-solvated $R_fZnI(R_2O)_2$ analogs. Hydrolysis processes were facile at low temperature and CF_3ZnI decomposed at low temperatures to yield $:CF_2$ which was trapped by an added alkene.[52]

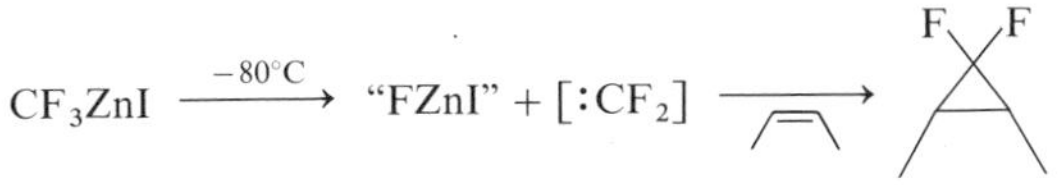

$$CF_3ZnI \xrightarrow{-80°C} \text{"FZnI"} + [:CF_2] \longrightarrow \text{(structure)}$$

$$(CF_3)_2CFI \xrightarrow[-50°C]{H_2O} (CF_3)_2CFH$$

4. SIMPLE ORBITAL MIXING PROCESSES

a. Ethylene. A series of matrix-isolated M–ethylenes (M = Cu, Ag, and Au) have been prepared.[53-55] For Cu, mono-, di-, and tri-C_2H_4 complexes were formed (cf. Table 6-3). As would be expected, the strongest π–C_2H_4–Cu interaction was found in the mono-complex, as shown by the lowest energy $v_{C=C}$. The strength of interaction decreased some with addition of a second ethylene molecule and even more with addition of a third. None of the complexes were found to be thermally stable above about 50°K, which is certainly quite different from the analogous Ni complexes (cf. Chapter 5).

It is interesting that the $v_{C=C}$ values for the mono-complexes of Cu, Ag, and Au are essentially identical. This is somewhat surprising, but it should be noted that the strength of each interaction is quite high. Perhaps there

TABLE 6-3

Matrix Isolated Ethylene Complexes with Cu, Ag, and Au Atoms

Complex[a]	$v_{C=C}(cm^{-1})$	Comments	Thermal stability	References
$(C_2H_4)_3Cu$	1517	D_{3h} planar, weakest interaction	40°K	53
$(C_2H_4)_2Cu$	1508	Intermediate strength of π–Cu interaction	10°–40°K	53
$(C_2H_4)Cu$	1475	Strongest interaction	10°–40°K	54, 55
$(C_2H_4)Ag$	1476	Only mono complex formed, purple	10°–40°K	53–55
$(C_2H_4)Au$	1475	Only mono complex formed, green	40°–70°K	54, 55

[a] All complexes were formed at 10°K in a high excess of argon.

is a limit regarding C_2H_4 bonding capabilities with these metals. For Ag and Au, the di- and tri-complexes were not formed, as they were in the Cu case.[53–55]

b. Hexafluoro-2-butyne and Acetylene. Metal atom–acetylene systems have been little investigated, mainly because of the complexity of product mixtures generally obtained. One acetylene, hexafluoro-2-butyne, (HFB), has been codeposited on macroscale with Cu, Ag, Au, or Zn atoms.[56] In the Ag and Zn cases, no complexes of notable stability were generated. With Cu and Au, however, moderately stable M–HFB complexes were formed. Little is known about their stoichiometries as yet. It would appear that for these metals HFB is superior to CO as a ligand since stable M–CO complexes are not known for these metals, and since the M–HFB complexes in question do not bond CO when exposed to it. The $\nu_{C=C}$ values showed that HFB was bound more strongly to Ni or Pd than Pt or Au.[56]

Kasai and McLeod have codeposited Cu atoms and acetylene on micro-scale with inert gases. Two complexes were formed, $Cu(C_2H_2)$ and $Cu(C_2H_2)_2$, as studied by ESR.[57] The hyperfine splittings observed indicate that these complexes are stabilized by dative bonding resulting from the interaction of the π-orbitals of acetylene and the valence orbitals 3d, 4s, and 4p of the Cu atom. Thus, a π-type bond is formed, in contrast to a similar study of Al–C_2H_2 (see Chapter 7).[57]

c. Arenes and Ethers. The Group IB and IIB metals form very unstable π–arene complexes at low temperature. There is evidence that electron-demanding arenes yield more thermally stable materials than electron-rich ones.[58,59] Copper atoms and hexafluorobenzene yielded a red complex at low temperature, the color of which may indicate that charge transfer took place. Further work is needed, especially with highly electron-demanding ligands.

Although the Cu–toluene and Ag–toluene complexes are very unstable, they have been used for the preparation of dispersed Cu and Ag/Al_2O_3 catalysts.[60] The dispersions were not nearly as fine as in the Ni–toluene

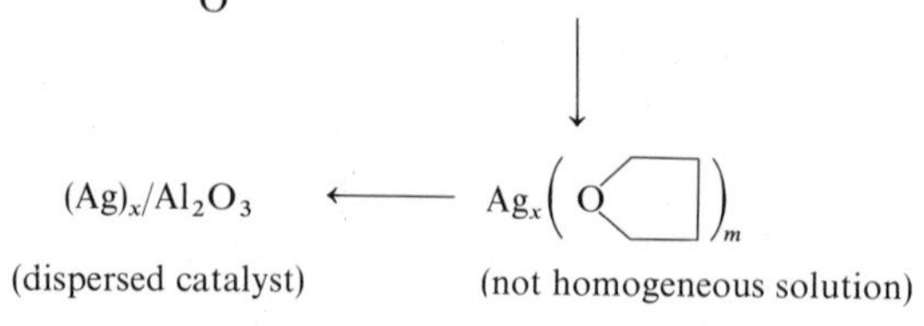

(Ag)$_x$/Al_2O_3 ← Ag$_x$$\left(O\!\!\begin{array}{c}\square\end{array} \right)_m$

(dispersed catalyst)   (not homogeneous solution)

system discussed in Chapter 5. M–etherates have also been employed for dispersion of Ag on catalyst supports. The Ag–THF dispersion worked better in this way than did the Ag–toluene dispersion.[61] However, it appeared that a true homogeneous solution ("solvated Ag atoms") was never formed, as with "solvated Ni atoms" described in Chapter 5.

d. Phosphines. Timms[39] reports that PN, generated as a reactive intermediate by pyrolysis of P_3N_5 at 850°C, reacted with Ag atoms on codeposition to yield $Ag(PN)_2$. These were microscale studies, and spectroscopic examination of the $Ag(PN)_2$ complex revealed that the Ag—P bonding rather than Ag—N bonding was involved.

This work beautifully illustrates one of the few examples wherein reactive metal atoms are generated and allowed to react with a relatively short-lived, reactive ligand.

e. Carbon Monoxide. Group IB metal atoms have been codeposited with CO on microscale. Table 6-4[62–68] summarizes the M–CO complexes prepared. In the Cu(CO), $Cu(CO)_2$, $Cu(CO)_3$ series, the $v_{C=O}$ values observed[62–64] are very strange. It would appear that backbonding to the π^* orbitals of CO is the most significant in the $Cu(CO)_2$ case, rather than in the expected Cu(CO) case. In fact the Cu(CO) system exhibits the highest $v_{C=O}$, which suggests the weakest backbonding interaction. These data are not easy to rationalize. The geometries of these complexes are linear in the Cu(CO) and $Cu(CO)_2$ systems and trigonal planar D_{3h} in the $Cu(CO)_3$ system. These are the same results found in the analogous Co, Ni, Pd, and Pt systems, and this geometry is therefore the unequivocally preferred bonding mode for these late transition metals. (Burdett, Poliakoff, Turner,

TABLE 6-4

Carbonyl Complexes of Group IB Metal Atoms Prepared by Matrix Isolation Spectroscopy Methods

Complex	$v_{C=O}(cm^{-1})$	Comments	References
Cu(CO)	2010	$C_{\infty v}$	62–64
$Cu(CO)_2$	1892	$D_{\infty h}$, linear	63, 64
$Cu(CO)_3$	1990, 1977	D_{3h}	63, 64
Ag(CO)	1958	$C_{\infty v}$	64, 65
$Ag(CO)_2$	1842, 1828	$D_{\infty h}$, linear	64, 65
$Ag(CO)_3$	1967, 1937	D_{3h} or C_2	64, 65
Au(CO)	2039	$C_{\infty v}$	66, 67
$Au(CO)_2$	1936	$D_{\infty h}$, linear	66–68

and Dubost[69] have reviewed metal carbonyl work as it relates to matrix isolation and structure determination by vibrational spectroscopy).

A comparison of $Cu(CO)_2$, $Ag(CO)_2$, and $Ag(CO)_2 \nu_{C=O}$ frequencies indicates that the strength of backbonding interaction is probably in the order $Ag > Cu > Au$ (Table 6-4).

f. Dinitrogen. No Group IB or IIB dinitrogen complexes have been reported.

g. Dioxygen. Copper, silver, and gold atoms have been codeposited with O_2 on microscale.[66–74] Little has been reported concerning these $\nu_{O=O}$ values or geometries. With Cu it is believed that $Cu(O_2)_2$ was formed, but in the case of Au, there was only the monocomplex $Au(O_2)$. This green Au complex exhibited $\nu_{O=O}$ at 1092 cm^{-1},[74] with apparent C_{2v} side-on bonding. This π-type bond contrasts the electron-transfer process for $Ag + O_2 \rightarrow Ag^+ O_2^-$.[34]

h. Carbon Dioxide. Silver atoms and CO_2 have been codeposited on microscale with the formation of a very weakly bound Ag–CO_2 complex. Spectra and CNINDO calculations favor a side-on bonded model of Cs symmetry as shown below.[73] An O–C–O angle of 170° was predicted by the calculations.

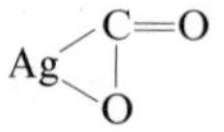

$$\text{Ag} \begin{smallmatrix} \diagup \\ \diagdown \end{smallmatrix} \begin{smallmatrix} C=O \\ | \\ O \end{smallmatrix}$$

i. Carbon Monoxide–Dioxygen Mixtures. Huber, McIntosh, and Ozin have carried out some very interesting work with Ag–CO/O_2 and Au–CO/O_2 mixtures.[35,70] These studies relate to Ag metal and Au metal as catalysts for oxidation processes. In the case of Au–CO–O_2 microscale codepositions, detectable intermediates, OC–Au–CO_3 and OC–Au–O, were formed. Apparently the following sequence took place on warming from 10°K to >40°K[70]:

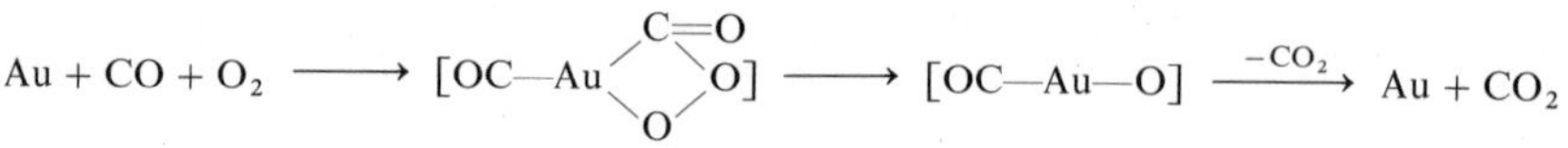

$$Au + CO + O_2 \longrightarrow \left[OC-Au \begin{smallmatrix} \diagup \\ \diagdown \end{smallmatrix} \begin{smallmatrix} C=O \\ | \\ O \end{smallmatrix} O \right] \longrightarrow [OC-Au-O] \xrightarrow{-CO_2} Au + CO_2$$

Note the four-membered ring intermediate. Could similar species be formed on bulk gold?

In the case of Ag–CO/O_2 mixtures, CO_2 was not formed, but apparently electron-transfer proceeded so that OC–$Ag^+ O_2^-$ was formed as an unstable but observable species (cf. Electron-Transfer Processes section earlier in this chapter).[35]

5. CLUSTER FORMATION PROCESSES

A huge effort has been made in industrial laboratories to understand the clustering processes of metal atoms. Silver has been extensively studied because of its importance in oxidation catalysis. Gold has been studied because it is reasonably amenable to study and understanding. For example, the classic chemical processes for production of metal sols have been extensively investigated in $HAuCl_4$ reduction systems.[75] It was determined that the first step in the gold precipitation involves formation of elementary nuclei (30–300 atoms), followed by further growth often with crystallographic twinning of particles. It was found that the gold sols are negatively charged which contributes to their dynamic equilibrium stabilization.[75,76] These sols form preferentially on the edges of crystalline surface supports, and it is believed that on edges the positive potential of the cations can overlap the negative potential of the gold particles. The basal faces of the crystal (kaolin) remain free of gold sols because negative OH sites exist in these areas.

Gold atoms on surfaces would probably act somewhat differently since the small clusters forming would presumably not be negatively charged. Neidermayer[77,78] has discussed the theoretical aspects of cluster growth on surfaces under equilibrium conditions. Parameters such as the heat of vaporization, the energy of desorption, and the energy of diffusion are involved, and these determine whether a cluster will grow in a two- or a three-dimensional array. In a general way, Neidermayer has mathematically treated the cluster growth problem as one would treat polymer growth. The process can be expressed as a set of differential equations which can be numerically solved if four values are known: (1) binding energy of the metal atoms, (2) lattice parameters, (3) substrate temperature, and (4) impingement rate. The growth process becomes governed by an equilibrium between impingement and desorption, which is indicated by a constant value of monomer concentration for a considerable time interval. What then is the critical nucleus size that might be maintained? Theoretically, an answer to this question is possible if the above values are known, but it is usually difficult to compare experimental with relevant theoretical results. However, at the average temperature normally employed (300°C), it is possible to establish equilibria with very small cluster sizes. However, low-temperature depositions to form clusters present other problems, and desorption equilibrium is not attained. Both the cluster concentration and the nucleus concentrations rise very steeply, and no nucleation period is observed. Generally, the cluster growth for gold under low-temperature conditions (80°K) on a clean surface involves an induction period of 10^{-5} sec, twin formation until 10^{-3} sec, constant growth to 10^{-1} sec, and then coalescence to a film.[77]

Thus, cluster formation in low-temperature matrices may follow the same mathematical model described by Niedermayer except that cluster formation would not be reversible. That is, the energy of diffusion is much too high for a low-temperature matrix. Some cluster studies in matrices have been carried out for Cu, Ag, and Au, and these will now be discussed under the following headings.

a. Metal Carbonyl Dimers and Higher Clusters.   As discussed in Chapter 5, kinetic studies of metal carbonyl dimers in low-temperature matrices have been carried out. An inert gas containing CO is a "reactive matrix." Silver vapor–CO codepositions were studied in some detail in order to demonstrate that metal atom matrix isolation techniques can be applied to kinetic analyses.[71–73] In the Ag–CO case, it was shown that $Ag(CO)_3$ was being converted to dimer $Ag_2(CO)_6$. The dimerization was kinetically studied first by carrying out experiments where metal flux was varied while CO deposition rate remained constant. In a second set of experiments, the Ag/CO metal ratio was held constant and the temperature varied (30°, 33°, 35°, and 37°K). Rate constants and diffusion coefficients were obtainable, the latter ranging from D $(cm^2/sec) = 1.8 \times 10^{-18}$ to 1.2×10^{-15}. The activation energy for this process was determined to be 1900 cal/mole.[72]

The $Ag_2(CO)_6$ species is believed to exist as an M–M bonded species.[65,72] Similarly, the $Cu_2(CO)_6$ adduct was prepared and also found to be M–M bonded, showing $v_{C=O}$ at 2039 and 2003 cm^{-1}, and probably D_3 symmetry.[63]

Moskovits and Hulse[79,80] have carried the M cluster–CO matrix isolation work even further. By using very low concentrations of carbon monoxide, CuCO, Cu_2CO, Cu_3CO, and Cu_4CO were spectroscopically detected in a low-temperature matrix. A localized bonding scheme was proposed,[79] and it was noted that $v_{C=O}$ rapidly approached the value for CO chemisorbed on Cu crystallites in moving from Cu–CO to Cu_4–CO.[80]

b. Matrix Isolated (Inert Gas) Bare Metal Dimers, Trimers, and Telomers. A significant breakthrough is represented by the work of Ozin and co-workers where photolysis was employed to induce matrix-isolated Ag atoms to begin cluster formation.[81,82] By irradiating isolated Ag atoms, light energy is absorbed by Ag → Ag* and the energy transferred to the matrix, thereby softening the matrix in the vicinity of the Ag atom. In this way Ag atoms can be moved more selectively and cluster formation controlled better than simply be external warming of the matrix as a whole.

Silver clusters $(Ag)_n$ ($n = 1-5$) were spectroscopically characterized, and, for the most part, these possessed molecular properties. Larger clusters ($n = 6-15$) were believed to represent intermediate stages and possessed optical properties of both molecular and bulk Ag.[81]

Silver atoms and disilver have also been matrix isolated, by Gruen and Bates.[83] Matrices of D_2, Ne, or N_2 were successfully employed. Likewise, Zn, Cd, and Cd_2 have been spectroscopically observed, trapped in rare gas matrices.[84–86]

Miller and Andrews[85] have recently prepared some Group IIA–IIB metal atom dimers on microscale by codepositing two different metals in a $10°K$ argon matrix. In this way absorption spectra for ZnHg, CdHg, MgZn, MgCd, MgHg, Hg_2, ZnCd, MgZn, and MgCd were obtained and detailed bonding analysis discussed.

By a similar method, Kasai and McLeod[86] have prepared AgM dimers (where M is a Group II metal) such as AgMg, AgCa, AgSr, AgBa, AgZn, AgCd, and AgHg. ESR studies of these matrix-isolated dimers indicated that coupling for all the complexes is essentially isotropic and that the unpaired electron resides in an orbital is given essentially by an antibonding combination of the valence s-orbitals of the Ag and M atoms.[86]

c. Small, Discrete Organometallic Clusters (Macroscale). No discrete organometallic M–M bonded clusters of Groups IB or IIB prepared by metal atom methods have yet been fully characterized. It is possible that for the hexafluoro-2-butyne (HFB) work mentioned earlier, where Cu–HFB and Au–HFB complexes were prepared, that these materials exist as M–M bonded clusters.[56]

The metal atom method has been used to prepare C_6F_5Cu and C_6F_5Ag.[46] The Cu analog,

$$C_6F_5Br + 2\,Cu(Ag) \longrightarrow (C_6F_5Cu)_n + CuBr$$

has also been prepared by other means and is known to exist as a tetramer.[40] This can be considered a cluster, although metal atom methods are not needed to obtain it.

d. Large Metal Clusters (Metal Slurries, Small Particles, or Crystallites). As briefly discussed under the Simple Orbital Mixing Section of this chapter, Ag–toluene and Ag–THF dispersions have been employed for deposition of small Ag crystallites on catalyst supports.[60] Copper can also be used. These studies qualitatively indicated that true "solvated metal atoms" were not formed with Cu and Ag as they were with Ni. Cluster growth of Ag in toluene or THF was rapid even before reaching the melting temperature of toluene or THF upon matrix warm-up. Thus, metal slurries form readily.

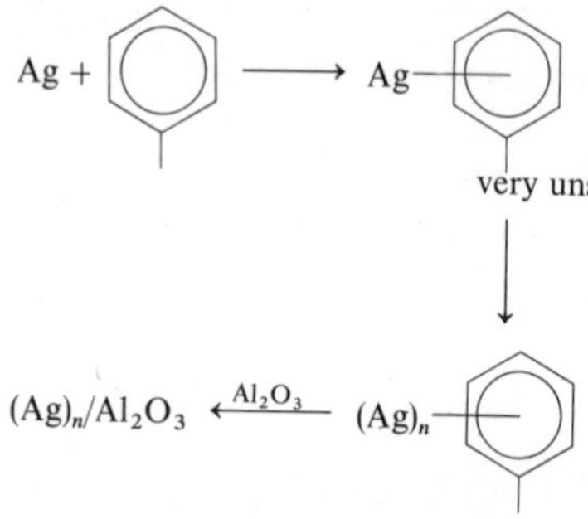

Copper slurries in diglyme, toluene, THF, and hexane have been prepared by codepositing Cu vapor with these solvents.[87] In the case of diglyme, a bright red-orange matrix was formed at $-196^\circ C$ which, on meltdown, turned to a black slurry. At reflux temperature this slurry reacted with iodobenzene to yield benzene in 80% yield. Apparently, abstraction of iodine by Cu particles caused the formation of C_6H_5 radicals that then scavanged H from solvent. No biphenyl was formed.

Zinc and cadmium slurries have proven to be quite useful. Active forms of Zn and Cd powders have been prepared in a wide variety of solvents, such as diglyme, dioxane, THF, toluene, and hexane. Slurries of Zn in these solvents reacted readily with alkyl bromides or iodides to give high yields of R_2Zn and ZnX_2.[88]

$$\text{Zn atoms} + \text{solvent} \xrightarrow{-196^\circ C} [\text{Zn} - \text{solvent}]$$

$$[\text{Zn} - \text{solvent}] \xrightarrow{\text{warm}} (\text{Zn})_n - \text{solvent}$$

$$R_2Zn + ZnI_2 \xleftarrow[\text{reflux}]{\text{RBr}} (\text{Zn})_n - \text{solvent}$$

R = Me, Et, Pr, etc.

A unique feature of these studies is that either polar or nonpolar solvents could be employed. Thus, it is very easy to prepare R_2Zn compounds in a solvent like hexane, and then to separate the soluble R_2Zn in a nonsolvated form simply by filtering off ZnI_2.[89]

The Zn slurries were also useful for Simmons–Smith reactions employing CH_2I_2 or CH_2Br_2[89]:

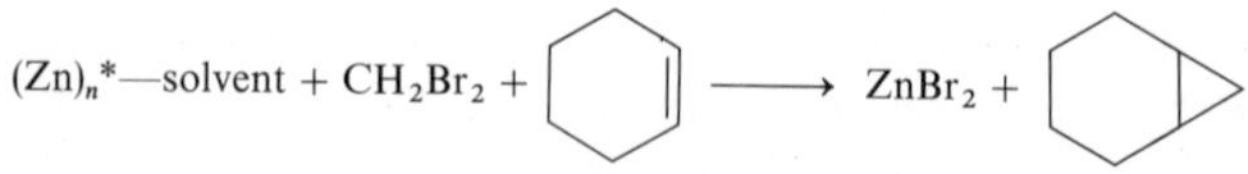

Another unique feature is that the Zn powders isolated by vacuum removal of solvent maintain their high activity even after >7 months storage under inert atmosphere.

Active cadmium slurries in either polar or nonpolar solvents were also prepared. These reacted with alkyl iodides to yield mainly RCdI, which could be prepared in nonsolvated form.[88,89]

$$(Cd)_n* \text{—solvent} + RI \longrightarrow RCdI$$

The active Cd powders were also storable for long periods with little loss in activity.

II. Groups IB and IIB Halide, Oxide, and Sulfide Vapors

A. Occurrence, Properties, and Techniques

The vapors of these materials are generally so difficult to form that they are found in nature only in the atmospheres of stars[90] or higher-temperature planets.[91] For example, the more volatile mercury salts have been detected as major cloud-forming species in the atmosphere of Venus. Detected were Hg_2Br_2, Hg_2I_2, HgS, Hg, and Hg_2Cl vapors.[91]

In the earth's atmosphere many of the more volatile species are oxidatively or hydrolytically unstable, and so are not found as vapors in appreciable amounts.

Table 6-5 summarizes available vaporization data in the literature, and includes vapor compositions when available.[92–136] Some comments on structure have also been included.

Hastie, Hauge, and Margrave have recorded the vibrational spectra of the first row transition metal difluorides isolated in rare gas matrices.[137] The bond angles and asymmetric stretch values for CuF_2 are $165 \pm 8°$, v_3 (argon) = 744 cm^{-1}, and for ZnF_2 $165 \pm 8°$, v_3 (argon) = 764 cm^{-1}. Similar studies on $ZnCl_2$, HgCl, and $CdCl_2$ showed that stable dimeric species were present in the Zn and Hg cases, but not with Cd.[138]

Buechler, Stauffer, and Klemperer[139] have studied electron deflections and mass spectrometry of molecular beams of a series of halide and oxide vapors. In this way they were able to determine whether the gaseous species possessed permanent dipoles or not. For the Group IIB halides, all of the vapor species were found to be linear, as were the dihalides of all the first row transition metals. Similar studies using electron diffraction have predicted a linear structure for Cd_2O and a bent (100° angle) structure for Cu_2O.[123]

Generally, it must be appreciated that, with this series of molecules, often large amounts of dimer, trimer, or tetramer can be present in the vapor, or that the vapor actually consists of different species. For example, HgCl vaporizes as both Hg and $HgCl_2$.[130,131] Likewise, CdS vaporizes as both Cd and S_2 molecules.[125,126,140] It is interesting to note that in the production of CdS photocells by evaporation of CdS onto a cool plate, the distance of the

TABLE 6-5
Vaporization Data for the Groups IB and IIB Halides, Oxides, and Sulfides

Compound[a]	mp (°C)	bp (°C)	Heat of vaporization or sublimation (kcal/mol)	Vapor composition (relative comp)	References[c]
CuBr	492	1345	18.1	$CuBr(1)$, $Cu_2Br_2(3)$,[b] $Cu_3Br_3(7)$ some Cu_6Br_6	92–95
CuBr$_2$	498				
CuCl	430	1490	17.0	$CuCl(1)$, $Cu_2Cl_2(3)$,[b] $Cu_3Cl_3(4)$, some Cu_4Cl_4	93, 96–99
CuCl$_2$	620 to CuCl	d 993	48.0		98
CuF	908	1100 subl			
CuF$_2$	d 950		63.9	Mainly CuF_2	100
CuI	605	1290	18.3	$CuI(1)$, $Cu_2I_2(1)$[b] $Cu_3I_3(3)$	93, 94
Cu$_2$O	1235			Cu, Cu_2O, CuO	101
CuO	1326	d 1800	63.0	Cu, Cu_2O, CuO	101, 102
Cu$_2$S	1100				
CuS	103	d 220			
AgBr	432	1505	44,41	AgBr, some trimer	94, 103–105
AgCl	455	1550	48,42.4	$AgCl$, Ag_3Cl_2, Ag_3Cl_3	98, 103, 104, 106–109
AgF	435	1159			

AgF$_2$	690	d 700			
Ag$_2$F	d 90				
AgI	558	1506	38.6	AgI, Ag$_3$I$_3$, I, I$_2$	94, 103, 105
AgO	d >100				
Ag$_2$O	d 300				
Ag$_2$S	825	d			
AuBr	d 115				
AuBr$_3$	−Br,97				
AuCl	d 170				
AuCl$_3$	d 254	265 subl			
AuI	d 120				
Au$_2$O$_3$	−0,160	−30,250			
Au$_2$S	d 240				
Au$_2$S$_3$	d 197				
ZnBr$_2$	394	650	28	Mainly ZnBr$_2$	104, 110, 111
ZnCl$_2$	283	732	26.1,30	Mainly ZnCl$_2$	104, 110–112
ZnF$_2$	872	1500	45.7		105
ZnI$_2$	446	d 624		Mainly ZnI$_2$	111
ZnO	1975				114, 115

[a] Vaporization is generally carried out by resistive heating of the materials in a crucible. ZnO has been vaporized by electron beam.[114]

[b] Calculated from Rosenstock et al.[93] from mass spectra ion intensities. Assume species are CuX, Cu$_2$X$_2$, and Cu$_3$X$_3$ before ionization.

[c] For general information and tabulation cf. Kelley and other authors[133–135] and "Handbook of Chemistry and Physics."[136]

TABLE 6-5 (*continued*)

Compound[a]	mp (°C)	bp (°C)	Heat of vaporization or sublimation (kcal/mol)	Vapor composition (relative comp)	References[c]
ZnS		1185		ZnS(1), decomp. prod. (10^4–10^5)	116–118
$CdBr_2$	567	863	28		104
$CdCl_2$	568	960	43.3,26.4	$CdCl_2(CdCl_2)_2$	98, 119–122
CdF_2	1100	1758	53.5		105
CdI_2	387	796	28		104
CdO	d 900	1559 subl	55.6		122–124
CdS		980 subl		Cd, S_2	125–127
		680–760 under UV			118
Hg_2Br_2	236	318 (345 subl)	52	Hg, $HgBr_2$	128, 129
$HgBr_2$	236	322			
$HgCl_2$	subl 400				
HgCl	276	302		Hg, $HgCl_2$	130, 131
Hg_2F_2	570	d			
HgF_2	d 645	650			
Hg_2I_2	254	351			
HgI_2	259	354		Mainly HgI_2	129
HgO	d 500			Hg	
HgS		446			132
Hg_2S	583 subl				118

vaporization source affects the properties of the deposit. It is believed that near the vaporization source the deposit is rich in S while further away it is rich in Cd.[127] Furthermore, the vaporization of CdS rich in S is greatly increased under the influence of UV light.[126] It is proposed that charge transfer is an important step in CdS vaporization, and UV light of greater than band-gap energy can affect this charge-transfer step.[126]

B. Chemistry

The Van Leirsburg–De Kock work describing CO, N_2, NO, and O_2 chemistry with NiF_2 and $NiCl_2$ (Chapter 5), also deals with CuF_2 and ZnF_2.[141] These salts were vaporized and codeposited with CO in a large excess of argon.

The interaction of CO with CuF_2 or ZnF_2 is considered to be essentially an interaction of the carbon with an ionic metal center. The $\nu_{C=O}$ values were found to increase upon complexation, as expected for an ionic interaction. Also Kasai, Whipple, and Weltner[142] have shown by ESR that matrix-isolated CuF_2 has ionic character. This conclusion is based on the fact that very little coupling between the free electrons on Cu and F was observed. It is assumed that other similar halides are also ionic.

Compound	$\nu_{C=O}$
$F_2Cu–CO$	2210
$F_2Zn–CO$	2186

A plot of $\nu_{C=O}$ vs the electric field at the carbon nucleus due to various fourth period dipositive metal ions showed that ZnF_2, CrF_2, MnF_2, and CaF_2 fit on a line, but that CuF_2 and NiF_2 fell well off the line. A simple electric field effect is not the entire explanation for the $\nu_{C=O}$ changes (all higher than $\nu_{C=O}$ for free CO), but it is an approximation.[141,143]

On macroscale, the vapors of CuCl and AgCl have been codeposited with unsaturated organic compounds such as norbornadiene, cyclooctadiene, and benzene.[144] These experiments were attempts to prepare new diene or arene complexes similar to those known for $AgNO_3$, or to prepare new CuCl complexes simply incorporating more organic ligands. Some known complexes are:

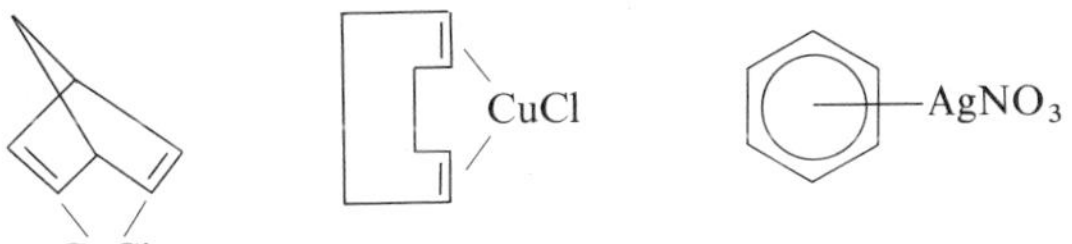

However, the vapor deposition experiments invariably yielded only the known complexes for CuCl, and with AgCl no organometallic product was formed with norboradiene or benzene. Since those known complexes are easily prepared by classical means, the vapor synthesis scheme for their preparation provided no advantage.

The apparent problem in these studies is the strong tendency of the metal halides to repolymerize through bridging with themselves.

No further work on the Chemistry of the vapors of the other Groups IB or IIB halides, oxides, or sulfides has been reported. There is much to do in this area.

References

1. C. S. Bull and O. Klemperer, *J. Sci. Instrum.* **20**, 179 (1943).
2. S. H. Williston, *J. Geophys. Res.* **73**, 7051 (1968).
3. V. Z. Fursov, N. B. Volfson, and A. G. Khvalouskii, *Dokl. Akad. Nauk SSSR* **179**, 1213 (1968).
4. W. A. Fowler, *Int. Astron. Union, Symp.* **26**, 335 (1966).
5. I. F. Danziger, *Mon. Not. R. Astron. Soc.* **131**, 57 (1965).
6. M. St. J. Burden and P. A. Walley, *Vacuum* **19**, 397 (1969).
7. N. F. Afanasev, S. N. Kapelyan, V. A. Morozov, L. P. Filippov, and Z. M. Yudovin, *Zh. Prikl. Spektrosk.* **11**, 883 (1969).
8. A. Guntherschulze, *Z. Phys.* **118**, 145 (1941).
9. A. Krupkowski and J. Golonka, *Bull. Akad. Pol. Sci. Ser. Sci. Tech.* **12**, 69 (1964).
10. S. Nagata, T. Nasu, and Y. Tomoda, *Oyo Butsuri* **27**, 459 (1958).
11. J. Van Audenhove, *Rev. Sci. Instrum.* **36**, 383 (1965).
12. E. B. Graper, *J. Vac. Sci. Techml.* **81**, 333 (1971).
13. G. M. Martynkevich, *Vestn. Mosk. Univ., Ser. Mat., Mekh., Astron., Fiz., Khim.* **13**, 151 (1958).
14. J. Drowart and R. E. Honig, *Mem. Soc. R. Sci. Liege* **18**, 536 (1957).
15. E. B. Owens and A. M. Sherman, *U.S., Dep. Commer., Off. Tech. Serv., AD* **275**, 468 (1962).
16. H. Kawano, *Bull. Chem. Soc. Jpn.* **37**, 697 (1964).
17. B. Siegel, *Q. Rev., Chem. Soc.* **19**, 77 (1965).
18. J. L. Dumas, *Rev. Phys. Appl.* **5**, 795 (1970).
19. A. Korunchikov and A. A. Yankovskii, *Zh. Prikl. Spektrosk.* **5**, 586 (1966).
20. S. I. Anisimov, A. M. Bonch-Bruevich, M. A. Elyashevich, Y. A. Imas, N. A. Pavlenko, and G. S. Romanov, *Zh. Tekh. Fiz.* **36**, 1273 (1966).
21. A. M. Bonch-Bruevich and Y. A. Imas, *Exp. Tech. Phys.* **15**, 323 (1967).
22. A. I. Chemenko, *Izv. Vyssh. Uchebn. Zaved. Fiz.* **1**, 140 (1958).
23. L. S. Nelson and N. A. Kuebler, *Spectrochim. Acta* **19**, 781 (1963).
24. M. G. Rossmann and J. Yarwood, *Br. J. Appl. Phys.* **5**, 7 (1954).
25. B. Vodar, S. Minn, and S. Offret, *J. Phys. Radium* **16**, 811 (1955).
26. P. C. Marx, E. T. Chang, and N. A. Gokeen, *High Temp. Sci.* **2**, 140 (1970).
27. M. N. Turko and I. I. Korshakevich, *Izv. Sib. Otd. Akad. Nauk SSSR, Ser. Tekh. Nauk* No. 3, p. 63 (1965).
28. G. M. Rosenblatt and C. E. Birchenall, *J. Chem. Phys.* **35**, 788 (1961).
29. G. Gattow and A. Schneider, *Angew. Chem.* **71**, 189 (1959).
30. R. F. Barrow, P. G. Dodsworth, A. R. Downie, E. A. N. S. Jeffries, A. C. P. Pugh, F. J. Smith, and J. M. Swinstead, *Trans. Faraday Soc.* **51**, 1354 (1955).
31. S. P. Choong and L. S. Wang, *Nature (London)* **204**, 276 (1964).

32. E. W. R. Steacie, *Can. J. Res. Sect B* **26**, 609 (1948).
33. E. W. R. Steacie, *Research (London)* 1, 541 (1948).
34. D. McIntosh and G. A. Ozin, *Inorg. Chem.* **16**, 59 (1977).
35. H. Huber and G. A. Ozin, *Inorg. Chem.* **16**, 64 (1977).
36. P. L. Timms, *Chem. Commun.* 1525 (1968).
37. P. L. Timms, *J. Chem. Soc., Dalton Trans.* p. 830 (1972).
38. P. L. Timms, *Angew. Chem., Int. Ed. Engl.* **14**, 273 (1975).
39. B. Mile, *Angew. Chem.* **80**, 519 (1968); *Angew. Chem., Int. Ed. Engl.* **7**, 507 (1968).
40. A. Cairncross and W. A. Sheppard, *J. Am. Chem. Soc.* **90**, 2186 (1968).
41. W. T. Miller and R. J. Burnard, *J. Am. Chem. Soc.* **90**, 7367 (1968).
42. W. T. Miller, R. H. Snider, and R. J. Hummel, *J. Am. Chem. Soc.* **91**, 6532 (1969).
43. K. K. Sun and W. T. Miller, *J. Am. Chem. Soc.* **92**, 6985 (1970).
44. S. S. Dua, A. E. Jukes, and H. Gilman, *J. Organomet. Chem.* **12**, 24 (1968).
45. R. J. DePasquale and C. Tamborski, *J. Org. Chem.* **34**, 1736 (1969).
46. K. J. Klabunde, *J. Fluorine Chem.* **7**, 95 (1976).
47. E. W. R. Steacie, "Atomic and Free Radical Reactions," Vol. I, pp. 413–416. Von Nostrand-Reinhold, Princeton, New Jersey.
48. P. Bender, *Phys. Rev.* **36**, 1535 (1930).
49. L. O. Olsen, *J. Chem. Phys.* **6**, 307 (1938).
50. H. W. Melville, *Trans. Faraday Soc.* **32**, 258 (1936).
51. D. J. LeRoy and E. W. R. Steacie, *J. Chem. Phys.* **12**, 369 (1944).
52. K. J. Klabunde, M. S. Key, and J. Y. F. Low, *J. Am. Chem. Soc.* **94**, 999 (1972).
53. H. Huber, D. McIntosh, and G. A. Ozin, *J. Organomet. Chem.* **112**, C50 (1976).
54. D. McIntosh and G. A. Ozin, *J. Organomet. Chem.* **121**, 127 (1976).
55. G. A. Ozin, H. Huber, and D. McIntosh, *Inorg. Chem.* **16**, 3070 (1977).
56. K. J. Klabunde, T. Groshens, M. Brezinski, and W. Kennelly, *J. Am. Chem. Soc.* **100**, 4437 (1978).
57. P. H. Kasai and D. McLeod, Jr., *J. Am. Chem. Soc.* **100**, 625 (1978).
58. K. J. Klabunde and H. F. Efner, *J. Fluorine Chem.* **4**, 114 (1974).
59. H. F. Efner, unpublished work from this laboratory.
60. K. J. Klabunde, D. Ralston, R. Zoellner, H. Hattori, and Y. Tanaka, *J. Catal.* **55**, 213 (1978).
61. T. Groshens, unpublished results from this laboratory.
62. L. Hanlan, H. Huber, and G. A. Ozin, *Inorg. Chem.* **15**, 2592 (1976).
63. H. Huber, E. P. Kundig, M. Moskovits, and G. A. Ozin, *J. Am. Chem. Soc.* **97**, 2097 (1975).
64. J. S. Ogden, *Chem. Commun.* p. 978 (1971).
65. D. McIntosh and G. A. Ozin, *J. Am. Chem. Soc.* **98**, 3167 (1976).
66. J. H. Darling, M. B. Garton-Sprenger, and J. S. Ogden, *J. Chem. Soc., Faraday Trans.* 2, *Symp.* 75 (1973).
67. D. McIntosh and G. A. Ozin, *Inorg. Chem.* **16**, 51 (1977).
68. D. McIntosh and G. A. Ozin, *Inorg. Chem.* **16**, 51 (1977).
69. J. K. Burdett, M. Poliakoff, J. J. Turner, and H. Dubost, *Adv. Infrared Raman Spectrosc.* 2, 1 (1976).
70. H. Huber, D. McIntosh, and G. A. Ozin, *Inorg. Chem.* **16**, 975 (1977).
71. M. Moskovits and G. A. Ozin, *in* "Cryochemistry" (M. Moskovits and G. A. Ozin, eds.), p. 261. Wiley (Interscience), New York, 1976.
72. D. McIntosh, M. Moskovits, and G. A. Ozin, *Inorg. Chem.* **15**, 1669 (1976).
73. G. A. Ozin, H. Huber, and D. McIntosh, *Inorg. Chem.* **17**, 1472 (1978).
74. D. McIntosh and G. A. Ozin, *Inorg. Chem.* **15**, 2869 (1976).
75. H. W. Kohlschutter, *Angew. Chem., Int. Ed. Engl.* **14**, 193 (1975).

76. M. Faraday, *Philos. Trans. R. Soc. London* **147**, 145 (1975).
77. R. Niedermayer, *Angew. Chem., Int. Ed. Engl.* **14**, 212 (1975).
78. K. J. Klabunde, *in* "Reactive Intermediates" (R. A. Abramovich, ed.), Plenum, New York, 1979.
79. M. Moskovits and J. E. Hulse, *J. Phys. Chem.* **81**, 2004 (1977).
80. M. Moskovits and J. E. Hulse, *Surf. Sci.* **61**, 302 (1976).
81. G. A. Ozin and H. Huber, *Inorg. Chem.* **17**, 155 (1978).
82. S. A. Mitchell and G. A. Ozin, *J. Am. Chem. Soc.* **100**, 6776 (1978).
83. D. Gruen and J. K. Bates, *Inorg. Chem.* **16**, 2450 (1977).
84. B. S. Ault and L. Andrews, *J. Mol. Spectrosc.* **65**, 102 (1977).
85. J. C. Miller and L. Andrews, *J. Chem. Phys.* **69**, 3034 (1978).
86. P. H. Kasai and D. McLeod, *J. Phys. Chem.* **82**, 1554 (1978).
87. T. Murdock, Ph.D. Thesis, University of North Dakota, Grand Forks (1977).
88. T. O. Murdock and K. J. Klabunde, *J. Org. Chem.* **41**, 1076 (1976).
89. K. J. Klabunde and T. O. Murdock, *J. Org. Chem.* **44**, 3901 (1979).
90. F. B. Hearnshaw, *Mem. R. Astron. Soc.* **77**, 55 (1972).
91. S. S. Lewis, *Icarus* **11**, 367 (1969).
92. P. I. Fedorov and M. N. Shakhova, *Izv. Vyssh. Uchebn. Zaved., Khim. Khim. Tekhnol.* **4**, 550 (1961).
93. H. M. Rosenstock, J. R. Sites, J. R. Walton, and R. Baldock, *J. Chem. Phys.* **23**, 2442 (1955).
94. K. Jellinck and A. Rudot, *Z. Phys. Chem., Abt. A* **143**, 55 (1929).
95. D. W. Schaaf and N. W. Gregory, *J. Phys. Chem.* **76**, 3271 (1972).
96. L. Brewer and N Lofgren, *J. Am. Chem. Soc.* **72**, 3038 (1950).
97. M. Guido, G. Balducci, G. Gigli, and M. Spoliti, *J. Chem. Phys.* **55**, 4566 (1971).
98. C. G. Maier, *U.S., Bur. Mines. Tech. Pap.* **360**, 1 (1925).
99. L. C. Wagner, P. Robert, Q. Grindstaff, and R. T. Grimley, *Int. J. Mass Spectrom. Ion Phys.* **15**, 255 (1974).
100. R. A. Kent, J. D. McDonald, and J. L. Margrave, *J. Phys. Chem.* **70**, 874 (1966).
101. E. Kazenas, D. Chizhikov, and Y. V. Tsvetkov, *Termodin. Kinet. Protsessov Vosstanov. Met., Mater. Konf., 1969* p. 14 (1972).
102. E. Mack, G. G. Osterhof, and H. M. Kraner, *J. Am. Chem. Soc.* **45**, 617 (1923).
103. H. M. Rosenstock, J. R. Walton, and L. K. Brice, Jr., *U.S. A. E. C. ORNL-2772* (1959).
104. H. Bloom, S. O'M. Bockris, N. E. Richards, and R. G. Taylor, *J. Am. Chem. Soc.* **80**, 2044 (1958).
105. O. Bernauer, *Ber. Bunsenges. Phys. Chem.* **28**, 1339 (1974).
106. A. Visnapuu and J. W. Jensen, *J. Less-Common Met.* **20**, 141 (1970).
107. P. Graeber and K. G. Weil, *Ber. Bunsenges. Phys. Chem.* **76**, 417 (1972).
108. L. C. Wagner and R. T. Grimley, *J. Phys. Chem.* **76**, 2819 (1972).
109. S. K. Chang and J. M. Toguri, *J. Less-Common Met.* **38**, 187 (1974).
110. F. J. Keneshea, Jr. and D. D. Cubicciotti, Jr., *J. Chem. Phys.* **40**, 191 (1964).
111. D. W. Rice and N. W. Gregory, *J. Phys. Chem.* **72**, 3361 (1968).
112. K. Sziklavari, *Banyasz. Kohasz. Lapok, Kohasz.* **104**, 274 (1971).
113. O. Ruff and L. L. Boucher, *Z. Anorg. Allg. Chem.* **219**, 376 (1934).
114. F. E. Dart, *Phys. Rev.* **78**, 761 (1950).
115. V. Bratanov, *Rudodobiv. Metal.* **23**, 38 (1968).
116. G. DeMaria, P. Goldfinger, L. Malaspina, and V. Piacente, *Trans. Faraday Soc.* **61**, 2146 (1965).
117. H. K. Pulker and E. Jung, *Thin Solid Films* **4**, 219 (1969).
118. W. Biltz, *Z. Anorg. Chem.* **59**, 273 (1908).
119. H. Bloom and B. J. Welch, *J. Phys. Chem.* **62**, 1594 (1958).

120. J. L. Barton and H. Bloom, *J. Phys. Chem.* **60**, 1413 (1956).
121. L. Topor and I. Moldoveanu, *Rev. Roum. Chim.* **17**, 1705 (1972).
122. M. Staerk, *Optik (Stuttgart)* **36**, 139 (1972).
123. A. I. Andreivskii and I. V. Kutovyi, *Nauchn. Zap. L'vov. Politekh. Inst., Ser. Khim. Tekhnol.* **29**, 7 (1955).
124. W. B. Hincke, *J. Am. Chem. Soc.* **55**, 1751 (1933).
125. F. Y. Pikus and G. N. Talnova, *Fiz. Tverd. Tela (Leningrad)* **12**, 1355 (1970).
126. G. A. Somorjai, *Surf. Sci.* **2**, 298 (1964).
127. L. Gombay and M. Zollei, *Acta Phys. Chem.* **2**, 28 (1956).
127a. E. I. Givargizov and P. A. Babasyan, *J. Cryst. Growth* **37**, 129 (1977).
128. G. Jung and W. Ziegler, *Z. Phys. Chem., Abt A* **150**, 139 (1930).
129. F. M. G. Johnson, *J. Am. Chem. Soc.* **33**, 777 (1911).
130. A. Smith and A. W. C. Menzies, *J. Am. Chem. Soc.* **32**, 1541 (1911).
131. A. Smith, *Z. Elektrochem.* **22**, 33 (1916).
132. A. Stock, F. Cucuel, F. Gerstner, H. Kohle, and H. Lux, *Z. Anorg. Allg. Chem.* **217**, 241 (1934).
133. K. K. Kelley, *U.S., Bur. Mines, Bull.* **383** (1935).
134. L. Brewer, "National Nuclear Energy Series," Vol. IV-19b, p. 193. McGraw-Hill, New York, 1950.
135. R. W. Kiser, J. G. Dillard, and D. L. Dugger, *Adv. Chem. Ser.* **72**, 153 (1968).
136. "Handbook of Chemistry and Physics" 56th ed., CRC Press, Cleveland, Ohio, 1975–1976.
137. J. W. Hastie, R. Hauge, and J. L. Margrave, *J. Chem. Soc., Chem. Commun.* 1452 (1969).
138. R. W. McNamee, Jr., *U.S. A. E. C. UCRL* **10**, 451, (1962).
139. A. Buechler, J. L. Stauffer, and W. Klemperer, *J. Am. Chem. Soc.* **86**, 4544 (1964).
140. G. A. Somorjai and D. W. Jepsen, *J. Chem. Phys.* **41**, 1389 and 1394 (1964).
141. D. A. Van Leirsburg and C. W. DeKock, *J. Phys. Chem.* **78**, 134 (1974).
142. P. H. Kasai, E. B. Whipple, and W. Weltner, Jr., *J. Chem. Phys.* **44**, 2581 (1966).
143. R. H. Hauge, S. E. Gransden, and J. L. Margrave, *J. Chem. Soc., Dalton Trans.* 745 (1979).
144. K. J. Klabunde, unpublished results.

CHAPTER 7

Boron, Aluminum, Gallium, Indium, and Thallium (Group IIIA)

I. Boron, Aluminum, Gallium, Indium, and Thallium Atoms (B, Al, Ga, In, Tl)

A. Occurrence, Properties, and Techniques

These elements occur as atoms in nature or as telemers in stars,[1] and the more volatile of the group (Ga, In, Tl) have been detected in the solar photosphere. And although the solar abundances of these and other elements have been reported,[2] whether their occurrence is as atoms, ions, or in chemical compounds is not clear. Pokhunkov[3] has reported the presence of Be, B, and MgO in small amounts in the upper atmosphere (100–210 km). Table 7-1[4–37a] brings together some of the information pertaining to vaporization of these elements, particularly that found in the original literature.

Vaporizations of these elements make for striking comparisons. Boron is distinctly different from the rest in that electron-beam vaporization is by far the preferred method, since a great deal of energy is needed to vaporize boron, and since molten boron is extremely corrosive. There have been numerous reports on the use of e-beam technology for boron vaporization[4,6,8,10] and an apparatus avoiding high-voltage vacuum feedthroughs or water-cooled connections has been developed.[8] Laser evaporations of B have also been successful[11] and mass spectrometry studies of the vapor formed showed the presence of B_1, B_2, and B_5 species. Arc vaporization and induction-heating vaporization of B have also been reported.[5,7] A graphite crucible at 2350°C has also been employed for B vaporization.[9] For chemical studies of B vapor, e-beam vaporizations have been successfully employed.[38,39] Boron vapor is essentially monatomic,[12] although small amounts of B_2[14] and B_5 (from laser evaporation)[8] have been observed as mentioned above.

The literature, is replete with reported methods for vaporizing aluminum. This is due to the great industrial importance of vapor-deposited Al films on materials ranging from plastics to metals to catalyst supports.[40,41] This

TABLE 7-1

Vaporization Data, Preferred Vaporization Techniques, and Vapor Compositions for B, Al, Ga, In and Tl (Group IIIA)

Element	mp (°C)	bp (°C)[a]	ΔH^{vap} (kcal/g atom)	Techniques	Vapor composition	References
B	2300	2550	133.8 128 101	e-beam, DC arc Induction, arc, laser Resistive heating–C crucible	B_1 from C arc[b] B_1, some B_2	4–15
Al	660	2467	73.4 73.5 79.8	Laser, resistive heating BN or TiB_2 crucible or C crucible Pulsed discharge, e-beam, levitation, Boride or AlN crucibles	Al_1	15–30
Ga	30	2403	59	C arc, resistive heating Al_2O_3 crucible	Ga Ga_2 (small)	13, 31–34
In	156	2080		C arc, resistive heating Al_2O_3 crucible	In In_2 (small)	15, 31, 32, 35
Tl	304	1457	39.8 43.1	Resistive heating Al_2O_3 crucible	Tl Tl_2 (small)	15, 36–38

[a] See "Handbook of Chemistry and Physics."[38a]
[b] ΔH_{diss} B_2 = 65.5 kcal/mole.[13]

metal is readily vaporized on energetic grounds but extremely corrosive when molten. Therefore, a great deal of work on the development of proper crucible liners for Al vaporization has been reported in the patent literature. Resistive heating in BN–TiB$_2$,[18,26] BN,[19] dense graphite,[22] B$_4$C,[24] AlN,[30] or metal boride crucibles and/or liners have been successfully employed. Focussed lasers,[16,17,23] e-beams,[21] induction or levitation heating,[25] and pulsed discharges[20] have also been employed. The composition of the Al vapor found is nearly all monatomic.[15,28]

The remaining elements of Group IIIA (Ga, In, Tl) are comparatively easy to vaporize. Resistive heating from ordinary crucibles such as W–Al$_2$O$_3$ works very well for these relatively volatile, low-melting metals. Their vapor compositions are essentially monatomic.[15,32,36]

Based on the prior literature, it appears that generation of the atoms of the Group IIIA elements is best done on a large scale using an e-beam apparatus for B, a BN-lined crucible for resistive heating of Al, and resistive heating in W–Al$_2$O$_3$ crucibles for Ga, In, and Tl. The vapors are essentially monatomic,[12–15,28,32,35,36] and the higher ΔH_{vap} for B and Al make them more interesting for metal atom chemistry than Ga, In, or Tl.

In this chapter a new technique is introduced—that of the study of metal

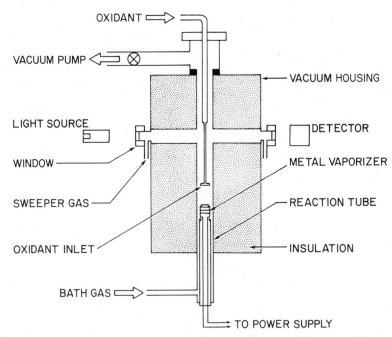

Figure 7-1. Schematic for a high temperature fast flow reactor (HTFFR) for the study of gas-phase metal atom reactions (after Fontijn and Felder[43]).

atom reactions in *high temperature fast flow reactors* (HTFFR). These reactors were developed primarily for the purpose of studying gas-phase metal atom oxidation reactions over a wide temperature range, in order to obtain detailed kinetic and thermodynamic information. Fontijn and Felder have published a series of papers describing these techniques.[43-48] Figure 7-1 illustrates the methodology. Metal atoms are evaporated from a high surface area source and swept rapidly by an inert bath gas into the region of the flow reactor where gaseous substrate is introduced. Concentrations of metal atoms, substrate, or products can be monitored spectroscopically along the reaction pathway. Also, chemiluminescence from excited-state intermediates can be monitored, and these studies have great importance in the chemical laser field. See the Abstraction Processes section under Chemistry for more information.

B. Chemistry

1. ELECTRON-TRANSFER PROCESSES

Only one brief report concerning electron-transfer has appeared. Aluminum vapor was deposited with TCNQ whereby electron transfer to TCNQ took place forming the Al^{3+} salt.[49]

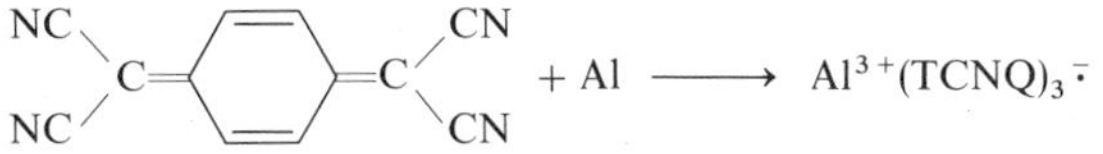

2. ABSTRACTION PROCESSES

Boron atoms, generated by e-beam vaporization of a boron rod (2500°–2800°C vaporization temperature), are odd electron species in the ^2P electronic state ($\cdot$B:). The odd electron causes the B atom to act as a radical-like species in many chemical reactions, thus causing polymerizations of unsaturated organic molecules. However, in the absence of facile radical reaction pathways, the B atom is a vigorous deoxygenating agent. Carbon dioxide reacts with B atoms vigorously, with light evolution, to yield carbon monoxide and boron oxides.[39] It is likely that an excited state "BO" molecule is initially formed.

$$B + CO_2 \longrightarrow \text{"BO"} + CO + \text{light}$$

Attempts to use B atoms to deoxygenate ketones has met with limited success, and about 2% of the boron atoms codeposited with 2-butanone yielded carbenoid rearrangement products[40,50-53]:

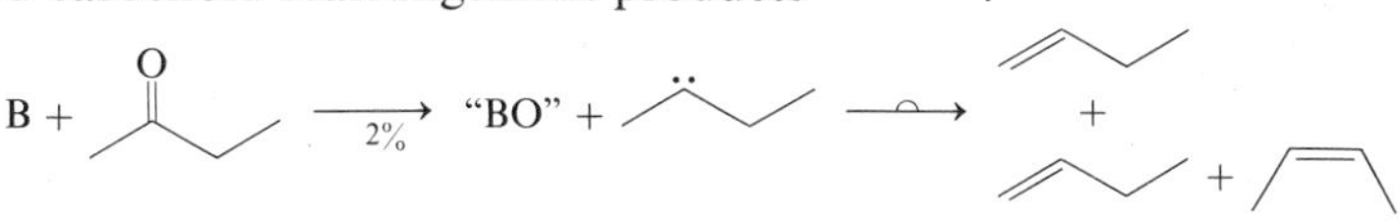

The main mode of reaction was reductive coupling of the ketone, with hydrogen abstraction from excess ketone substrate to form 2-(2-butoxy)-4,5-diethyl-4,5-dimethyl-1,3,2-dioxaborolane.[39,49–52]

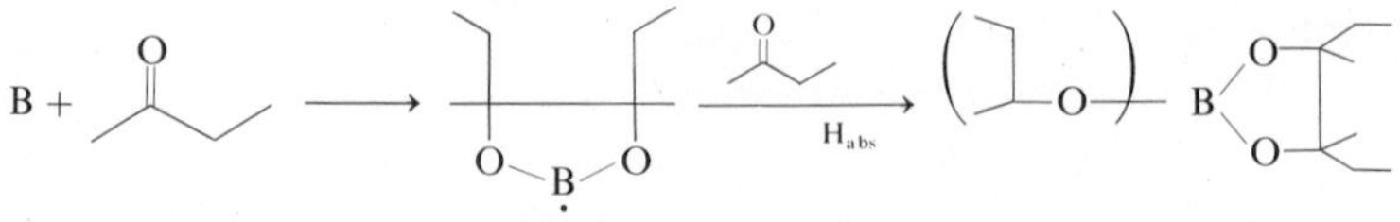

Deoxygenation of ketones, aldehydes, and epoxides by Al atoms has been studied in detail.[43] Aluminum was vaporized by resistive heating (W crucible), and Table 7-2 summarizes the results of a wide variety of deoxygenation reactions. The deoxygenation yields were not high in the ketone and aldehyde studies, and Al-induced coupling products were observed as with B atom reactions. However, epoxides were more interesting in that *either* efficient deoxygenation took place *or* the epoxide was catalytically polymerized by the Al atoms *or* the Al oxide products derived from Al deoxygenation pathways. In the case of cyclohexene oxide, a clean, soluble polymer of MW 1250 was produced.[43]

It should be noted that even ethers are deoxygenated by Al atoms, demonstrating the tremendous oxygen affinity of Al atoms.

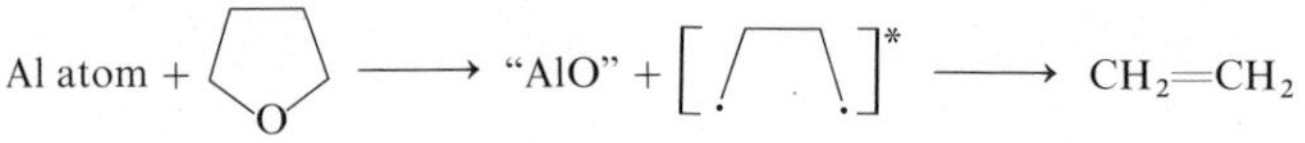

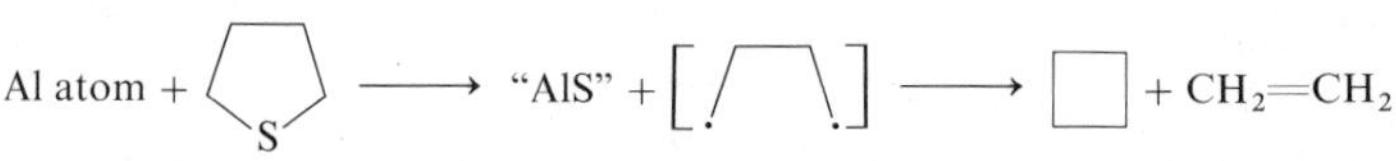

The Al atom desulfurization reaction is less exothermic, and this difference manifests itself in a change in product distributions for tetrahydrafuran vs tetrahydrothiophene. Lower exothermicity may allow a lower energy 1,4-diradical to be formed that will have a higher coupling efficiency. Similar studies, in carbon atom chemistry are discussed in Chapter 8.[54]

Preliminary results indicate that Al atoms also deoxygenate CO_2 accompanied by chemiluminescence.[52] It is difficult, however, to rationalize the light emission, since the primary reaction $Al + CO_2 \rightarrow AlO + CO$ should be almost thermoneutrol.[43–48] Further work is obviously necessary in this area.

A word of caution is in order regarding this deoxygenation work. The low yields encountered may indicate that these processes actually take place

TABLE 7-2

Aluminum Atom Deoxygenation and Desulfurization Reactions[a]

Substrate	Volatile products	Comments
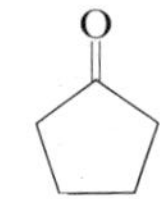 cyclopentanone	 cyclopentene (80)[c]	13%[b] Overall yield
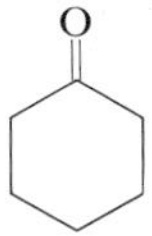 cyclohexanone	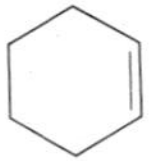 cyclohexene (75), bicyclic (20)	26% Overall, strained cyclic hydrocarbon also formed
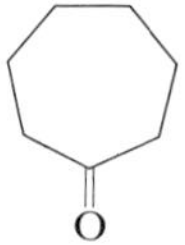 cycloheptanone	 cycloheptene (75), 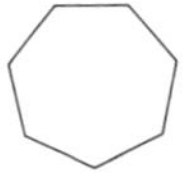 (trace), bicyclic (25)	17% Overall, strained cyclic hydrocarbon also formed
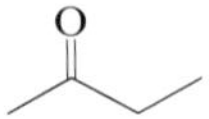 2-butanone	 (37),  (35),  (12),  (7)	13% Overall, cis and trans in equal amounts. Some propane and propene also formed
 diethyl ether	(73),  (27)	7.9% Overall

(continued)

TABLE 7-2 (*continued*)

Substrate	Volatile products				Comments
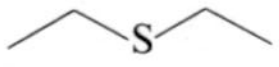 (S, ethyl sulfide)	 (41)	(55)	(4)		11% Overall
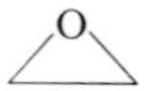 (epoxide)	(23)	(77)			58% Overall, no polymer formed
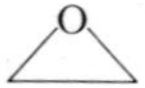 (epoxide)	 (93)	(4)	(2)	(2)	30% Overall, no polymer formed
(cyclohexene oxide)	—				Polymer formed (catalytic)
 (tetrahydrofuran)	 (7)	(4)	(19)	(70)	4.8% Overall

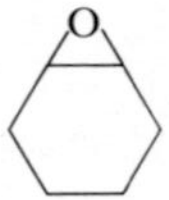

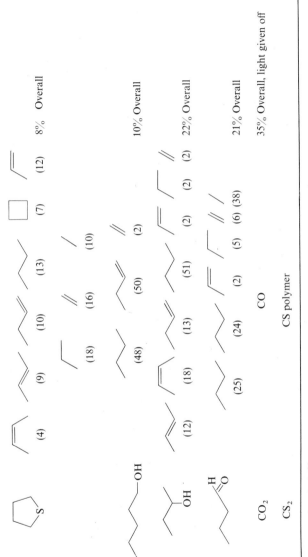

(4) (9) (10) (13) (7) (12)   8% Overall

(18) (16) (10)   10% Overall

(48) (50) (2)   22% Overall

OH

OH

(12) (18) (13) (51) (2) (2)   21% Overall

H
O

(25) (24) (2) (5) (6) (38)   35% Overall, light given off

CO_2

CS_2

CO

CS polymer

[a] From Murdock.[51]
[b] Based on Al vaporized.
[c] Numbers in parens show relative distribution of products.

161

in the gas phase just prior to entering the low-temperature matrix. Preliminary experiments have suggested this to be the case with Mg atom–ketone reactions.[53]

Using the HTFFR technique described earlier in this chapter, a variety of gas-phase Al atom abstraction–oxidation reactions have been studied kinetically. Some of these reactions are shown below:

$$Al + O_2 \xrightarrow[727-1427°C]{1-50 \text{ torr}} AlO + O$$

1127°C 3.1×10^{-13} ml/molecule sec[44,46,47]
(laser-induced fluorescence used to measure [AlO])

$$Al + CO_2 \longrightarrow AlO + CO$$

37°C 1.5×10^{-13} ml/molecule·sec[45]

217°C 6.9×10^{-13}

477°C 1.6×10^{-12}

(2.6 kcal/mole Ea)

1227°C 9×10^{-12}

1607°C 3.8×10^{-11}

Fluorine atom abstractions by gas-phase Al atoms, with chemiluminescence, have also been studied.[48] In these studies, Al atoms were allowed to contact halogens (F_2, Cl_2, Br_2, and I_2) or nitrogen or sulfur fluorides (NF_3 and SF_6). Chemiluminescent flames resulted, and aluminum subhalides, nitrogen, and sulfur subhalides were produced in the process.

$$Al + NF_3 \longrightarrow NH + N_2 + AlF_{1-3}$$

$$Al + F_2 \longrightarrow AlF + F$$

It would be of great interest to extend these studies to other types of reactions, such as oxidative addition or simple orbital mixing processes (see one example in the Simple Orbital Mixing Processes section of this chapter).

3. OXIDATIVE ADDITION PROCESSES

Boron atoms are capable of combined oxidative addition and abstraction reactions. For example, bromobenzene codeposited with B atoms leads to $C_6H_5BBr_2$, probably through a $C_6H_5B–Br$ intermediate. Alcohols and ethers undergo similar reactions.[40,50] Boron atoms and water yield boric acid $(HO)_3B$ and H_2 as final products, and it has been proposed that this reaction proceeds through an oxidative addition type of intermediate.[40]

Similar studies where HCl and HBr were codeposited with B atoms yielded BCl_3, (BBr_3), and H_2.[39]

$$B \text{ atoms} + C_6H_5Br \longrightarrow [C_6H_5\dot{B}Br] \longrightarrow C_6H_5{}^{\cdot} + C_6H_5BBr_2$$

$$B \text{ atoms} + CH_3OCH_3 \longrightarrow [CH_3O\dot{B}CH_3] \xrightarrow{CH_3OCH_3} {}^{\cdot}CH_3 + (CH_3O)_2BCH_3$$

$$B \text{ atoms} + CH_3OH \longrightarrow [CH_3O\dot{B}H] \xrightarrow{CH_3OH} (CH_3O)_2BH + H^{\cdot}$$

$$(CH_3O)_3B + H_2 \xleftarrow{} {}^{CH_3OH}$$

$$B \text{ atoms} + H_2O \longrightarrow [HO\dot{B}H] \xrightarrow{H_2O} HO-\overset{-}{\underset{\underset{H \nearrow \overset{+}{O} \nwarrow H}{|}}{B}}-H$$

$$H_2 + B(OH)_3 \longleftarrow (HO)_2\overset{-}{\underset{\underset{H \nearrow \overset{+}{O} \nwarrow H}{|}}{B}}-H \xleftarrow{H_2O} HOB\underset{OH}{-}H + 1/2\,H_2 \xleftarrow{}$$

$$B \text{ atoms} + HX \longrightarrow BX_3 + H_2$$

$$B \text{ atoms} + BX_3 \longrightarrow B_2X_4$$

Reactions with BX_3 or PX_3 gave low yields of B_2X_4 and P_2X_4.[39]

Benzene and ammonia codeposited with B atoms yielded only intractable polymers.[39] The benzene polymerization is probably radical-like, but the ammonia polymerization must be more complex.

Aluminum atoms react with alkyl and aryl halides upon codeposition and matrix warm-up to yield aluminum sesquihalides, $R_3Al_2X_3$ or $Ar_3Al_2X_3$, although they have only been detected by analysis of their hydrolysis products.[40] Aluminum metal itself reacts with alkyl halides,[55] and later in this chapter we will see that activated aluminum metal yields $Ar_3Al_2X_3$ with halides.

Aluminum atoms codeposited with ammonia yield $Al(NH_2)_3$ (presumably) and $\frac{3}{2}H_2$ at very low temperatures. Matrix warm-up led to AlN, and the sequence is suggested below. It is probable that a σ-complex is initially formed $[H_3N \rightarrow Al]$ followed rapidly by oxidative addition to yield $[H_2NAlH]$.[40]

$$Al + 3\,NH_3 \longrightarrow Al(NH_2)_3 + 3/2\,H_2$$

$$Al(NH_2)_3 \xrightarrow[\text{up}]{\text{warm}} (AlN)_n + 2\,NH_3$$

Aluminum atoms and water yield aluminum hydroxides and H_2 during deposition and more H_2 upon matrix warm-up, suggesting two different modes of H_2 formation.[40]

Oxidative addition of acetylene to Al atoms is covered in the next section.

4. SIMPLE ORBITAL MIXING PROCESSES

It is difficult to determine if Al atom–alkene reactions should be placed under the Oxidative Addition or the Orbital Mixing section, as this is a controversial area. Skell and co-workers[56–58] codeposited Al atoms with propene, and the resultant organoaluminum mixture was hydrolyzed with D_2O to yield deuterated propane, 2,3-dimethybutane, 2-methylpentane, and traces of n-hexane. Aluminacyclopropane radicals have been proposed

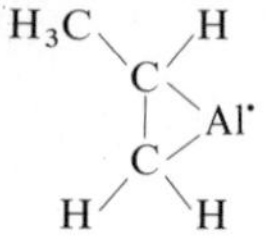

as intermediates in the eventual formation of the necessary mixture of coupled $RR'R''Al$ alkyls. There seems little doubt that aluminum alkyls and dialuminoalkanes are formed in this process, and matrix-isolation studies at $-196°C$ support this claim.[58] Thus, σ-C—Al bonds are formed at $-196°C$ and above. Also, some C—C coupling is induced. Kasai and McLeod[59] carried out matrix-isolation studies at 4°K with Al atoms and ethylene–neon matrices. A green matrix was initially formed whose ESR spectrum revealed a unique sextet attributed to hyperfine interaction with the ^{27}Al nucleus (100% abundance, $I = \frac{5}{2}$). Of the three proposed possible bonding modes shown below, Kasai and McLeod prefer the third where the C=C bond is

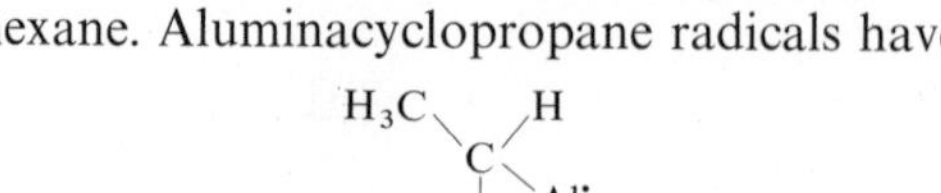

Skell Model   Kasai Model

intact, but donating to an empty orbital on Al and with some backbonding from Al to C=C. The ESR data indicates this type of bond since the Al coupling tensor indicates a low spin density in the 3s orbital, but a high density in the 3p.

Kasai and co-workers have also studied the Al–acetylene reaction in a low-temperature matrix.[60] The product, according to ESR splitting parameters, was a vinyl radical.

This is in contrast to the Kasai work with ethylene, where a π-type bond was implicated. Thus, major bonding differences between C_2H_2 and C_2H_4 were found, which is indeed interesting.

The Kasai model does differ from the Skell model, but the difference in temperatures and the high dilutions used by Kasai must be noted.

A very limited amount of matrix-isolation spectroscopy work has been reported employing Group IIIA metals. Aluminum and gallium carbonyls have been formed in a matrix, but the stoichiometries of these species are not clearly defined. For example, formulations of $Al_x(CO)_2$ and $Ga_x(CO)_2$ have been discussed.[61-64] One interesting outcome of this and related work on Ge and Sn is that it appears that although the d-level is filled in these metals, the p-orbitals are capable of $p\pi-\pi^*$ backbonding (with minimal $d\pi-\pi^*$ contributions). This, of course, is consistent with the Kasai model of the Al–ethylene bonding interaction. And, in comparing the Al and Ga spectra, even though the structures are not well defined, it is evident that Al "backbonds" better than Ga.

Complex	$\nu_{C=O}$ (cm^{-1})
$Al_x(CO)_2$	1890, 1939
$Ga_x(CO)_2$	2006, 1912

A gas-phase Al atom reaction (cf. Fig. 7-1) with NO has been reported.[65] An Al–NO gas-phase adduct was formed in argon bath gas with a rate coefficient of 2.5×10^{-31} ml^2/molecule^2 sec.

5. CLUSTER FORMATION PROCESSES

As we have seen in previous chapters, small metal particles of extreme chemical reactivity can be produced by dispersion of metal atoms in organic solvents followed by warming and partial reagglomeration of the metal atoms. Coordinated solvent or solvent fragments bind to the metal particles or clusters, protecting their surface and discouraging further crystal growth.

Aluminum vapor codeposited at $-196°C$ with xylene or toluene yields colored π-arene–Al complexes.[66] Upon warm-up, Al—Al bond formations occur and fine black slurries are obtained. In a similar way Al–hexane slurries can be obtained. These slurries are extremely pyrophoric and are very reactive with alkyl and aryl halides.[66] For example, iodobenzene reacted in an Al–toluene slurry on refluxing several hours to yield phenyl-aluminum sesquiiodide.

$$Al/toluene + C_6H_5I \longrightarrow (C_6H_5)_3Al_2I_3$$

Bromobenzene reacted similarly, although chlorobenzene was fairly unreactive under these conditions. The slurry reactivities were found to vary in the order Al/xylene > Al/toluene > Al/hexane.[66]

Indium slurries in diglyme, dioxane, toluene, and hexane have been prepared as well.[66] These slurries react with alkyl iodides upon reflux to yield a mixture of R_2InI, $RInI_2$, and InI.

In both the Al and In cases, the dry metal powders obtained by evaporation of excess solvent were storable under aneroebic conditions for many months without appreciable loss in activity. After the storage period under pure N_2 or Ar, deoxygenated solvent was added and the slurries easily regenerated. These storage data are very significant as they imply that the activated powders could be useful commercial chemical reagents. Their large-scale production is quite feasible as industrial scale Al vaporization is already known.[42] Indium vaporization is even more trivial than Al vaporization, so a large-scale supply of metal vapor-activated In powders is also feasible.

II.   Boron, Aluminum, Gallium, Indium, and Thallium Halide, Oxide, and Sulfide Vapors

A. Occurrence, Properties, and Techniques

A long series of low-valent molecules have been spectrally detected in the cooler vapors of the sun, including BH, AlH, BO, AlO, AlS, and BN.[67-69] Similarly, AlH and AlO have been detected in sunspots at temperatures near 4900°K.[70] In 1965 a Ga(I) species was also added to the list of solar atmospheric compounds.[71] Of course, it is not surprising in view of these data that AlH and AlCl have been reported to exist in the outer layers of K and M type stars.[72] It should be emphasized, however, that these binary species exist only in the outer, cooler reaches of the sun and stars, and ternary or larger molecules were not detected.

The generation of these and similar reaction species in the laboratory usually involves a high-temperature disproportionation reaction, or the high-temperature decomposition of the normal form of a salt or oxide to form a metastable vapor species. Two examples are shown below:

$$AlF_3 + Al \longrightarrow AlF_{(g)}$$

$$Al_2O_3 \longrightarrow AlO + Al_2O + O_2 \text{ (cf. Table 7-3)}$$

Since there are many examples of these procedures, before discussing techniques in detail, we will summarize vaporization procedures, properties, disproportionation procedures, and vapor compositions. Table 7-3[73-147] summarizes the properties, vaporization, vapor compositions, and where appropriate, disproportionation data on the species of interest. Note the inclusion of some data on GaAs and GaP which were included because of the high interest in these materials in solar cell and semiconductor applications.

Simple vaporization of many of these materials leads to the generation of telomeric vapor species. This is particularly true of boron sulfides, although boron oxides such as B_2O_3 have only a slight tendency to telomerize.

TABLE 7-3
Properties, Vaporization, and Disproportionation Data for the Group IIIA Subhalides, Halides, Oxides, Sulfides, and Arsenides

Species	bp (°C)	mp (°C)	Method of formation (°C)	Vapor composition	ΔH_{vap} (kcal/mole)[a]	References
BF			$BF_3 + B$ (2000)	BF, BF_3		73
BCl			B_2Cl_4 (1100)	BCl, B_2Cl_4, BCl_3		74
BF_3	−127	−100				75
BCl_3	−107	12				
BBr_3	−46	91				
BI_3	50	210				
BO						76, 77
B_2O_3	460	1860	Vap of B_2O_3 (2352)	B_2O_3 mostly, small amounts B_2O_2, BO, B, O		78–81
BS_2			Vap of BS_2 (550–1100)	BS_2, $(BS_2)_2$, $(BS_2)_4$		82, 83
B_2S_3	310		Vap of B_2S_3 (300–600)	B_2S_3 polymers, BS_2, B_2S_2, B_6S_{12}, B_7S_{14}, B_8S_{16}, B_9S_{18}, $B_{10}S_{20}$	36	81, 84–86
			$H_2S + B$			87
AlF			$AlF_3 + Al$ (800–1000)	AlF, $(AlF)_2$		88–92
AlCl						93, 94
AlBr						93
AlF_3		1291 subl				92
$AlCl_3$		183 subl				95
$AlBr_3$	97	263				93
AlI_3	191	360				

(continued)

TABLE 7-3 (continued)

Species	bp (°C)	mp (°C)	Method of formation (°C)	Vapor composition	ΔH_{vap} (kcal/mole)[a]	References
Al₂O				Matrix-isolated		96, 97
Al₂O₃	2020	3996	Vap of Al₂O₃	Al₂O₃, AlO, Al₂O, Al, O, O₂		98–101
		2980	(2000–4000)			
		2210	(2309–2605)	Mainly AlO and Al₂O		102–111
			(1690 under vac)			
Al₂S						112, 113
Al₂S₃	1100	1500 subl				112, 113
GaF			CaF₂ + Ga	GaF		89, 92
			(800–1300)			
			(AlF₃ + Ga)			
			(700–900)			
GaF₃		800 subl				92
GaCl₃	77.9	201	Vap or Ga + F₂	GaCl₃, Ga₂Cl₆, GaCl₂		95, 114–116
			Ga₂Cl₆ → GaCl₃			
			(258–516)			
GaBr₃	121.5	279	Ga₂Br₆ → GaBr₃	GaBr₃, Ga₂Br₆		114
			(223–430)			
GaI₃	212	346	Ga₂I₆ → GaI₃	GaI₃, Ga₂I₆	23.0	114, 117
			(356–527)			
Ga₂O	1900		Ga₂O₃ + Ga	Ga₂O only		96, 97, 118
Ga₂O₃	970			Ga₂O, Ga₄O₂		119
Ga₂S₂	800d			Ga₄S₅, Ga₂S	86.9	120, 121
Ga₂S		1530	Vap of Ga₂S₂	Ga₂S₂, Ga₂S	39.6	97, 120, 122
GaS	965		Vap of Ga₂S	Ga₂S₂, Ga		
			(950–1200)			

Ga₂S₃	1255		Vap of Ga₂S₃ (950)	Ga₄S₅, S		123, 124
GaAs	1238		Vap of GaAs (700–900)	Ga, As, As₂, As₄	90	123, 124
GaP			Vap of GaP (741–953)	Ga, P₂	84	124, 125
InF			CaF₂ + In	InF		89
InCl	225	608		InCl	27.8	117, 126–128
InBr	220	662				126, 127
InI						126–128
InCl₂	235	550	Vap of InCl₂ (500–700)	InCl₂, InCl, InCl₃		126, 127
InBr₂	235	632	Vap of InBr₂ (500–700)	InBr₂		126, 127
InF₃	1170	>1200			38.4	117
InCl₃	586	600		InCl, InCl₃	24.3	116, 129
In₂Cl₃		641	Vap of In₂Cl₃ (340–450)			117
InBr₃	436				34.1	117
InI₃	210				28.2	117
In₂O		600 subl	In₂O₃ + In (600–950)	In₂O only		96, 97, 118, 130
In₂O₃		850 subl	Vap of In₂O₃ (1290–1490)	In₂O, In₄O₂	118	119, 131
InS	692	850 subl				122
In₂S	653					132
In₂S₃	1050	850 subl	Vap of In₂S₃ (600–800) (960–1360)	InS, In₅S₆	59.3	133, 134

(continued)

TABLE 7-3 (*continued*)

Species	bp (°C)	mp (°C)	Method of formation (°C)	Vapor composition	ΔH_{vap} (kcal/mole)[a]	References
TlF	327	826 655		TlF, Tl$_2$F$_2$	23.7 monomer 33.7	135, 136, 137
TlCl	430	720		TlCl, Tl$_2$Cl$_2$	24.4 monomer 32.3	137, 138
TlBr	480	815		TlBr, Tl$_2$Br$_2$	24.3 monomer 32.3	137, 139
TlI	440	823	Vap of TlI	TlI	25.2 33.4	137, 140
Tl$_2$O$_3$	717	875 d				
Tl$_2$O	300		Vap of Tl$_2$O, or Tl$_2$O$_3$, or Tl$_2$O$_3$ + Tl	Tl$_2$O	42 30.2	141–144
Tl$_2$S	449				20.4	122, 143, 144
Tl$_2$S$_3$	260					

[a] Dissociation energies[145-147] in kcal/mole: BF = 185 (180; BS = 143; BCl = 127; BBr = 103; AlF = 156 (159); AlCl = 117; AlBr = 105; AlI = 87; GaF = 143 (138); GaCl = 99; GaBr = 80, InF = 125 (121); InCl = 102; InBr = 92; InI = 79; TlF = 109 (101); TlCl = 90; TlBr = 76; TlI = 65.

Aluminum oxide, however, yields a very complex vapor mixture, with Al_2O and AlO the predominant species.

Mechanistic studies of the vaporization of Al_2O_3 crystals has indicated that the 0001 face vaporizes faster than the $01\bar{1}0$ face, which is faster than the $1\bar{0}12$ face.[100] Apparently, the least dense face most readily vaporizes, indicating that diffusion range is quite important. It is interesting that pure Al_2O, Ga_2O, and In_2O are accessible by use of the $M_2O_3 + 4M$ disproportionation reaction.[102,118] The chemistry of these species can be readily studied. No work has been reported yet, although the IR spectra of the gaseous species and matrix isolated species have been reported.[96,97] In the case of Tl_2O, the molecule is stable, and gaseous Tl_2O can be prepared by simple vaporizations of solid Tl_2O.[98,142]

The techniques involved in the generation of some of these species merits further discussion. Direct vaporization of these molecules is generally straightforward, and the same methods employed for metals are used. For Al_2O_3, for example, e-beams, induction furnaces, lasers, and arc-image furnaces have been employed.[108,109,111–113] However, when disproportionation reactions are involved, special methods are used that have not previously been discussed. We shall look at one such system in detail, that of BF generation.

When BF_3 is passed over granular B at 2000°C, high yields of gaseous BF can be generated, and if this BF is allowed to pass to a low-temperature wall, its reaction chemistry can be investigated. The apparatus employed by Timms is shown in Fig. 7-2. The heater for the system was a 30 Kc induction heater. A mixture of BF and BF_3 emerged from the bottom of the graphite container through a $\frac{1}{8}$ inch hole and condensed on a liquid nitrogen-cooled wall just below the graphite container.

B. Chemistry

1. ELECTRON-TRANSFER PROCESSES

No literature has been located on this process with these reactive particles.

2. ABSTRACTION PROCESSES

No literature has been located concerning this reaction with these reactive particles.

3. OXIDATIVE ADDITION PROCESSES

Timms[73] has found that the codeposition of BF with excess BF_3 yields some B_2F_4. Subsequent reaction of B_2F_4 with BF yields B_3F_5:

$$BF + BF_3 \longrightarrow B_2F_4$$

$$BF + B_2F_4 \longrightarrow B_3F_5$$

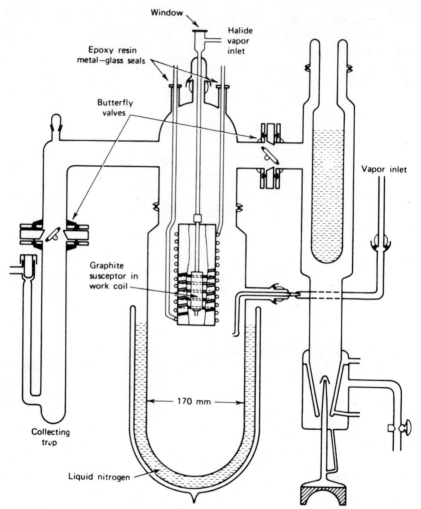

Figure 7-2. Apparatus for BF generation from BF_3 and hot granular boron (after Timms[73]).

In the presence of CO or PF_3, BF telomers that apparently originate from the $BF-B_2F_4$ reaction can be trapped:

$$BF + B_2F_4 \longrightarrow FB{-}B\overset{F}{\underset{B-F}{\diagdown}}$$

The final products $(BF_2)_3BCO$ and $(BF_2)_3BPF_3$ are isolable crystalline compounds.

Generally, BF was found to polymerize readily, and often only pyrophoric $(BF)_n$ polymers were found. For example, BF codeposited with SiF_4 yielded only $(BF)_n$ polymer.[147a] However, the generation of SiF_2 simultaneously with BF (also in the presence of BF_3) yielded a $F_2Si–B_2F_4$ adduct:

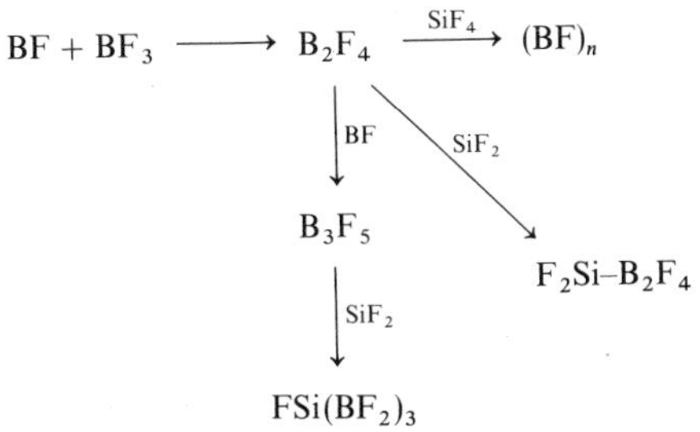

$$BF + BF_3 \longrightarrow B_2F_4 \xrightarrow{\ SiF_4\ } (BF)_n$$

with B_2F_4 reacting via BF to form B_3F_5, and via SiF_2 to form $F_2Si–B_2F_4$; B_3F_5 reacting via SiF_2 to form $FSi(BF_2)_3$.

Oxidative addition of BF to C≡C triple bonds readily takes place when acetylene and BF are codeposited at $-196°C$.[148] Initial formation of a $(F_2BCH{=}CH)_2BF$ adduct was proposed, which eliminated BF_3 on warming to form, a diboron cyclohexadiene derivative quantitatively from the primary adduct:

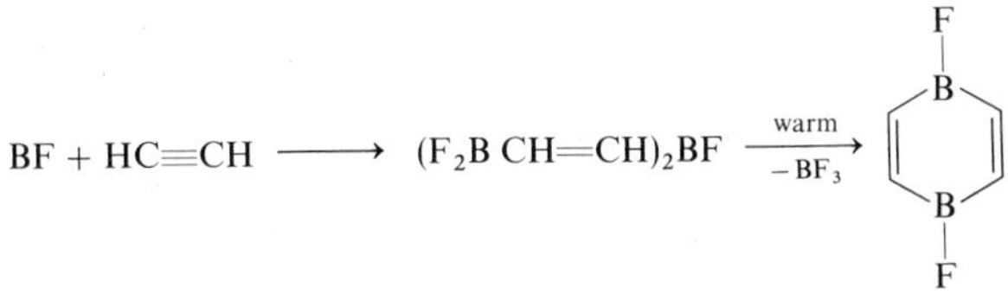

$$BF + HC{\equiv}CH \longrightarrow (F_2B\,CH{=}CH)_2BF \xrightarrow[-BF_3]{warm} \text{[diboron cyclohexadiene]}$$

Similar products were formed when BF and propyne or 2-butyne were cocondensed. In the case of propyne two isomers are possible, and both were observed:

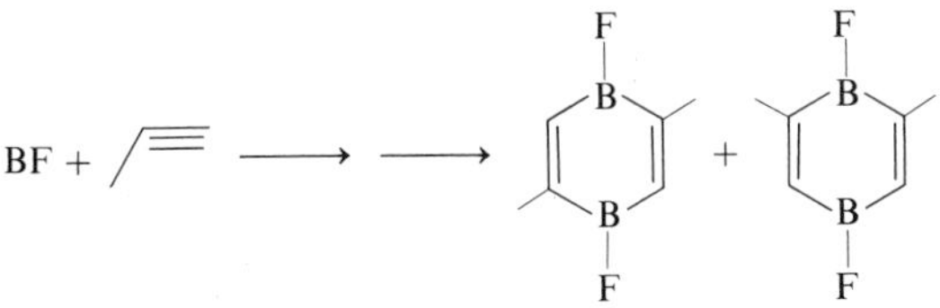

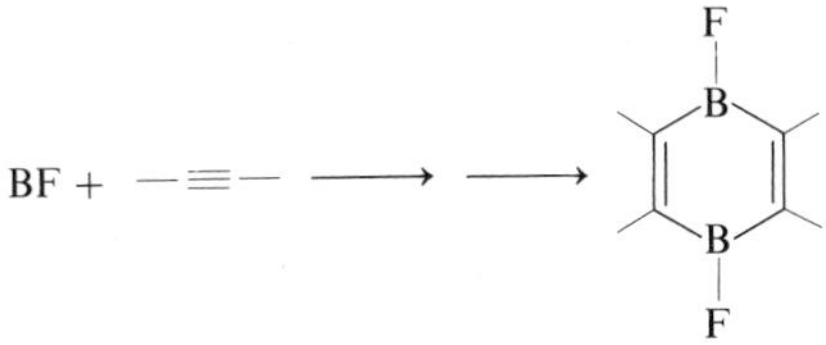

$$BF + \text{(propyne)} \longrightarrow \longrightarrow \text{[isomer 1]} + \text{[isomer 2]}$$

$$BF + \text{(2-butyne)} \longrightarrow \longrightarrow \text{[diboron product]}$$

A very similar species, BCl, can be prepared by pyrolysis of B_2Cl_4; the chemistry of this species is much like that of BF, especially with acetylenes.[74] However, with alkenes some differences are encountered. For example, with BCl, cyclic double addition products were formed, as shown below. However, with BF, F-transfer from excess BF_3 (also present in the matrix) apparently allowed the formation of a double BF_2 adduct[74,149]:

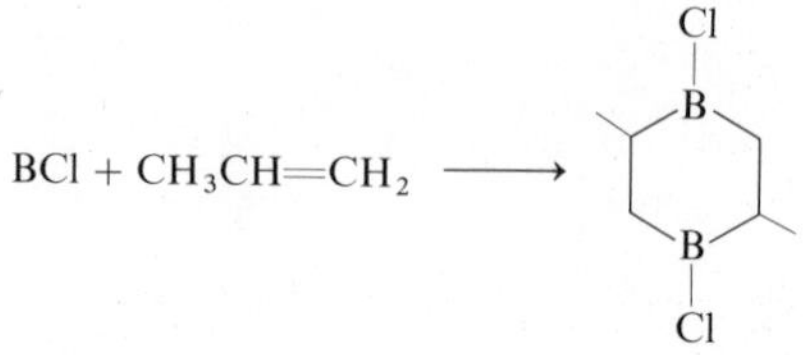

$$BF + CH_3CH=CH_2 \longrightarrow CH_3-CH-CH_2$$
$$\qquad\qquad\qquad\qquad\qquad\quad | \qquad |$$
$$\qquad\qquad\qquad\qquad\qquad BF_2 \quad BF_2$$

4. SIMPLE ORBITAL MIXING PROCESS

No literature has been located concerning this type of reaction.

5. CUSTER FORMATION PROCESSES

The polymerization of BF could be considered under this heading, but these studies have been covered under the Oxidative Addition section.

6. DISPROPORTIONATION PROCESSES

Semenkovich[94] has generated AlF and AlCl in the presence of various small, gaseous molecules, metal oxides, or metal carbides at very high temperature. In the $1150°-1200°C$ range, AlX interacts with MC by disproportionation as follows:

$$6\,AlX + 3\,MC \xrightleftharpoons{1150-1200°C} Al_4C_3 + 3\,MX_2 + 2\,Al$$

References

1. H. Reeves, *Mem. Soc. R. Sci. Liege, Collect. 8°* **19**, 235 (1970).
2. O. Hauge and O. Engvold, *U.S. C. F. S. T. I., AD Rep.* **723633** (1970); *U.S. A. E. C.* **NP-18857** (1970).
3. A. A. Pokhunkov, *Iskusstv. Sputniki Zemli* No. 13, p. 110 (1962).
4. M. Burden and P. A. Walley, *Vacuum* **19**, 397 (1969).
5. Y. V. Kathavate and G. D. Rihani, *Curr. Sci.* **32**, 158 (1963).
6. K. L. Erdman, D. Axén, J. R. MacDonald, and L. P. Robertson, *Rev. Sci. Instrum.* **35**, 122 (1964).
7. P. N. Walsh, AEC Accession No. 46956, Rep. No. NP-15514 (Vol. 1) (1965).
8. M. Aubecq, M. Brabers, M. Heuset, and M. Meullemans, *Mem. Sci. Rev. Metall.* **62**, 373 (1965).

9. A. Hashizume, *J. Sci. Res. Inst., Tokyo* **51**, 211 (1957).
10. H. M. O'Bryan, *Rev. Sci. Instrum.* **5**, 125 (1934).
11. J. Berkowitz and W. A. Chupka, *J. Chem. Phys.* **40**, 2735 (1964).
12. R. P. Burns, A. J. Jason, and M. G. Inghram, *J. Chem. Phys.* **46**, 394 (1967).
13. G. Verhaegen and J. Drowart, *J. Chem. Phys.* **37**, 1367 (1962).
14. Y. A. Priselkov, Y. A. Sapozhnikov, and A. V. Tseplyaeva, *Izv. Akad. Nauk SSSR, Otd. Tekh. Nauk, Metall. Topl.* No. 1, p. 134 (1960).
15. E. B. Owens and A. M. Sherman, *U.S., Dep. Commer., Off. Tech. Serv.*, AD **275,468** (1962).
16. A. I. Korunchikov and A. A. Yankovskii, *Zh. Prikl. Spektrosk.* **5**, 586 (1966).
17. J. L. Dumas, *Rev. Phys. Appl.* **5**, 795 (1970).
18. I. Ames, L. H. Kaplan, and P. A. Roland, *Rev. Sci. Instrum.* **37**, 1737 (1966).
19. J. C. Meaders and M. D. Carithers, *Rev. Sci. Instrum.* **37**, 1612 (1966).
20. N. V. Afanas'ev, S. N. Kapel'van, V. A. Morozov, L. P. Filippov, and Z. M. Yudovin, *Zh. Prikl. Spektrosk.* **11**, 883 (1969).
21. E. B. Graper, *J. Vac. Sci. Technol.* **8**, 333 (1971).
22. V. P. Perevezentsev, A. N. Zhunda, A. G. Zeberin, and L. V. Sinel'nikova, *Tsvetn. Met.* **44**, 40 (1971).
23. A. M. Bonch-Bruevich and Y. A. Imas, *Exp. Tech. Phys.* **15**, 323 (1967).
24. T. Kraus, German Patent 1,078,401 (Cl. 48b) to Balzers Vakuum G.m.b.H. (1955).
25. J. Van Auderhove, *Rev. Sci. Instrum.* **36**, 383 (1965).
26. V. Mandorf, Jr., British Patent 943, 698 (Cl. C23c) to Union Carbide Corp. (1963).
27. W. Reichelt, German Patent 1,085,743, to W. C. Heneaus G.m.b.H. (1960).
28. V. K. Kulifeev and G. A. Ukhlinov, *Izv. Vyssh. Uchebn. Zaved., Tsvetn. Metall.* **11**, 43 (1968).
29. Y. A. Priselhov, U. A. Sapozhnikov, and A. V. Tseplyaeva, *Izv. Akad. Nauk SSSR, Otd. Tekh. Nauk., Netall. Topl.* No. 1, p. 106 (1959).
30. G. Long and L. M. Foster, *Am. Ceram. Soc., Bull.* **40**, 423 (1961).
31. A. I. Chernenko, *Izv. Vyssh. Uchebn. Zaved., Fiz.* No. 1, p. 140 (1958).
32. J. Drowart and R. E. Honig, *Bull. Soc. Chim. Belg.* **66**, 411 (1957).
33. G. Matern, Y. A. Sapozhinikov, S. Khardzhosukanto, and Y. A. Priselkov, *Izv. Akad. Nauk SSSR, Met.* No. 3, p. 210 (1969).
34. S. K. Haynes, *Phys. Rev.* **71**, 832 (1947).
35. Y. A. Priselkov, Y A. Sapozhnikov, A. V. Tseplyaeva, and V. V. Karelin, *Izv. Vyssh. Uchebn. Zaved., Khim. Khim. Tekhnol.* **3**, 447 (1960).
36. S. A. Shchukarev, G. A. Semenov, and I. A. Rat'kovskii, *Zh. Neorg. Khim.* **7**, 469 (1962).
37. J. Bohdansky and H. E. J. Schims, *J. Phys. Chem.* **71**, 215 (1967).
38. L. K. Genov, A. N. Nesmeyanov, and Y. A. Priselkov, *Dokl. Akad. Nauk SSSR* **140**, 159 (1961).
38a. "Handbook of Chemistry and Physics," 56th ed. CRC Press, Cleveland, Ohio, 1975–1976.
39. P. L. Timms, *Chem. Commun.*, p. 258 (1968).
40. M. J. McGlinchey and P. A. Skell, *in* "Cryochemistry" (M. Moskovits and G. A. Ozin, eds.), p. 153. Wiley (Interscience), New York, 1976.
41. K. J. Klabunde, *in* "Reactive Intermediates" (R. A. Abramovich, ed.). Plenum, New York, 1979 (in press).
42. C. Baer, *Res. Dev.* p. 51 (1974)
43. A. Fontijn and W. Felder, *J. Phys. Chem.* **83**, 24 (1979).
44. A. Fontijn, W. Felder, and J. J. Houghton, *Symp. (Int.) on Combust.* **15**, 775 (1974).
45. A. Fontijn and W. Felder, *J. Chem. Phys.* **67**, 1561 (1977).
46. W. Felder and A. Fontijn, *J. Chem. Phys.* **64**, 1977 (1976).
47. A. Fontijn, W. Felder, and J. J. Houghton, *Symp. (Int.) Combust.* [*Proc.*] **16**, 871 (1977).

48. S. Rosenwaks, *J. Chem. Phys.* **65**, 3668 (1976).
49. F. R. Gamble and H. M. McConnell, *Phys. Lett. A*, **26**, 162 (1968).
50. W. N. Brent, Ph.D. Thesis, Pennsylvania State University, University Park (1974).
51. T. O. Murdock, Ph.D. Thesis, University of North Dakota, Grand Forks (1977).
52. T. O. Murdock and T. Groshens, unpublished work from this laboratory.
53. Private communications with P. S. Skell.
54. P. S. Skell, K. J. Klabunde, J. H. Plonka, J. S. Roberts, and D. L. Williams-Smith, *J. Am. Chem. Soc.* **95**, 1547 (1973).
55. H. Adkins and C. Scanley, *J. Am. Chem. Soc.* **73**, 2854 (1951).
56. P. S. Skell and M. J. McGlinchey, *Angew. Chem., Int. Ed. Engl.* **14**, 195 (1975).
57. P. S. Skell, D. L. Williams-Smith, and M. J. McGlinchey, *J. Am. Chem. Soc.* **95**, 3337 (1973).
58. P. S. Skell and L. R. Wold, *J. Am. Chem. Soc.* **94**, 7919 (1972).
59. P. Kasai and D. McLeod, Jr., *J. Am. Chem. Soc.* **97**, 5609 (1975).
60. P. H. Kasai, D. McLeod, Jr., and T. Watanabe, *J. Am. Chem. Soc.* **99**, 3521 (1977).
61. M. Moskovits and G. A. Ozin, *in* "Cryochemistry" (M. Moskovits and G. A. Ozin, eds.), p. 261. Wiley (Interscience), New York, 1976.
62. A. J. Hinchcliffe, D. D. Oswald, and J. S. Ogden, *Chem. Commun.* p. 338 (1972).
63. A. J. Hinchcliffe and D. D. Oswald, unpublished results.
64. S. J. Ogden, *in* "Cryochemistry" (M. Moskovits and G. A. Ozin, eds.), p. 231. Wiley (Interscience), New York, 1976.
65. A. Fontijn, *Chem. Phys. Lett.* **47**, 142 (1977).
66. K. J. Klabunde and T. O. Murdock, *J. Org. Chem.*, **44**, 3901 (1979).
67. A. Schadee, *Int. Astron. Union, Symp.* **26**, 92 (1964).
68. H. D. Babcock, *Astrophys. J.* **102**, 154 (1945).
69. G. F. Gahm, B. Lindgren, and K. P. Lindrous, *Astron. Astrophys., Suppl. Ser.* **27**, 277 (1977).
70. R. S. Richardson, *Astrophys. J.* **73**, 216 (1931).
71. L. H. Aller, *Adv. Astron. Astrophys.* **3**, 1 (1965).
72. M. S. Vardya, *Mon. Not. R. Astron. Soc.* **134**, 347 (1966).
73. P. L. Timms, *J. Am. Chem. Soc.* **89**, 1629 (1967).
74. P. L. Timms, *Acc. Chem. Res.* **6**, 118 (1973).
75. W. H. Beattie, *Appl. Spectrosc.* **29**, 334 (1975).
76. W. Weltner, Jr., *Proc. Meet. Interagency Chem. Rocket Propulsion Group Thermochem., 1st, 1963*, Vol. 1, p. 27 (1964).
77. J. Drowart, A. Pattoret, and S. Smoes, *Proc. Br. Ceram. Soc.* No. 8, p. 67 (1967).
78. A. Sommer, P. N. Walsh, and H. W. Goldstein, *Adv. Mass. Spectrom.* **2**, 110 (1963).
79. J. R. Soulen and J. L. Margrave, *J. Am. Chem. Soc.* **78**, 2911 (1956).
80. S. C. Cole and N. W. Taylor, *J. Am. Chem. Soc.* **18**, 82 (1935).
81. J. Y. Shen and P. W. Gilles, *J. Phys. Chem.* **76**, 2035 (1972).
82. Y. K. Grinberg, E. G. Zhukov, and V. A. Koryazhkin, *Dokl. Akad. Nauk SSSR* **190**, 589 (1970).
83. J. M. Brom, Jr. and W. Weltner, Jr., *J. Mol. Spectrosc.* **45**, 82 (1973).
84. H. Y. Chen and P. W. Gilles, *J. Am. Chem. Soc.* **92**, 2309 (1970).
85. F. T. Greene and P. W. Gilles, *J. Am. Chem. Soc.* **86**, 3964 (1964).
86. Y. K. Grinberg, E. G. Zhukov, and V. A. Koryazhkin, *Dokl. Akad. Nauk SSSR* **184**, 847 (1969).
87. F. T. Greene and P. W. Gilles, *J. Am. Chem. Soc.* **84**, 3598 (1962).
88. M. H. Boyer, E. Murad, Y. H. Inami, and D. L. Hildenbrand, *Rev. Sci. Instrum.* **39**, 26 (1968).
89. J. Hoeft, F. J. Lovas, E. Tiemann, and T. Toering, *Z. Naturforsch., Teil B* **25**, 901 (1970).

References

177

90.　A. Y. Baimakov, *Tr. Leningr. Politekh. Inst.* **188**, p. 156 (1957).

91.　A. Schneider and W. Schmidt, *Z. Metall.* **42**, 43 (1951).

92.　J. W. Hastie, R. H. Hauge, and J. L. Margrave, *J. Fluorine Chem.* **3**, 285 (1973).

93.　H. Schnoeckel, *Anorg. Chem., Org. Chem.* **31B** (9), 1291 (1976).

94.　S. A. Semenkovich, *Zh. Prikl. Khim.* **33**, 552 (1960).

95.　H. Preiss, *Z. Anorg. Allg. Chem.* **389**, 280 (1972).

96.　A. A. Mal'tsev and V. F. Shevel'kov, *Kolebatel'nye Specktry Neorg. Khim.* p. (1971).

97.　P. A. Perov. V. F. Shevel'kov, and A. A. Maltsev, *Vestn. Mosk. Univ., Khim.* **16**, 109 (1975).

98.　B. F. Yudin and A. K. Karklit, *Zh. Prikl. Khim.* **39**, 537 (1966).

99.　R. C. Paule, *High Temp. Sci.* **8**, 257 (1976).

100.　M. Peleg and C. B. Alcock, *High Temp. Sci.* **6**, 52 (1974).

101.　M. Farber, R. D. Srivastava, and O. M. Uy, *J. Chem. Soc., Faraday Trans. 1* **68**, 249 (1972).

102.　L. Brewer and A. Scarey, *J. Am. Chem. Soc.* **73**, 5308 (1951).

103.　A. A. Hasapis, A. J. Melveger, M. B. Panish, L. Reif, and C. L. Rosen, *U.S. A. E. C.* **WADD-TR-60-463** (Pt. II) (1961).

104.　R. P Burns, *J. Chem. Phys.* **44**, 3307 (1966).

105.　O. Ruff and O. Goecke, *Z. Angew. Chem.* **24**, 1459 (1910).

106.　O. Ruff and P. Schmidt, *Z. Anorg. Allg. Chem.* **117**, 172 (1921).

107.　G. W. Sears and L. Navias, *J. Chem. Phys.* **30**, 111 (1959).

108.　R. P Burns, A. J. Jason, and M. G. Inghram, *J. Chem. Phys.* **40**, 2739 (1964).

109.　R. F. Walker, J. Efimenko, and N. L. Lofgren, *Planet. Space Sci.* **3**, 24 (1961).

110.　R. J. Ackermann and R. J. Thorn, *J. Am. Chem. Soc.* **78**, 4169 (1956).

111.　D. W. Moore, *Natl. Symp. Vac. Tech.* Vol. 6, p. 181 (1959).

112.　P. S. P. Wei, D. J. Nelson, and R. B. Hall, *J. Chem. Phys.* **62**, 3050 (1975).

113.　A. Hejduk, Z. Marchwicki, T. Ohly, and M. Szreter, *Pr. Nauk. Inst. Technol. Elektron. Politech. Wroclaw*, **10**, 133 (1973).

114.　W. Fischer and O. Jubermann, *Z. Anorg. Allg. Chem.* **227**, 227 (1936).

115.　A. W. Laubengayer and F. B. Schirmer, *J. Am. Chem. Soc.* **62**, 1578 (1940).

116.　H. Schaefer and M. Binnewies, *Rev. Chim. Miner.* **13**, 24 (1976).

117.　F. J. Smith and R. F. Barrow, *Trans. Faraday Soc.* **54**, 826 (1958).

118.　V. F. Shevel'kov and A. A. Mal'tsev, *Teplofiz. Vys. Temp.* **3**, 486 (1965).

119.　A. J. Hinchcliffe and H. H. Ogden, *J. Phys. Chem.* **77**, 2537 (1973).

120.　H. Spandu and F. Klanberg, *Z, Anorg. Allg. Chem.* **295**, 300 (1958).

121.　V. Piacente, G. Bandi, V. DiPaolo, and D. Ferro, *J. Chem. Thermodyn.* **8**, 391 (1976).

122.　V. F. Shevel'kov, Y. S. Ryabov, and A. A. Mal'tsev, *Vestn. Mosk. Univ., Khim.* **13**, 645 (1972).

123.　C. Y. Lou, *U.S. A. E. C.* **UCRL-19685** (1970).

124.　O. G. Folberth, *Phys. Chem. Solids* **7**, 295 (1958).

125.　P. K. Lee and R. C. Schoonmaker, *Condens. Evaporation Solids, Proc. Int. Symp., 1962* p. 379 (1964).

126.　C. Robert, *Helv. Phys. Acta* **9**, 405 (1936).

127.　C. Robert and M. Wehrli, *Helv. Phys. Acta* **8**, 322 (1935).

128.　Y. Kuniya, S. Hosako, and M. Hosaka, *Denki Kagaku* **42**, 20 (1974).

129.　V. N. Fadeev and P. I. Fedorov, *Zh. Neorg. Khim.* **8**, 2007 (1963).

130.　N. J. Valderrama and K. T. Jacob, *Thermochim. Acta* **21**, 215 (1977).

131.　S. A. Shchukarev, G. A. Semenov, I. A. Rat'kovskii, and V. A. Perevoshchikov, *Zh. Obshch. Khim.* **31**, 2090 (1961).

132.　A. R. Miller and A. W. Searcy, *J. Phys. Chem.* **69**, 3826 (1965).

133.　R. A. Isakova, V. N. Nesternov; and A. S. Shendyapin, *Zh. Neorg. Khim.* **8** (1), 18 (1963).

134.   Y. V. Rumyantsev, G. M. Zhiteneva, and V. P. Kochkin, *Tr. Vost.-Sib. Fil., Akad. Nauk SSSR* **25**, 110 (1960).

135.   F. J. Kenshea, Jr. and D. Cubicciotti, *J. Phys. Chem.* **69**, 3910 (1965).

136.   M. L. Lesiecki and J. Nibler, *J. Chem. Phys.* **63**, 3452 (1975).

137.   R. F. Barrow, E. A. Jeffries, and J. M. Swinstead, *Trans. Faraday Soc.* **51**, 1650 (1955).

138.   D. Cubicciotti, *J. Phys. Chem.* **68**, 1528 (1964).

139.   D. Cubicciotti, *J. Phys. Chem.* **68**, 3835 (1964).

140.   D. Cubicciotti, *J. Phys. Chem.* **64**, 1410 (1965).

141.   I. A. Ralkovskii and G. A. Semenov, *Izv. Vyssh. Uchebn. Zaved., Khim., Khim. Tekhnol.* **13**, 168 (1970).

142.   D. Cubicciotti, *High Temp. Sci.* **2**, 213 (1970).

143.   M. G. Shakhtakhtinskii, *Tr. Inst. Fiz., Akad. Nauk Az. SSR* **11**, 52 (1963).

144.   M. G. Shakhtakhtinskii, *Dokl. Akad. Nauk SSSR* **123**, 1071 (1958).

145.   E. Murad, D. L. Hildenbrand, and R. P. Main, *J. Chem. Phys.* **45**, 263 (1966).

146.   R. F. Barrow, *Trans. Faraday. Soc.* **56**, 952 (1960).

147.   K. A. Gingerich, *J. Chem. Soc. D* No. 10p. 580 (1970).

147a. R. W. Kirk and P. L. Timms, *J. Am. Chem. Soc.* **91**, 6315 (1969).

148.   P. L. Timms, *J. Am. Chem. Soc.* **90**, 4585 (1968).

149.   P. L. Timms, *Endeavour* **27**, 133 (1968).

CHAPTER **8**

Carbon, Silicon, Germanium, Tin, and Lead (Group IVA)

I. Carbon, Silicon, Germanium, Tin, and Lead Atoms and Vapors (C, Si, Ge, Sn, Pb)

A. Occurrence, Properties, and Techniques

As with many of the other elements, atoms or telomers of these elements occur in stars, comets, and flames.[1,2] Of particular interest is carbon, C_3 having been detected in the tail of comets by the prominent Swings bands.[3-5] Also, Goldfinger believes, because of the following thermodynamically favorable processes, that C atoms must be important in flames.

$$CH \longrightarrow :C: + :CH_2 \text{ (in flames and stars)}$$

$$C_x + CH \longrightarrow C_{x+1} + H \text{ (in soot formation)}$$

These and other low-valent carbon species are very likely of importance in flames and stars.

Atomic carbon reactions with O_2 may occur in the upper atmosphere forming CO_2, and this may be of importance in the natural ^{14}C radiocarbon cycle.[6] Table 8-1 brings together data on vaporization properties and compositions, and some of the original literature references.[7-45]

Carbon is perhaps the most intriguing element with regard to vaporization properties. The carbon arc has been used as an intense heat and light source for over 70 years. The temperatures generated in a carbon arc are believed to be as high as $6000°K$.[46] However, sublimation of C allows for much lower temperatures at the surface, of the order of $3800°K$.[47]

Knudsen cell measurements indicate a relative composition of C(1), C_2(2.8), C_3(4.5), C_4(0.35), and C_5(0.5)[7-9] for carbon vapor in thermal equilibrium. And, indeed, straight thermal vaporization of carbon from resistively heated graphite electrodes is richest in C_3.[14,15] Unfortunately, it is difficult to determine precisely the vapor composition from a carbon arc, the most convenient source of carbon vapor. Chemical studies indicate a higher proportion of C_1, perhaps as high as 60%, with the remainder being C_2 and C_3.[14-16]

179

TABLE 8-1

Vaporization Data, Preferred Vaporization Techniques, and Vapor Compositions for C, Si, Ge, Sn, and Pb (Group IVA)

Element	mp (°C)	bp (°C)	ΔH vap (kcal/g atom)	Techniques	Vapor composition	References
C	3632	4827	173–195[a]	Thermal vap	Knudsen Cell $C_1(1)$, $C_2(2.8)$, $C_3(4.5)$, $C_4(0.35)$, $C_5(0.5)$	7–13
				Arc vap	$C_1(1)$, $C_2(0.4)$, $C_3(0.1)$	14–19
				Laser	C_1–C_{10}, C_2 and C_3 predominant	20–24
				e-beam		24–27
Si	1410	2355	90.6	Resistive heating-Si rod		25, 28
				Knudsen cell		29
				Arc		30
				Induction heating-levitation		26, 31–35
				e-beam		27
				Laser		
Ge	937	2830		Resistive heating, W-Al_2O_3		24, 36
				Laser		20, 36
				Induction heating-levitation		25, 30
				e-beam		26, 27, 37
Sn	232	2260		Resistive heating, W-Al_2O_3 crucible		21, 34
				Laser		24, 26, 27, 38–40
				e-beam		
Pb	328	1740		Resistive heating, W-Al_2O_3 crucible		21, 22, 34
				Arc discharge	monatomic	38, 41
				Laser		39, 40, 42
				e-beam		24, 26, 27, 43–45

[a] These values for C_2 and C_3 approach 200 kcal/mole. Also, the bond energy in $C_2 \approx 140$ kcal/mole and $C_3 \approx 149$ kcal/mole[11]

Electron-beam and laser evaporative methods have also been employed for graphite.[10,17-23,25,26] A 3J laser generated a surface temperature of 4060°-4100°K,[20] with C_2 and C_3 species predominating in the vapor.

To summarize, graphite can be vaporized by a variety of means, including e-beam, laser, resistive heating, and arc. Generally, C_2 and to a greater extent, C_3 species predominate, and theoretical considerations do predict that C_3 should predominate.[48] The arc is both the most inexpensive and the most convenient way to generate carbon vapor, but as we will see later, the high-energy conditions of the arc cause a complex mixture of excited electronic states to form as well as a mixture of C_1, C_2, and C_3 species.

Silicon acts much more like a metal during vaporization. Silicon can be melted before vaporization. However, molten Si is so corrosive that there are no good crucible materials to contain it, and so e-beam methods are generally used ("containerless method"), although Knudsen cell, resistive heating, and arc vaporizations have also been reported.[29-35]

Germanium, tin, and lead are much more easily vaporized than either carbon or silicon. Resistive heating from $W-Al_2O_3$ crucibles is quite satisfactory. However, e-beam and laser methods have also been employed extensively.[37-42]

Summarizing, Si, Ge, Sn, and Pb, are readily vaporized, Si best by e-beam methods and the others by resistive heating of $W-Al_2O_3$ crucibles. All of them form mainly monatomic vapors. Kant,[26] however, believes that the higher polymers (M_2, M_3, etc) will increase in the vapor with higher temperatures in the bulk. Thus, it would be expected that higher pressure vaporizations (inert atmosphere) might lead to more $(M)_n$ vapor species.

B. Chemistry

1. ELECTRON-TRANSFER PROCESSES

No electron-transfer studies have been reported. It is interesting to note, however, that e-beam evaporization of Si and cocondensation with a variety of organic substrates, such as CH_3OH, yields a brilliant purple matrix,[50] probably due to solvated electrons. Their source is probably the e-beam–Si interface where stray electrons are given off rather than from Si atom electron transfer to the matrix.

2. ABSTRACTION PROCESSES

Oxygen abstraction by carbon atoms has been studied extensively because this process leads to the production of a variety of interesting reactive fragments in the matrix. In these abstraction processes "ylide-like" intermediates are likely.[51-57]

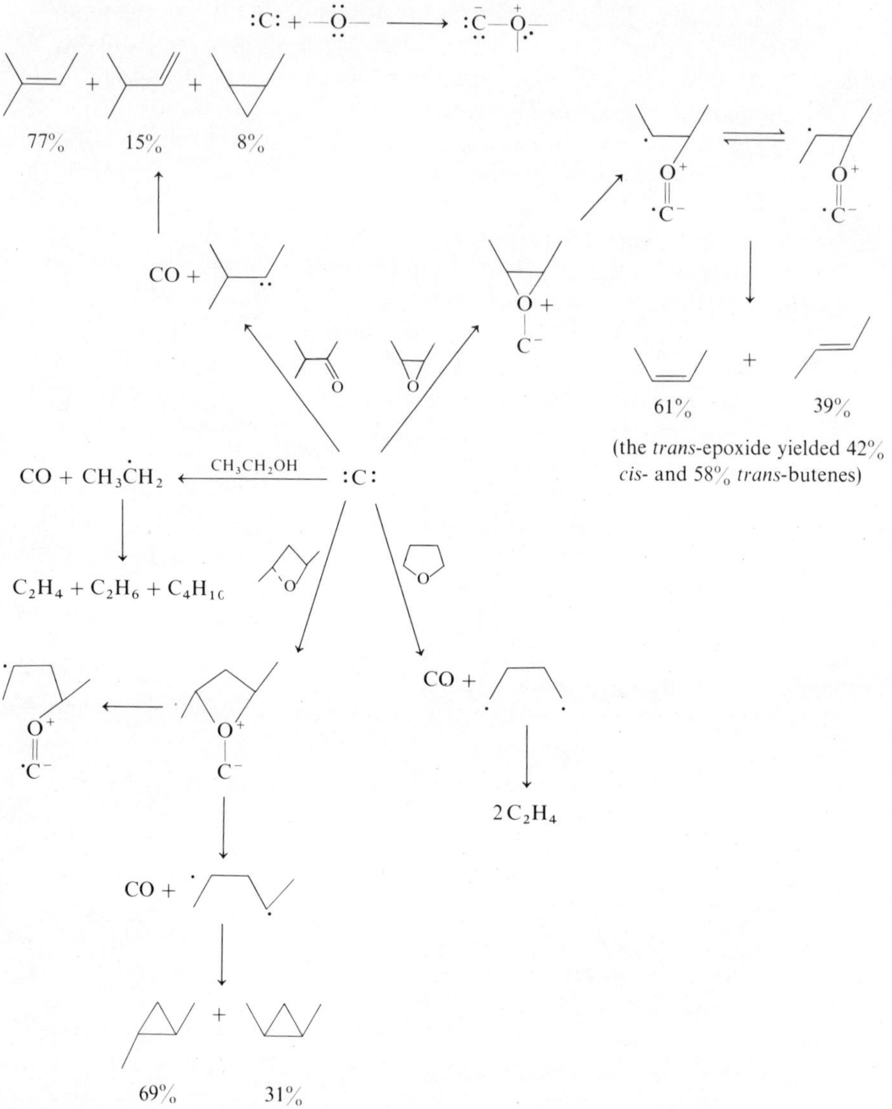

$$:C: + \cdot\ddot{O}- \longrightarrow \cdot\bar{\ddot{C}}-\overset{+}{\underset{..}{O}}-$$

77%   15%   8%

$$:\overset{+}{\underset{\cdot C^-}{O}}: \rightleftharpoons \cdot\overset{+}{\underset{\cdot C^-}{O}}$$

$$\overset{+}{\underset{C^-}{O}}$$

$$CO +$$

61%   39%

(the *trans*-epoxide yielded 42%
cis- and 58% *trans*-butenes)

$$CO + CH_3\dot{C}H_2 \xleftarrow{CH_3CH_2OH} :C:$$

$$C_2H_4 + C_2H_6 + C_4H_{10}$$

$$CO +$$

$$2C_2H_4$$

$$\overset{+}{\underset{\cdot C^-}{O}} \longleftarrow \overset{+}{\underset{C^-}{O}}$$

$$CO +$$

69%   31%

Deoxygenation of aldehydes and ketones yielded carbenes as intermediates. For alkyl-substituted carbenes, rearrangement to olefins was very fast. However, with $Cl_2C{=}O$ and $CH_3O(H)C{=}O$, the intermediates $Cl_2C:$ and $CH_3O(H)C:$ could be trapped by added 2-butene.[53,54] Epoxides and other cyclic and noncyclic ethers were also deoxygenated by C_1.[53,56,57] The deoxygenation proceeded nonstereospecifically with epoxides and oxetanes, suggesting the intermediacy of reasonably long-lived diradical species. Furthermore, larger cyclic ethers such as THF yielded almost exclusively diradical cleavage products. These types of nonstereospecific reactions and cleavages suggested that the intermediate species possessed excess vibrational and/or electronic energy in the matrix.[58] Thus, it was of interest to study a similar but less exothermic abstraction process, that of desulfurization by C_1 (C_1 deoxygenation ~85–100 kcal excess, C_1 desulfurization ~45 kcal excess[58]). Indeed, analogous desulfurization of episulfides yielded much more stereospecific reactions, and in the case of tetrahydrothiophene some

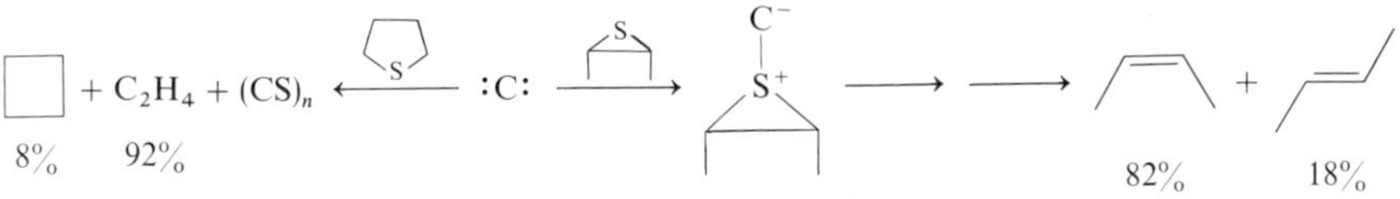

cyclobutane coupling product was observed.[58,59] The intermediacy of CS in a similar $C_1 + CS_2$ reaction was demonstrated by O_2 trapping to form COS. This species existence in the matrix was also implied by the formation of $(C\text{–}S)_n$ polymer upon matrix warm-up.

Hydrogen abstraction by C_1, C_2, and C_3 has been observed, and is especially important with C_2. Acetylene is the major product from C_2 when any hydrogen-containing substrate is employed. Plonka and Skell[60,61] have carried out some elegant deuterium-labeling experiments that strongly suggest that both triplet and singlet C_2 are formed in the arc and carried to the matrix. Thus, triplet C_2 was believed to react by two separate abstraction processes whereas singlet C_2 reacted by an intramolecular double abstraction followed by rearrangement.[60,61]

$$\cdot C{\equiv}C\cdot(t) + CD_3CHO \longrightarrow CD_3{-}\dot{C}{=}O + HC{\equiv}C\cdot$$

$$\cdot C{\equiv}CH + CD_3CHO \longrightarrow CD_3\dot{C}{=}O + HC{\equiv}CH\ (28\%)$$

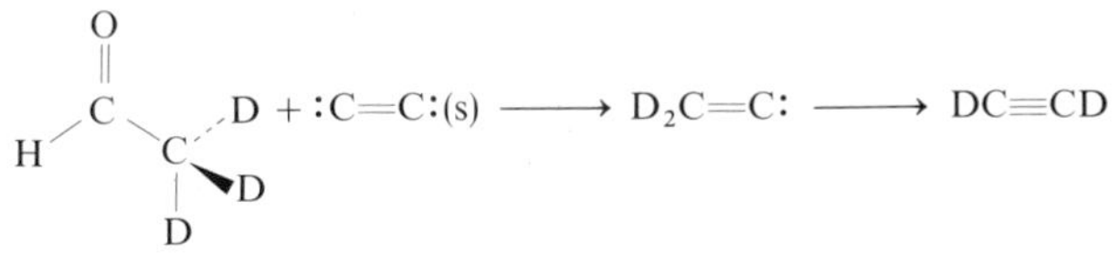

Further evidence for the intramolecular step was obtained from C_2 + acetone experiments where CH_3COCH was trapped.[62]

$$C_2 + CH_3\overset{O}{\overset{\|}{C}}{-}CH_3 \longrightarrow C_2H_2 + CH_3\overset{O}{\overset{\|}{C}}{-}\ddot{C}H$$

$$CH_3\overset{O}{\overset{\|}{C}}{-}\ddot{C}H \xrightarrow[\text{rearr}]{\text{wolff}} CH_3CH{=}C{=}O \xrightarrow[CH_3OH]{} CH_3CH_2CO_2CH_3$$

C_3 also reacts with organic substrates by hydrogen abstraction. Abstraction products from arc-generated C_3 are propyne, allene, propene, and traces of propane (^{14}C-enriched carbon was employed to determine the proportion of these products coming from C_3).[63] However, if C_3 was generated by resistive heating, these abstraction products were not observed, implying that electronically excited C_3 formed in the arc was responsible.

In a similar way, arc-generated C_4 ($\sim 2\%$ by weight of species produced in the arc) forms ethylacetylene, methylallene, 1,3-butadiene, vinylacetylene, and traces of diacetylene, by hydrogen abstraction processes (again ^{14}C labeling was employed to prove C_4 intermediacy.)[64] And again thermally vaporized carbon did not produce C_4 capable of these abstraction reactions.

Fontijn and co-workers have employed a very interesting and different approach to the study of metal atom abstraction reactions.[65,66] These high temperature fast flow reactor systems (described in Chapter 7) allow the study of metal atom reactions in the gas phase at temperatures ranging from $\sim 25°$ to $1400°C$. These studies so far have been mainly concerned with metal atom abstraction–oxidation processes. For example, the $Sn + N_2O$ reaction has been carefully studied using kinetic methods.[65,66] Chemiluminescence from the SnO^* formed was employed as a reaction monitor as studies were carried out in the $25°–680°C$ range and $4–110$ torr range. At the higher temperatures,

$$Sn + N_2O \longrightarrow SnO(a^3\Sigma^+(1) - X^1\Sigma^+) + N_2$$

$$Sn + O_2 \longrightarrow SnO(c-X^1\Sigma^+) + O$$

it was proposed that the actual reaction was $Sn + O_2$, the O_2 being formed from N_2O thermal decomposition.[65,66]

The importance of these studies relates to the possible generation of new chemical lasers. Kinetics and chemiluminescence are the important features rather than the production of new compounds. However, crossover of these high temperature flow methods with the normal cryochemical methods for studying the chemistry of high-temperature species should be of great future value.

3. OXIDATIVE ADDITION–ABSTRACTION PROCESSES

If we consider the species of carbon vapor to be in low valencies (0 for C_1, $+2$ for C_2, and $+2$ for the terminal carbons of C_3), then insertion and addition reactions can be considered oxidative addition to C_1, C_2, or C_3. In every case, C is not variable in its oxidation state, and a valency of $+4$ is always attained.

The ground state for C_1 is ^3P and a long-lived excited state is ^1S. Both of these species are apparently formed and undergo chemical reactions in the matrix. These species react with saturated hydrocarbons in moderate to low yields [based on 40% of $(C)_n \rightarrow C_1$] to give insertion followed by rearrangement for ^1S or double abstractions for ^3P[67-69]:

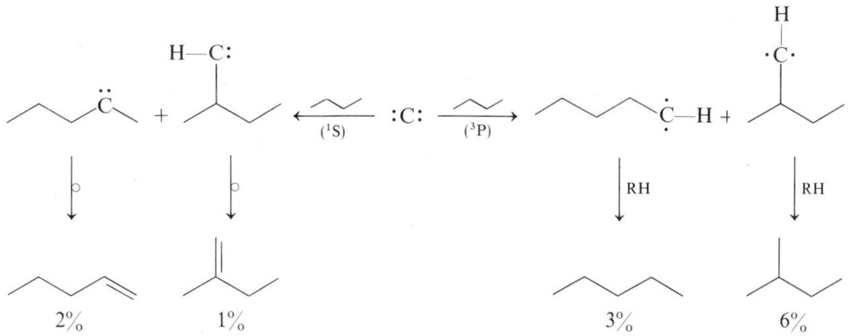

Similar results were obtained for alcohols; however, double O—H insertion was the favored process.[52]

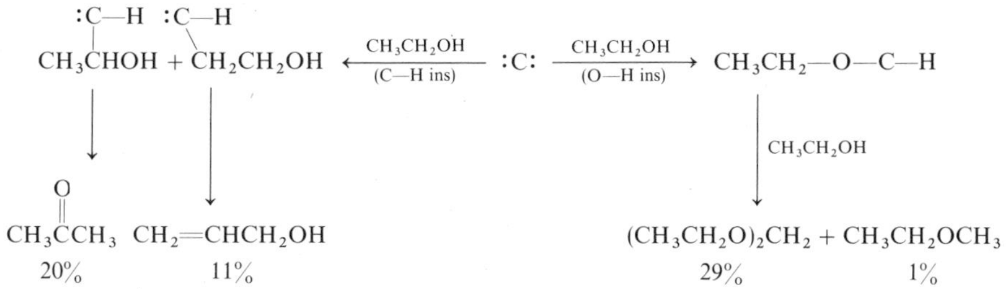

The reaction of carbon vapor with water vapor has also been examined.[68] In this case, acetylene, CO, and H_2 were the major products, especially acetylene. It is possible that the acetylene results from C_2 and the CO and H_2 from C_1. However, when a similar experiment involving the codeposition

$$C \text{ vapor} + H_2O \longrightarrow HC\equiv CH + CO + H_2$$

of H_2O–C vapor at $-196°C$, H_2 was not formed, but instead formaldehyde polymers and related products.[69] For alkyl halides, C_1 undergoes C—X insertion (predominant) as well as C—H insertion, sometimes with subsequent insertion to yield cyclopropanes, or with rearrangement to yield alkenes[70,71]:

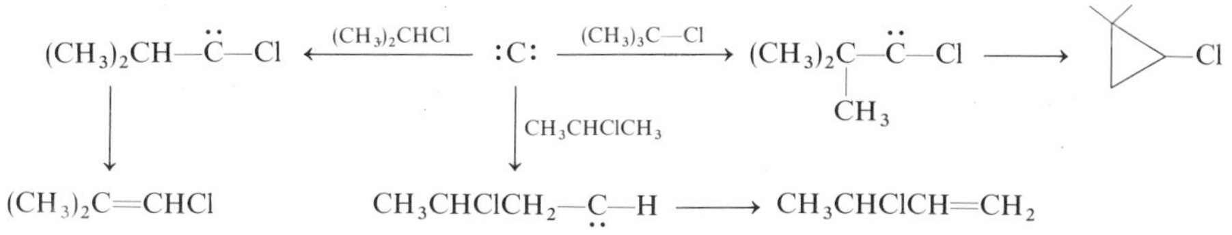

The major reaction of C_1 with alkenes is the formation of allene and some propyne, and a small amount of a double alkene addition product was isolated.[72] Spiropentanes are apparently not formed in these reactions:

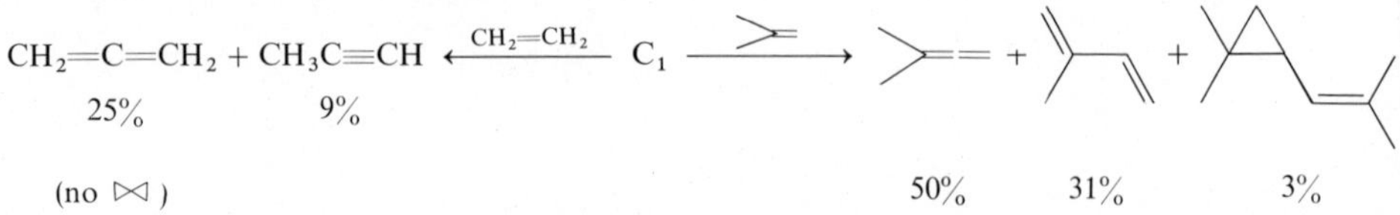

$$CH_2{=}C{=}CH_2 + CH_3C{\equiv}CH \xleftarrow{CH_2{=}CH_2} C_1 \longrightarrow \quad + \quad + $$

25%   9%

(no ⋈)   50%   31%   3%

Boron–boron and boron–hydrogen bonds are also susceptible to insertion by C_1.[73] Pentaborane-9 reacted to yield carbahexaborane-7 whereas B_2F_4 and C_1 yielded $(BF_2)_4C$.[74]

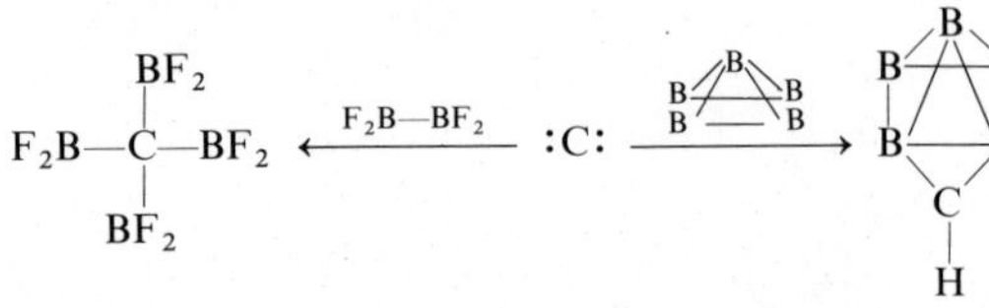

$$F_2B{-}\underset{\underset{BF_2}{|}}{\overset{\overset{BF_2}{|}}{C}}{-}BF_2 \xleftarrow{F_2B{-}BF_2} {:}C{:} \longrightarrow$$

Trimethylsilane reacts with C_1 mainly by C—H insertion rather than Si—H insertion, which probably reflects the statistical preference for C—H and the nonselectivity of a carbon atom.[75] However, C_2 and C_3 insert the more labile Si—H bond selectively:

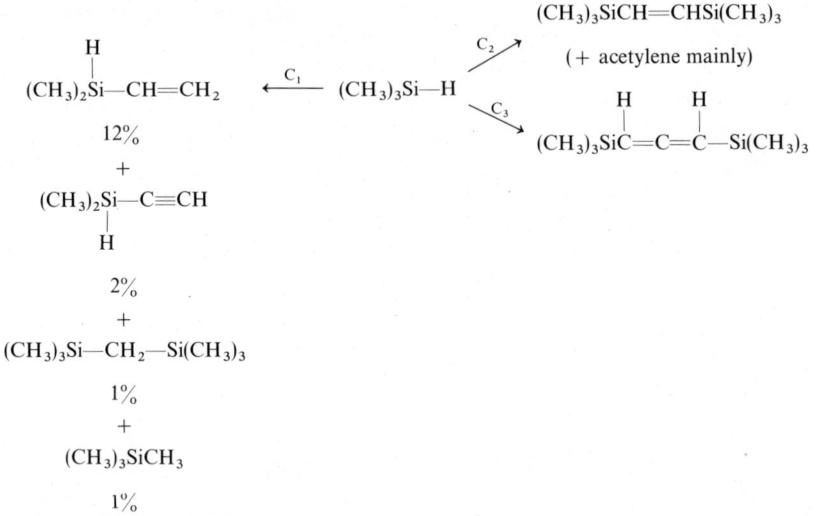

$$(CH_3)_3SiCH{=}CHSi(CH_3)_3$$

(+ acetylene mainly)

$$\overset{\overset{H}{|}}{(CH_3)_2Si}{-}CH{=}CH_2 \xleftarrow{C_1} (CH_3)_3Si{-}H \quad\overset{C_2}{\nearrow}\quad \overset{C_3}{\searrow}\quad (CH_3)_3Si\overset{\overset{H}{|}}{C}{=}C{=}\overset{\overset{H}{|}}{C}{-}Si(CH_3)_3$$

12%

+

$$(CH_3)_2\underset{\underset{H}{|}}{Si}{-}C{\equiv}CH$$

2%

+

$$(CH_3)_3Si{-}CH_2{-}Si(CH_3)_3$$

1%

+

$$(CH_3)_3SiCH_3$$

1%

C_2 also undergoes addition and insertion reactions, although hydrogen abstraction to form acetylene is by far the predominant process. Notably, isobutane and dimethyl ether yielded allenic and acetylenic products, the ether being especially reactive in this process.[76]

$$CH_3OCH=C=CH_2 + CH_3OCH_2C\equiv CH \xleftarrow{\ CH_3OCH_3\ } C_2 \xrightarrow{\ (CH_3)_3CH\ } (CH_3)_2CH\ CH=C=CH_2$$

$$30\% \qquad\qquad\qquad small \qquad\qquad\qquad \Bigg\downarrow {\scriptstyle CH_3CH_2CH_3} \qquad\qquad\qquad 10\%$$

$$CH_3CH_2CH=C=CH_2$$

$$1\%$$

It is possible that these C_2 insertion reactions occur via singlet C_2 by intramolecular hydrogen abstraction followed by carbene coupling.

$$C_2 + RCH_3 \longrightarrow [R\overset{..}{C}H + H_2C=C:] \xrightarrow[\text{coupling}]{\text{rotation and}} RCH=C=CH_2$$

R = alkyl or alkoxy

The chemistry of C_3 probably remains the most interesting aspect of carbon vapor chemistry. This dicarbene reacts efficiently with alkenes to yield bisethanoallenes.[49,77] Ground state singlet C_3 adds stereospecifically whereas excited state triplet C_3 adds nonstereospecifically in the second step (the proportion of triplet C_3 formed can be increased by increasing the arc voltage).

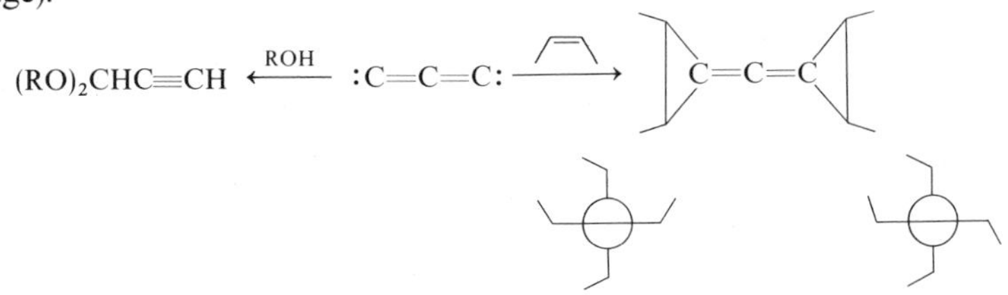

$$(RO)_2CHC\equiv CH \xleftarrow{\ ROH\ } :C=C=C: \longrightarrow$$

93%   (from singlet)   7%

50%   (from triplet)   50%

Alcohols react with C_3 to give 1,1-dialkoxypropynes as products, apparently by two O—H insertion reactions.[63]

Shevlin and his co-workers[78-84] have devised clever means of generating carbon atoms from simple thermal decomposition of chemical reagents. The compounds shown below have both been employed, the 5-tetrazoyldiazonium chloride being the most useful.

The chemistry of these chemically generated carbon atoms is summarized below. Note that epoxide deoxygenation is more stereoselective than with arc generated C atoms. This result indicates that higher energy species are involved in the arc studies. Also, it is possible that the chemically generated atoms react by lower energy stepwise processes. The following reactions are referenced as: epoxides,[78,82] cyclopropane,[80] CO,[81] C_2H_4,[82] and propane.[83]

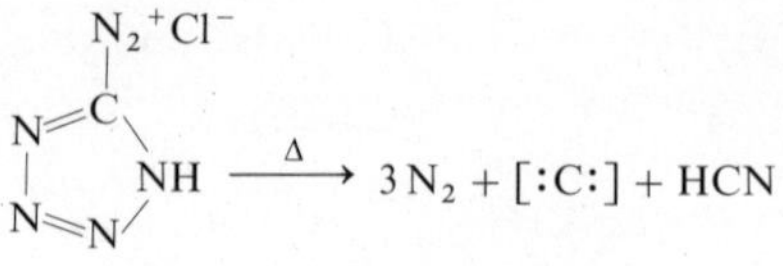

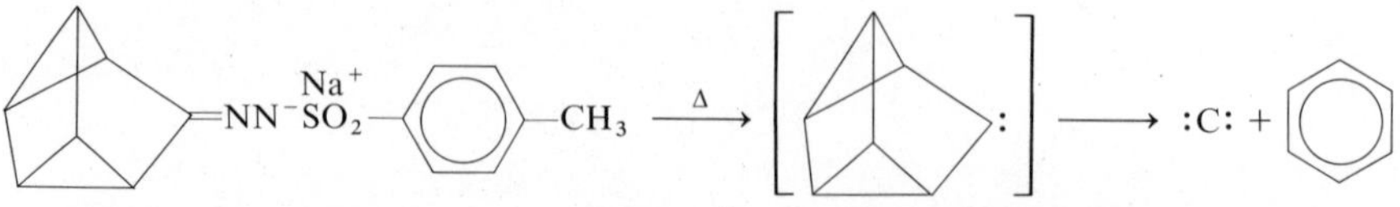

$$3\,N_2 + [:C:] + HCN$$

$$:C: + \text{(benzene)}$$

Theoretical studies of the C + epoxide → CO + olefin reaction predict a concerted process where the C strips the O along a reaction coordinate leading directly to products without a local energy minimum.[84]

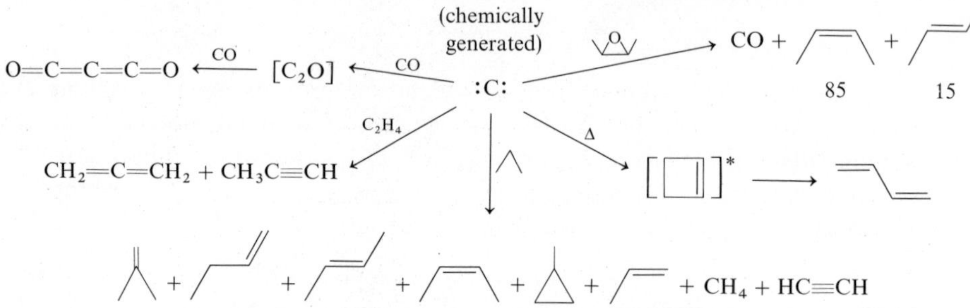

Silicon and germanium atoms also undergo oxidative addition reactions. However, very reactive substrates must be employed since these atoms are not nearly as reactive as carbon atoms in a low-temperature matrix. For example, codeposition of alkyl halides or aryl halides with Si atoms yielded no volatile products. It was believed that [RSiX] or [ArSiX] species formed, and that these polymerized in the cold matrix.[50] However, more reactive substrates such as Cl_2, HCl, HBr, CH_3OH, and $(CH_3)_3SiH$ yielded double insertion products [85,86]:

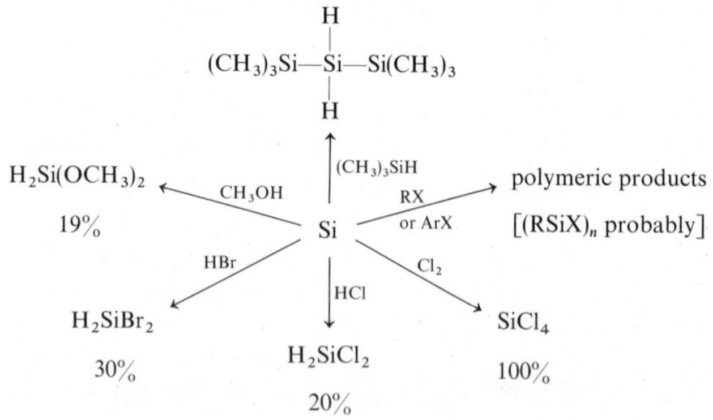

These results are typical of atom chemistry in general in that multiple oxidative addition reactions are not very favorable in low-temperature matrices, and only occur with exceedingly reactive materials.

Germanium atom chemistry has not been developed to any great extent. One report has appeared which deals with reactions of systems containing labile C—X bonds such as CCl_4, $SiCl_4$, and $CHCl_3$. These substrates allow an initial oxidative addition followed by halogen abstraction. These reactions are reminiscent of the chemistry of $(^3P)C_1$, and indeed McGlinchey and Tan believe triplet Ge atoms are involved.[87]

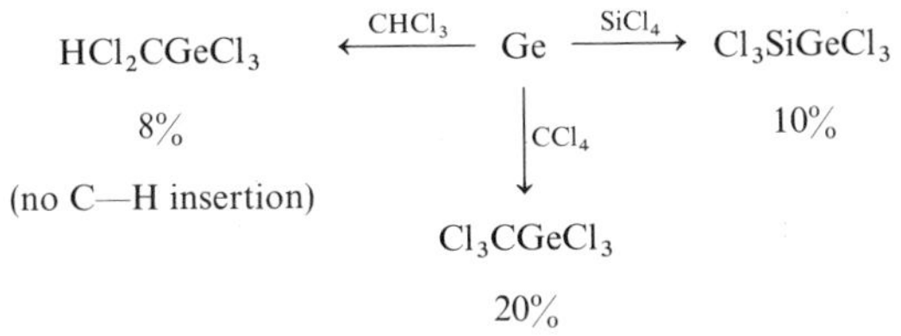

$$HCl_2CGeCl_3 \xleftarrow{\ CHCl_3\ } Ge \xrightarrow{\ SiCl_4\ } Cl_3SiGeCl_3$$

$$8\% \qquad\qquad\qquad \Big\downarrow CCl_4 \qquad\qquad 10\%$$

(no C—H insertion)

$$Cl_3CGeCl_3$$

$$20\%$$

Tin and lead atom reactions with alkyl halides has not lead to the production of R_2MX_2 or similar compounds. In fact, Sn atoms did not react with normal alkyl halides under codeposition reaction conditions.[88]

Matrix-isolation spectroscopy has been employed for the study of Ge and Sn atom reactions with O_2. These studies, carried out at near liquid helium temperatures, indicated that oxidative addition of O_2 to Ge and Sn atoms had taken place.[89-91] This reaction seems quite surprising considering the

$$Ge + O_2 \longrightarrow O\text{—}Ge\text{—}O$$

$$D_{\infty h} \text{ point group}$$

$$Sn + O_2 \longrightarrow O\text{—}Sn\text{—}O$$

$$D_{\infty h} \text{ point group}$$

low temperatures involved, and it might be expected that simple complexation (Simple Orbital Mixing) would be the preferred reaction mode.

4. SIMPLE ORBITAL MIXING PROCESSES

Apart from the "ylide-like" intermediates proposed in C_1–ether reactions, there are no examples of Simple Orbital mixing in C, Si, Ge, Sn, and Pb atom chemistry.

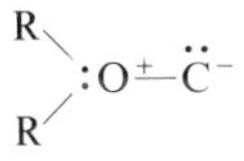

$$\begin{matrix} R\diagdown \\ \diagup \\ R \end{matrix} :O^\pm\text{—}\ddot{C}^-$$

5. CLUSTER FORMATION PROCESSES

The codeposition of Sn or Pb vapors with toluene, THF, or diglyme followed by warm-up yielded black slurries of Sn and Pb with high reactivities.[92]

The reaction of the Sn–THF slurry with CH_3I yielded a mixture of CH_3SnI_3, $(CH_3)_2SnI_2$, and $(CH_3)_3SnI$. The relative composition of this mixture was somewhat solvent dependent. In a similar way, CF_3Br yielded $(CF_3)_nSn(Br)_{4-n}$. However, with CF_3I, only one organometallic product was observed, as with C_6F_5Br.[92] In some cases, such as with C_6H_5I, no organometallics were isolated, and SnI_4 was the only characterizable product.

$$CF_3SnBr_3$$

$$+$$

$$(CF_3)_2SnBr_2 \xleftarrow{CF_3Br} (Sn)_n-THF \xrightarrow{CH_3I} (CH_3)_3SnI$$

$$(slurry)$$

$$+$$

$$(CF_3)_3SnBr \qquad\qquad \swarrow C_6F_5Br \qquad \searrow CF_3I \qquad (CH_3)_2SnI_2$$

$$(C_6F_5)_3SnBr \qquad (CF_3)_2SnI_2$$

Lead slurries in THF or diglyme were reactive with CH_3I yielding small amounts of $(CH_3)_3PbI$.[92] This is the first example of direct Pb–RX reaction in this way (Pb–Na alloy not needed).

II. Vapors of Carbon, Silicon, Germanium, Tin, and Lead Subhalides, Oxides, and Sulfides (Excluding Carbenes, CO, and CO_2)

A. Occurrence, Properties, and Techniques

A wide variety of species would, by definition, fit under this category, including carbenes, CO, and CO_2. However, because of the wide literature coverage of carbene reactions, they will not be considered here, nor will CO or CO_2 chemistry since these are obviously not "high-temperature" species.

The chemistry of a host of other reactive species will be covered, however, including CS, SiO, SiO_2, SiS, SiX_2, GeX_2.

Natural occurrence of species such as these is rare, except perhaps in certain stars and other high-temperature bodies.[2] Dickinson and Gottlieb believe SiO, which exhibits a characteristic absorption at 86846.9 MHz,[93] exists in small amounts in interstellar space. Parson[94] believes SiO_2 vapors may have been involved in the formation of the earth. Carbon monosulfide has been observed in significant amounts in the upper atmosphere,[95,96] and as an intermediate in the photolysis and combustion of carbon disulfide.[97-101]

The generation of these species usually involves discharge or high-tempera-ture disproportionation processes. For example, CS formation has been carried out by radiofrequency glow discharge or by a high voltage vacuum discharge on CS_2[95-103] Figure 8-1 illustrates the methodology employed.

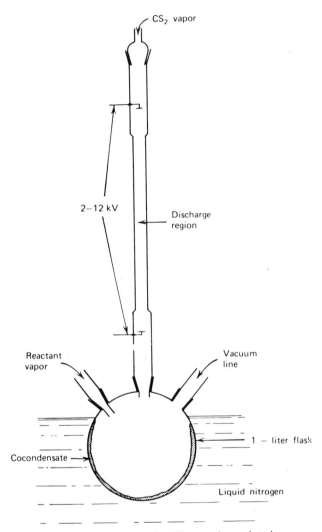

Figure 8-1. Apparatus for CS generation and study.

$$CS_2 \xrightarrow[\text{discharge}]{\text{glow}} \quad CS \quad + (S)_n$$

moveable in
a vacuum system

In a similar way, employing a high-temperature disproportionation process, SiF_2 has been prepared from SiF_4 and Si metal.[104,105] Figure 7-2 illustrates this methodology.

TABLE 8-2

Properties, Vapor Composition, and Methods of Preparation of the Group IVA Subhalides, Oxides, and Sulfides (Excluding Carbenes, CO, and CO$_2$)

Species	mp (°C)	bp (°C)	Method of formation	Vapor composition or other species present; comments	References
CS[a]	−130	Polymerizes	Discharges	CS, CS$_2$, S	94–103, 106–114
			Sulfur fed C arc,	C vapor, CS,	115–118
			CS$_2$ phot,	CS$_2$, CS, S	119
			Shockwaves,		
			K vap + Cl$_2$C=S	CS, KCl	120
			CS$_2$ + O;		121, 122
			MnS + C,		123
			:C: + CS$_2$		58
SiF$_2$[b]			SiF$_2$H$_2$ photo	HSiF$_2$, SiF$_2$H$_2$	124
			SiF$_4$ + Si[c]	SiF$_4$	108–114
			Si$_2$F$_6$ pyrolysis		125
			Li + SiF$_4$	Monomolecular, diamagnetic SiF$_2$ in matrix	126a
SiCl$_2$[d]			SiCl$_4$ + Si(1150°C)	SiCl$_4$, SiCl$_2$	127–129
			HCl + Si		130
SiBr$_2$[e]			SiBr$_4$ + Si(950°C)		128
SiO[f]	>1702	1880	e-beam, evap		24, 131, 132
			Si + SiO$_2$		133, 134
			Evap. from C cloth		135, 136
			Laser evap.		137
SiO$_2$[g]	1700	3050 2590	e-beam evap		24, 138, 139
GeF$_2$[h,i]	d350	subl	Laser evap		137
			GeF$_2$ subl	GeF$_2$, (GeF$_2$)$_2$	24, 140–142
			GeF$_4$ + Ge		

Compound	mp	bp/subl	Method	Products	References
$GeCl_2$	d		$GeCl_4$ + Ge; $GeHCl_3$ pyrol		24, 128–129a, 143, 144
$GeBr_2$	122	d	$GeBr_2$ evap; GeH_2Br_2 photolysis in matrix at 12°K		24, 145
GeO		710 subl	evap; SiO + Ge		24, 146; 136
$GeO_2{}^j$	1115			GeO, GeO_2, O	24, 147, 148
GeS^k	530	430 subl	Ge + H_2S + H_2; GeS subl	GeS, $(GeS)_2$	24, 149; 150–156
$GeS_2{}^l$	800	600 subl	Ge + S; GeO_2 + H_2S; GeS_2 subl; 650°–780°C	GeS, S_2, Ge	24, 149; 157; 151, 157; 152
$SnF_2{}^m$	246		Evap	Dimers, trimers	158
$SnCl_2{}^n$	216	652	Evap		24, 144, 159, 160
$SnBr_2$	d1080	620	Evap		24, 105, 160a
SnO^o		1080	SnO_2 evap; SnO evap	SnO, O_2; SnO	24, 161; 162–164
SnO_2			SnO_2 evap		
SnS^p	880	1230	SnS evap.; 1000°–1200°C	SnS; $(SnS)_2$	24, 156, 165–167
$PbF_2{}^q$	855	1290	PbF_2 evap	Pb, PbF_4	24, 158
$PbCl_2{}^r$	501	954	$PbCl_2$ evap; 530–660°C	$PbCl_2$	24, 144; 168–170

(continued)

TABLE 8-2 (continued)

Species	mp (°C)	bp (°C)	Method of formation	Vapor composition or other species present; comments	References
PbBr₂[s]	373	1166	PbBr₂ evap 530–660°C	PbBr₂	24, 168, 169 171
PbI₂[t]	402	916 954	PbI₂ evap	PbI, PbI₂ (PbI₂ bent)	24, 172, 173
PbO[u]	888	1132	Evap	Pb_nO_n n = 1, 2, 3, 4, 5, 6 (Pb_4O_4 favored) some Pb and O_2	24, 163, 164, 174, 175 176
PbS[v]				Pbs mainly, some Pb, S_2 at 1000°C $(PbS)_2$	156, 177

[a] See Dyne and other authors[178-188] for specific properties.

[b] See Margrave et al.[189-191]

[c] See later discussion for further references.

[d] For more details see Hastie et al.[192], from matrix studies Cl—Si—Cl angle is 105°.[128]

[e] From matrix IR studies, Br—Si—Br angle is 109°.[128]

[f] Heat of formation of SiO = 23.3 kcal/mole[133]; dissociation energy = 188 kcal/mole.[136] See Margrave et al.[190]

[g] Heat of evaporation = 167 kcal/mole.[139]

[h] Heat of sublimation = 20.5 kcal/mole.[140]

[i] For specific properties see Margrave et al.[190] and Takeo et al.[193]; dimer is C_{2h} nonplanar species according to matrix IR and Raman studies.[142]

[j] Heat of sublimation = 100–120 kcal/mole.[147]

[k] Heat of sublimation = 27 kcal/mole[151,152]; dimer D_{2h} planar.[156]

[l] Heat of sublimation = 71 kcal/mole.[151,152]

[m] See Margrave et al.[190] for specific properties.

[n] Heat of vaporization = 21 kcal/mole.[144,160,168-170,172]

[o] Heat of vaporization = 70 kcal/mole.[162]

[p] Heat of vaporization = 45.7 kcal/mole[165]; dimer D_{2h} planar.[156]

[q] For specific studies see Margrave et al.[190]

[r] Heat of vaporization = 30–34 kcal/mole.[144,160,168-170,172]

[s] Heat of vaporization = 30–32 kcal/mole.[168,171]

[t] Molecular beam study.[173]

[u] Heat of vaporization = 64 kcal/mole[175,176]; PbO dissociation energy = 88.4 kcal/mole.[177,194]

[v] Dimer D_{2h} planar as shown by matrix isolation studies.[156]

$$Si + SiF_4 \xrightarrow{1200°C} \quad :SiF_2$$

moveable in a
vacuum system

Table VIII-2 summarizes the properties, vapor compositions, and methods of preparation of the species of interest in this section.[24,58,94–103,106–194]

As can be ascertained from Table VIII-2, a variety of methods are available for production of interesting chemical species, such as CS, SiF_2, SiO, SiS, and GeS. Many of these species can be prepared in monomolecular forms by simple vaporization alone (e.g., GeS and SnS). On the other hand, high-energy discharge procedures or very high-temperature disproportionation reactions are often required (e.g., CS and SiF_2 preparation) and involve some extra equipment. However, even these methods are not difficult. Thus, it seems surprising that only one of these species, SiF_2, has actually been studied very much as a chemical reagent.

The properties of SiF_2 and CS are unique in that these molecules are long-lived (seconds) in the gas phase at moderately low pressures. Moreover, CS can be condensed as a white solid at $-196°C$, which is stable for hours at this temperature.[195] However, upon warming to ca $-160°C$, violent polymerization takes place giving off light, heat, and crackling sounds, with the formation of $(CS)_n$.[102,103,109,112,195] Much vacuum glassware has been destroyed in this way, and metal reactors or traps are preferred.[102,103] A note of caution is also in order, since gram quantities of CS could be dangerous.[102,103]

SiF_2 is similar to CS in that it has a strong tendency to telomerize in a low-temperature matrix. It is different, however, in that this telomerization will occur at $-196°C$. Further polymerization is possible but does not occur explosively. Another difference is that $(SiF_2)_n$ $(n = 1, 2, 3)$ is very reactive with other added molecules at low temperature whereas CS is not. This brings us to the next section, where the chemistry of CS and SiF_2 (mainly) will be summarized. Essentially, only CS and SiF_2 will be discussed since other species of interest in Group IVA have not been used as chemical reagents yet (note a couple of rare exceptions later).

B. Chemistry (*Excluding Carbenes, CO and CO_2*)

1. ELECTRON–TRANSFER PROCESSES

No reports have been located for this reaction type with these species.

2. ABSTRACTION AND OXIDATION PROCESSES

Steudel[195–197] generated CS from CS_2 by electric discharge. When the walls of the discharge tubes were coated with Se or Te, CS abstracted atoms from these films to form SCSe and SCTe.

Oxygen atom abstraction by CS has also been observed.[97,198,199] These gas-phase studies were also carried out with SO, but in this case an O–S

$$CS + O_2 \longrightarrow COS + O$$
$$CS + SO \longrightarrow CO + S$$
$$CS + O \longrightarrow CO + S$$

exchange occurred, with the formation of S_2 and CO.[121] A similar reaction was encountered when CS was allowed to react with O atoms.[121,122,200] These studies are, of course, related to combustion processes for CS_2 and CS.

It has been reported that CS is a mild deoxygenation agent at low temperature, capable of removing oxygen from propylene oxide but not from CO_2.[59]

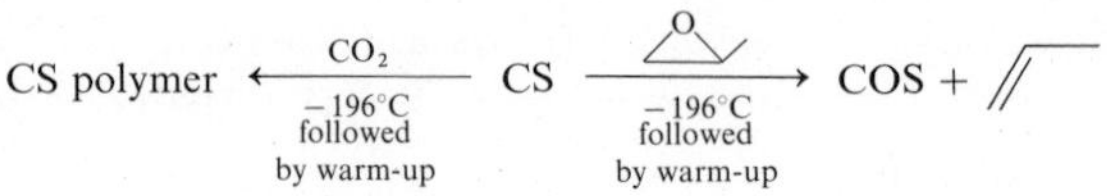

The reaction of metal oxides with gaseous CS yielded metal sulfide and (presumably) CO.[110] Iron oxide reacted more efficiently than mercury or silver oxides.[110] In these studies, it was found that $AgNO_3$, $Pb(OAc)_2$, and Na_2O_2 did not react with CS.

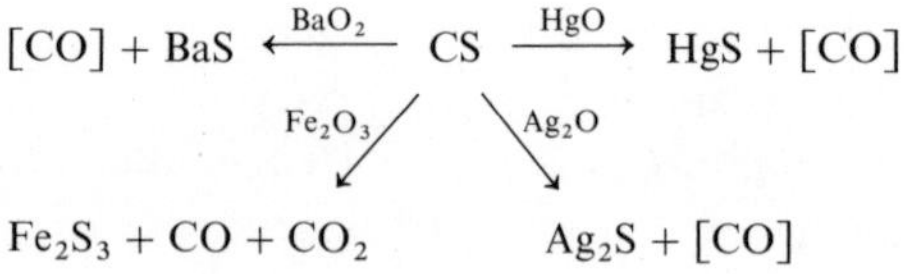

[CO] indicates a presumed product

Dewar and Jones[110] also reported the reaction of CS with sulfuric acid as "the most striking reaction we have observed." The CS was passed over the surface of H_2SO_4 that had been placed in a wide U-tube. This acid acquired a yellow color, then deep orange-red, and then sulfur precipitated. The gaseous products were SO_2, CO, and CO_2

$$CS + H_2SO_4 \longrightarrow SO_2 + CO + CO_2 + S$$

3. OXIDATIVE ADDITION PROCESSES

a. CS. Steudel[197] added gaseous halogens just after the flowing discharge zone for CS production. The CS–X_2 mixture was swept to a cold wall ($-196°C$) and the condensate studied by IR spectroscopy. In this way CSX_2 derivatives were detected (but not isolated).

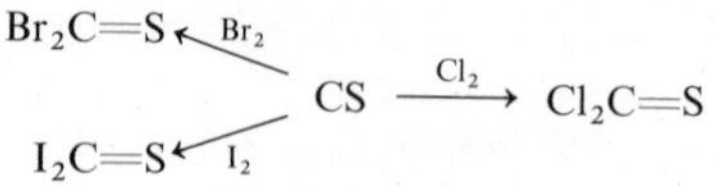

We have reported the first synthetic-scale studies where CS was used as a chemical reagent.[102,103] Employing a high voltage AC discharge through CS_2, CS was prepared. Sulfur plated out quickly in the discharge tube, whereas the CS could be moved through the vacuum flow system where it was either cocondensed with other reagents or swept into a cool stirred solution. Excesses of halogens and mixed halogens allowed the preparation of CSX_4 compounds. Unfortunately, iodo derivatives could not be prepared in this way due, apparently, either to facile halogen-exchange processes or simply to poor stability.

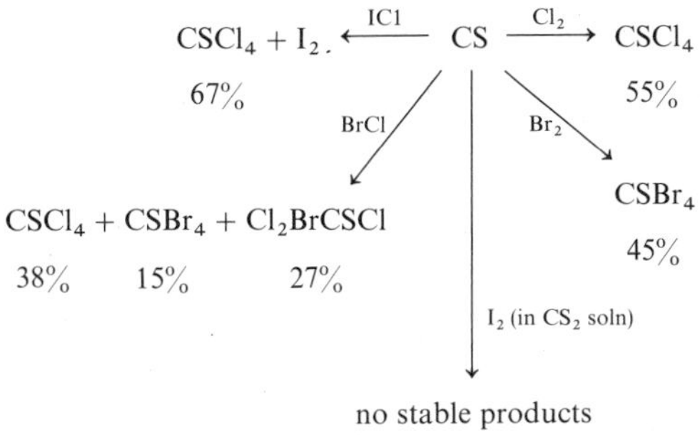

$$CSCl_4 + I_2 . \xleftarrow{\text{ICl}} CS \xrightarrow{\text{Cl}_2} CSCl_4$$

67%   55%

BrCl   Br₂

$CSCl_4 + CSBr_4 + Cl_2BrCSCl$

38%   15%   27%

$CSBr_4$

45%

I_2 (in CS_2 soln)

no stable products

Although two molecules of Cl_2 or Br_2 readily added twice to CS, HCl and HBr only added once. Using these hydrogen halides, HXC=S derivatives, to our knowledge previously unknown, have been prepared. These thioformyl halides trimerized on warming, but could be trapped by Cl_2 to form HCl_2CSCl.

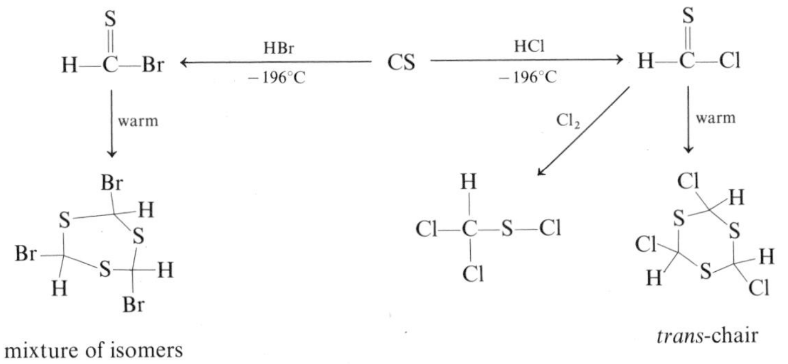

mixture of isomers

trans-chair

It was interesting although somewhat disappointing that CS only reacted under these conditions with vigorous reagents such as halogens and hydrogen halides. It was determined that CS behaves as a Lewis base, not as an electrophilic carbene.[102,103] However, in spite of this, Lewis acids such as BCl_3[102,103,182] RCO_2[102,103] and $RCOCl$[102,103] did not react by oxidative addition processes, but only served to complex the CS weakly at low temperature and to delay the eventual CS polymerization reaction.[102,103] Similar results were obtained when CS was allowed to cocondense with weak Lewis bases, such as ethylene,[109,198] butene,[102,103,198] acetylene,[108] benzene,[108] ammonia,[182] or alkanes.[201] Furthermore, it was found that CS would not directly displace CO as a ligand from $Ni(CO)_4$, $Fe(CO)_5$, or $P\phi_3$ from $RhCl(P\phi_3)_3$.

b. SiF_2. This carbeneoid species reacts readily with unsaturated organic compounds upon codeposition at $-196°C$. However, since Si=C bonds and silacyclopropanes and silacyclopropenes are not stable entities, dimers or molecular fragments are commonly observed as final products. Thus, the dimerization and trimerization of SiF_2 is quite facile and has been observed in a variety of SiF_2 studies. It is difficult to know in many cases whether the dimerization of SiF_2 occurred before or after initial reaction with substrate. The latter possibility seems likely in many instances as it would be anticipated that the mono-adducts of SiF_2 would not be stable with a variety of substrates, e.g.,

Likewise, these mono-adducts could rearrange by C—X insertion, which has also been observed.

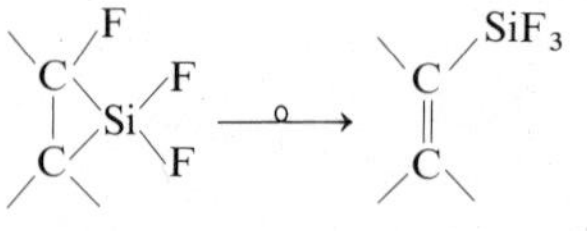

With these reaction modes in mind, the following reaction scheme for SiF_2 oxidative addition reactions with unsaturated organic compounds is outlined (CF_2=CFH,[202-205] CF_2=CF_2,[202-205] C_6H_6,[206,207] C_6F_6,[206,207] CH_2=CH_2,[208] CH_2=CHCH=CH_2,[202-205] HC≡CH,[202-205,209] CH_3C≡CH,[202-205,210] CH_3C≡CCH_3,[202-205] CF_3C≡CH,[211] CH_3CH=CH_2,[212] CH_2=C=CH_2,[213] CF_3COCl,[214] CF_3COF[214a]).

A variety of other substrates have also been allowed to react with SiF_2. Reactions only took place, however, in the condensed phase. As mentioned before, in the gas phase, SiF_2 has an extraordinarily long half-life, nearly 150 seconds at less than 1 torr.[104,202-205,215,216] Even after addition of BF_3, CO, or PF_3, in the gas phase *SiF_2 did not react.* Oxygen did react with SiF_2 under these conditions, however, yielding a mixture of silicon oxyfluorides.

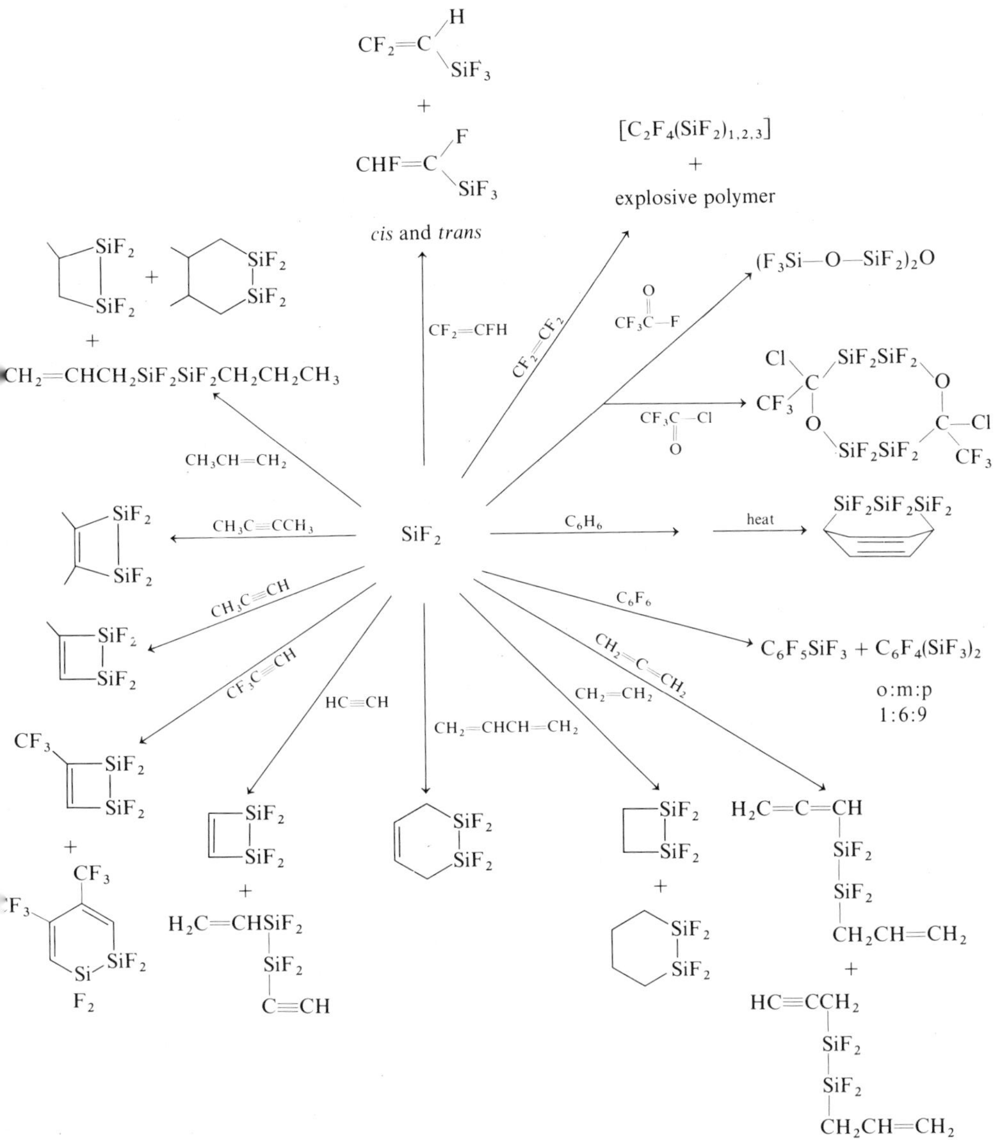

$$CF_2=C{\overset{H}{\underset{SiF_3}{\diagup}}}$$

+

$$CHF=C{\overset{F}{\underset{SiF_3}{\diagup}}}$$

cis and trans

$$[C_2F_4(SiF_2)_{1,2,3}]$$

+

explosive polymer

$$(F_3Si—O—SiF_2)_2O$$

$$CF_3C{\overset{O}{\diagup}}F$$

$$CF_3C{\underset{O}{\diagup}}Cl$$

$$CH_2=CHCH_2SiF_2SiF_2CH_2CH_2CH_3$$

$$CH_3CH=CH_2$$

$$CH_3C{\equiv}CCH_3$$

$$SiF_2$$

$$C_6H_6$$ heat $$SiF_2SiF_2SiF_2$$

$$C_6F_6$$

$$C_6F_5SiF_3 + C_6F_4(SiF_3)_2$$
o:m:p
1:6:9

$$CH_3C{\equiv}CH$$

$$CF_3C{\equiv}CH$$

$$HC{\equiv}CH$$

$$CH_2=CHCH=CH_2$$

$$CH_2=C=CH_2$$

$$CH_2=CH_2$$

$$H_2C=CHSiF_2$$
$$SiF_2$$
$$C{\equiv}CH$$

$$H_2C=C=CH$$
$$SiF_2$$
$$SiF_2$$
$$CH_2CH=CH_2$$
+
$$HC{\equiv}CCH_2$$
$$SiF_2$$
$$SiF_2$$
$$CH_2CH=CH_2$$

Upon condensation of SiF_2, a red-yellow deposit forms that turns white on warming, forming an $(SiF_2)_x$ polymer. This material is plastic-like and tough, but pyrophoric in air and reactive with water. Pyrolysis of this material at about 200°C yielded a mixture of perfluorosilanes from SiF_4 through $Si_{16}F_{34}$ (which, parenthetically, is a viable preparative method for some of these compounds).

When SiF_2–SiF_4 mixtures were added to chemical reagents, and this mixture condensed at $-196°C$, a variety of Oxidative Addition reactions took place. With BF_3, products were isolated that contained two and three SiF_2 moieties.[202–205,215,216]

$$SiF_2 + BF_3 \longrightarrow \underset{\underset{F}{|}}{\overset{\overset{F}{|}}{F-Si}}-\underset{\underset{F}{|}}{\overset{\overset{F}{|}}{Si}}-BF_2 + \underset{\underset{F}{|}}{\overset{\overset{F}{|}}{F-Si}}-\underset{\underset{F}{|}}{\overset{\overset{F}{|}}{Si}}-\underset{\underset{F}{|}}{\overset{\overset{F}{|}}{Si}}-BF_2$$

From the reactions outlined, it appears that the SiF_2–SiF_2 species is important and that this species and higher telomers exist as diradicals in low-temperature matrices, and indeed may be the reacting species.[217–222] In fact, many of the organic substrate–SiF_2 reactions may proceed through free radical intermediates formed by initial attack by one end of the SiF_2 moiety.[202–205,212]

A number of inorganic and saturated organic substrates also react with SiF_2, as outlined below (H_2O,[202–205,223] CF_3I,[224] I_2,[225] H_2S,[226] GeH_4,[227,228] $(CH_3)_3SiOCH_3$,[202–205,229] $(CH_3)_2SO$,[202–205] $BF–BF_3$,[230] BF_3,[217,231] CH_3OH,[202–205,232] SOF_2,[233] CS_2,[234] S_2Cl_2,[234] HBr,[235] B_2H_6,[236] PH_3[237]). The most interesting product of the SiF_2–H_2O reaction is $HF_2SiOSiF_2H$, which is probably formed from the initial insertion product HF_2SiOH.

Somewhat similarly $(CH_3)_3Si\,OCH_3$ reacted with SiF_2 to yield apparently first an Si—O insertion product which decomposed to the products indicated.[202–205] Methanol reacted analagously apparently to yield unstable CH_3OSiF_2H, which then reacted with excess CH_3OH to form the final product $(CH_3O)_2SiF_2$.[202–205,232]

Simple oxidative addition reactions took place with excesses of CF_3I, I_2, H_2S, GeH_4, and BF_3, with SiF_2, $(SiF_2)_2$, and/or $(SiF_2)_3$. In the BF_3 reaction, there is evidence that an excited state SiF_2^* species is responsible for the production of F_3SiBF_2.[212]

Reactions of SiF_2 with $Si(CH_3)_4$, $Si(CH_3)_3Cl$, $Ge(CH_3)_4$ and $CHCl_3$ yielded no products other than SiF_2 polymer. Attempted reactions with CF_3CN, CH_3CN, $CNCl$, and $(CN)_2$ yielded only a complex black mixture that on strong heating gave perfluorosilanes and triazine derivatives.[202–205] Additionally, Margrave and co-workers[202–205] have cocondensed the vapors of SiF_2 and NaF or LiF yielding upon warming Na_2SiF_6 or Li_2SiF_6 and Si metal. It seems likely that $Na^+SiF_3^-$ and $Li^+SiF_3^-$ are unstable intermediates. And finally, SiF_2 reactions with GeF_4, CF_4, C_2F_6, c—C_4F_8, and C_2F_4 yielded explosive matrices. This is rationalized by noting that Si—F bonds are considerably stronger than Ge—F or C—F bonds, and so

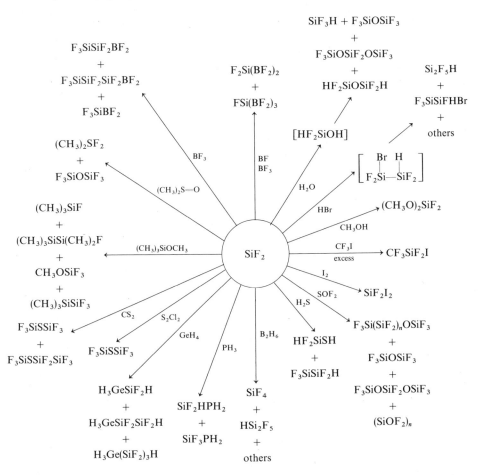

convertion to SiF_4 and Ge (or C) can take place vigorously. This is an oxidation–reduction process with fluorine as oxidant.

In conclusion, it can be stated that SiF_2 behaves somewhat like metal atoms in that two competing processes are always operating: (1) SiF_2 polymerization, which is a very low activation energy process, and (2) oxidative insertion and/or abstraction reactions by SiF_2 or its small clusters $(SiF_2)_n$. In addition, $(SiF_2)_n$ is quite capable of radical-type reactions.

It also should be noted that SiF_2 and related species apparently form multiply bonded Si compounds in low-temperature matrices. Thus, there is considerable theoretical and mechanistic evidence for $X_2Si{=}SiX_2$, $X_2Si{=}CX_2$, and $X_2Si{=}O$.[238] However, there is also direct spectroscopic evidence for $F_2Si{-}SiF_2$ triplet. It is evident that divalent silicon species are interesting reactive entities that fully exploit the possible alternatives for chemical reactions.[238]

The generation of SiF_2 by nuclear recoil techniques has allowed the production of truly monomeric SiF_2 in both singlet and triplet states.[239,240] Thus, ^{31}Si recoil atoms react with PF_3 to yield SiF_2 in a singlet/triplet ratio of 1:3.5.[239] The singlet species reacts with 1,3-butadiene to yield a metallo-cycle, while the triplet reacts by indirect means through radical pathways.

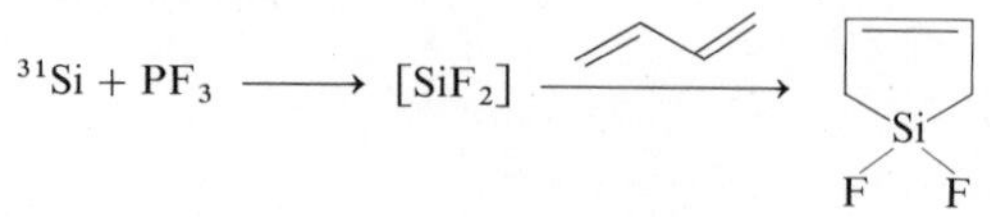

$$^{31}Si + PF_3 \longrightarrow [SiF_2]$$

c. SiCl$_2$. The reagent $SiCl_2$ can be prepared analogously to SiF_2. However, $SiCl_2$ is much shorter-lived in the gas phase. The chemistry of $SiCl_2$ under cocondensation conditions is similar to SiF_2, although with $SiCl_2$, SiX_2 often reacts instead of $(SiX_2)_n$, so often the case with SiF_2. For example, $SiCl_2$ reacted with C_2H_2 to yield 1,4-metalloid-substituted cyclo-hexadienes, as shown below.[212] It is likely that a silacyclopropene inter-mediate precedes formation of this product. Carbenoid character of $SiCl_2$ is also observed in alkene reactions, propene yielding 1,4-metalloid cyclo-hexane derivatives.[212] Silacyclopropane is a likely intermediate. A further $SiCl_2$ insertion product was also observed in the propene reaction. Likewise, oxidative addition processes were observed in the reactions of BCl_3, PCl_3, and CCl_4.[241] Polymer was also formed in the case of CCl_4, and in studies of $SiCl_2$ with $SnCl_4$, C_6H_6, or C_2H_6 only polymers were formed[241] ($HC{\equiv}CH$,[212] $CH_3CH{=}CH_2$,[212] BCl_3,[241] PCl_3,[241] CCl_4[241]).

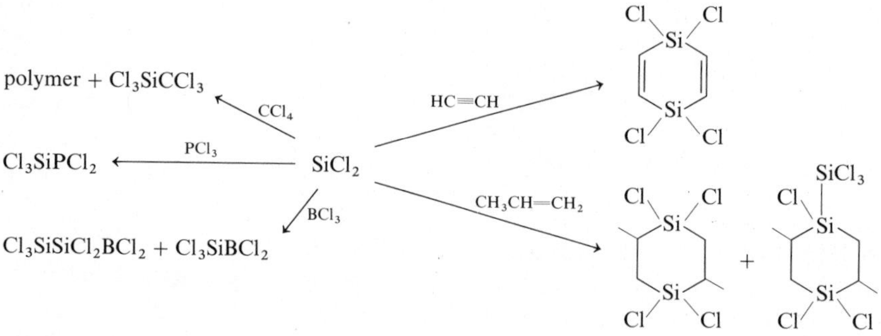

d. SiO. Vaporization of $(SiO)_n$ yields SiO. Condensation of SiO at $-196°C$ with a variety of organic compounds generally yields oxygen-bridged polymers. It has not yet proven possible to prepare compounds containing one SiO moiety,[212,242] although the possibility of preparing new silicone polymers exists. Unfortunately, these polymers also have some

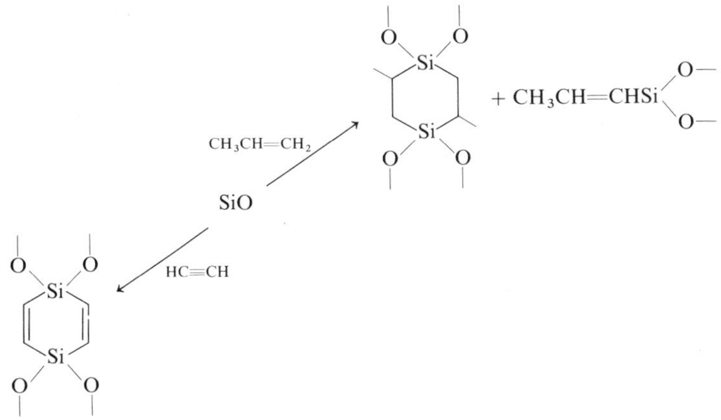

$$\text{SiO} \xrightarrow[\text{HC}\equiv\text{CH}]{\text{CH}_3\text{CH}=\text{CH}_2} \quad + \text{ CH}_3\text{CH}=\text{CHSi}$$

Si—H bonds, which adversely affect their air sensitivity properties.

Note that acetylene[212] and propene[212] both reacted somewhat analogously to $SiCl_2$, yielding 1,4-metalloid substituted cyclohexadiene or cyclohexane polymers.

e. GeF$_2$. This has not been treated as a "high-temperature species," but it should be mentioned that GeF_2 shows "carbene-like" character. $(GeF_2)_n$ can be dissolved in solvents such as dioxane, and allowed to react with alkyl halides[243] or dienes.[244] For example, 2,3-dimethyl-1,3-butadiene adds to GeF_2 to form a metallocyclopentene derivative.

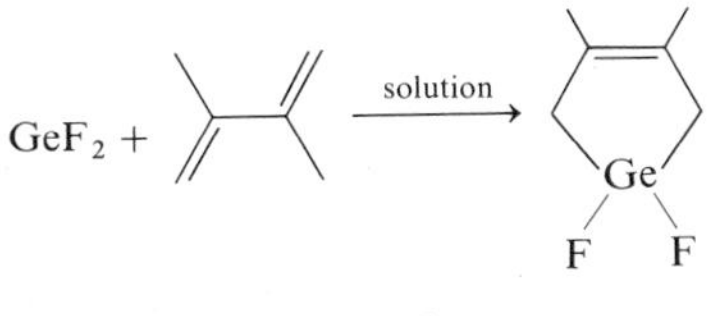

$$\text{GeF}_2 + \quad\xrightarrow{\text{solution}}$$

4. SIMPLE ORBITAL MIXING PROCESSES

Almost all of the chemistry that CS, SiF_2, $SiCl_2$, etc. undergo fits best under the Oxidative Addition or Oxidation Processes sections. However, a few matrix-isolation studies have been carried out that suggest the formation of molecular complexes.

Cocondensation of Ni atoms with CS yielded what Verkade and co-workers believe is $Ni(CS)_4$.[245] However, if this compound is formed, it is quite unstable, decomposing apparently to CS polymer, Ni, and perhaps NiS. Ozin[246] and co-workers have also carried out Ni–CS codepositions, preparing unstable $Ni(CS)_n$ complexes.

Deposition of SiF_2 on a cold window with inert gas has demonstrated how readily SiF_2 telomers can form.[222] In fact, matrix isolation studies of SiF_2–CO mixtures provided evidence for $(SiF_2)_2(CO)_2$. It is interesting that ν_{CO}

shifts down to 1880 cm^{-1} upon complexation of CO and SiF_2, and it seems likely that these CO molecules are bridging SiF_2–SiF_2 moieties.

Other compounds, at $20°$–$50°K$, were deposited with SiF_2–SiF_4 mixtures. Complex products resulted from O_2 and NO reactions.[222] In the case of BF_3, oxidative addition took place to yield $F_3SiSiF_2BF_2$ even as low as $35°K$ (SiF_2–SiF_2 the apparent reacting species).

5. CLUSTER FORMATION PROCESSES

There is a strong tendency for CS to cluster (polymerize) explosively. A high polymer is formed of very low reactivity. Small clusters of CS have not been observed spectroscopically or in chemical reactions.

There is also a strong tendency for SiF_2 to form clusters. In this case, the small clusters (telomers) are extremely reactive, and most often the final products of SiF_2 reactions contain $(SiF_2)_n$ ($n = 2, 3$) moieties. This chemistry has been covered under the Oxidative Addition and Oxidation Processes and Simple Orbital Mixing sections of this chapter.

There is less tendency for $SiCl_2$ or SiO to cluster during a reaction with an added substrate (in excess). However, these, as well as CS and SiF_2 polymerize very vigorously in the absence of other available reaction pathways.

References

1. D. L. Lambert, *Highlights Astron.* **3**, 237 (1974).
2. F. Querci and M. Querci, *Astron. Astrophys.* **39**, 113 (1975).
3. B. Rosen and P. Swings, *Ann. Astrophys.* **16**, 82 (1953).
4. A. E. Douglas, *Astrophys. J.* **114**, 466 (1951).
5. A. C. Danks, D. L. Lambert, and C. Arpigny, *Astrophys. J.* **194**, 745 (1974).
6. M. Pandow, C. MacKay, and R. Wolfgang, *J. Inorg. Nucl. Chem.* **14**, 153 (1960).
7. W. A. Chupka and M. G. Inghram, *J. Phys. Chem.* **59**, 100 (1955).
8. J. Drowart, R. P. Burns, G. DeMaria, and M. G. Inghram, *J. Chem. Phys.* **31**, 1131 (1959).
9. W. A. Chupka and M. G. Inghram, *J. Chem. Phys.* **21**, 1313 (1953).
10. I. V. Golubtosv, *Simp. Ispol'z. Metodov Mechenykh At. Soversh. Tekhnol. Protsessov Proizvod. Primen. Yad.-Fiz. Metodov Anal. Sostava Veshchestva, 1968* p. 38 (1969).
11. G. Glockler, *J. Chem. Phys.* **22**, 159 (1954).
12. T. Doehard, P. Goldfinger, and F. Waelbroeck, *J. Chem. Phys.* **20**, 757 (1952).
13. R. Honig, *J. Chem. Phys.* **22**, 126 (1954).
14. R. F. Harris, Ph.D. Thesis, Pennsylvania State University, University Park (1968).
15. J. J. Havel, Ph.D. Thesis, Pennsylvania State University, University Park (1972).
16. J. H. Plonka, K. J. Klabunde, and P. S. Skell, unpublished results.
17. P. Guillery, *Z. Naturforsch., Terl A* **10**, 248 (1955).
18. H. Gadacz and L. Reimer, *Naturwissenschaften* **47**, 104 (1960).
19. B. Vodar, S. Minn, and S. Offret, *J. Phys. Radium.* **16**, 811 (1955).
20. P. D. Zavitsanos, L. E. Brewer, and W. E. Sauer, *Proc. Natl. Electron Conf.* **24**, 864 (1968).
21. S. I. Anisimov, A. M. Bonch-Bruevich, M. A. Elyashevich, Y. A. Imas, N. A. Paulenko, and G. S. Romanov, *Zh. Tekh. Fiz.* **36**, 1273 (1966).

22. A. I. Korunchikov and A. A. Yankovshii, *Zh. Prikl. Spektrosk.* **5**(5), 586 (1966).
23. A. M. Covington, G. N. Liu, and K. A. Lincoln, *AIAA J.* **15**(8), 1174 (1977).
24. "Handbook of Chemistry and Physics," 56 ed. CRC Press, Cleveland, Ohio, 1975–1976.
25. M. Burden and P. A. Walley, *Vacuum* **19**(9), 397 (1969).
26. A. Kant, *U.S. Dep. Commer., Off. Tech. Serv., AD* **283**, 1342 (1962); *U.S. Gov. Res. & Dev. Rep.* **37**, 16 (1962).
27. E. B. Owens and A. M. Sherman, *U.S. Dep. Commer., Off. Tech. Serv., AD* **275**, 466 (1962).
28. A. V. Tseplyaeva, Y. A. Prisekov, and V. V. Karelin, *Vestn. Mosk. Univ., Ser.* **15**, 36 (1960).
29. M. Naoe and S. Yamanaka, *Jpn. J. Appl. Phys.* **8**, 287 (1969).
30. E. A. Roth, E. A. Margerum, and J. A. Amick, *Rev. Sci. Instrum.* **33**, 686 (1962).
31. R. Thun and J. B. Ramsey, *Natl. Symp. Vac. Technol.* p. 196 (1959).
32. H. A. Hill, *Rev. Sci. Instrum.* **27**, 1086 (1956).
33. S. Namba, *Proc. Symp. Electron Beam Technol., 4th, 1962* p. 304 (1962).
34. E. B. Graper, *J. Vac. Sci. Technol.* **8** (1), 333 (1971).
35. L. Ernst, *Surf. Sci.* **21**, 193 (1970).
36. V. A. Batanov, I. A. Bufetov, S. G. Lukishova, and V. B. Ferorov, *Kvantovaya Elektron (Moscow)* No. 2, p. 436 (1974).
37. L. N. Nemirovskii, *Prib. Tekh. Eksp.* p. 192 (1968).
38. M. A. Luzhnova and Y. D. Raikhbaum, *Teplofiz. Vys. Temp.* **7**, 313 (1969).
39. A. M. Bonch-Bruevich and Y. A. Imas, *Exp. Tech. Phys.* **15**, 323 (1967).
40. A. Korunchikov and A. A. Yankovskii, *Zh. Prikl. Spektrosk.* **5**, (5) 586 (1966).
41. N. V. Afanasev, S. N. Kapelyan, V. A. Morozov, L. P. Filippov, and Z. M. Yudovin, *Zh. Prikl. Spektrosk.* **11**, 883 (1969).
42. J. L. Dumas, *Rev. Phys. Appl.* **5**, 795 (1970).
43. J. Bohdansky and H. E. Schins, *J. Phys. Chem.* **71**, 215 (1967).
44. J. H. Kim and A. Cosgarea, Jr., *J. Chem. Soc.* **44**, 806 (1966).
45. E. V. Bolshin, I. A. Myasnikov, and D. G. Tabatadze, *Zh. Fiz. Khim.* **45**, 2499 (1971).
46. H. G. MacPherson, *J. Appl. Physiol.* **13**, 97 (1942).
47. N. K. Chaney, V. C. Hamister, and S. W. Glass, *Trans. Electrochem. Soc.* **67**, 107 (1935).
48. K. S. Pitzer and E. Clementi, *J. Am. Chem. Soc.* **81**, 4447 (1959); also cf. Skell *et al.* [49]
49. P. S. Skell, L. D. Wescott, Jr., J. P. Goldstein, and R. R. Engel, *J. Am. Chem. Soc.* **87**, 2829 (1965).
50. P. Owens and P. S. Skell, private communications.
51. P. S. Skell, J. J. Havel, and M. J. McGlinchey, *Acc. Chem. Res.* **6**, 97 (1973).
52. P. S. Skell and R. F. Harris, *J. Am. Chem. Soc.* **91**, 4440 (1969).
53. P. S. Skell, J. H. Plonka, and R. R. Engel, *J. Am. Chem. Soc.* **89**, 1748 (1967).
54. P. S. Skell and J. H. Plonka, *J. Am. Chem. Soc.* **92**, 2160 (1970).
55. J. H. Plonka and P. S. Skell, *Tetrahedron Lett.* p. 4557 (1970).
56. P. S. Skell and J. H. Plonka, *Chem. Commun.* p. 1108 (1970).
57. P. S. Skell, J. H. Plonka, and K. J. Klabunde, *Chem Commun.* p. 1109 (1970).
58. P. S. Skell, K. J. Klabunde, J. H. Plonka, D. L. Williams-Smith, and J. S. Roberts, *J. Am. Chem. Soc.* **95**, 1547 (1973).
59. K. J. Klabunde and P. S. Skell, *J. Am. Chem. Soc.* **93**, 3807 (1971).
60. P. S. Skell and J. H. Plonka, *J. Am. Chem. Soc.* **92**, 5620 (1970).
61. P. S. Skell, J. H. Plonka, and R. F. Harris, *Chem. Commun.* p. 689 (1970).
62. P. S. Skell, J. H. Plonka, and L. S. Wood, *Chem. Commun.* p. 710, (1970).
63. P. S. Skell and R. F. Harris, *J. Am. Chem. Soc.* **91**, 699 (1969).
64. R. F. Harris and P. S. Skell, *J. Am. Chem. Soc.* **90**, 4172 (1968).
65. W. Felder and A. Fontijn, *J. Chem. Phys.* **69**, 1112 (1978).

66. A. Fontijn and W. Felder, *Chem. Phys. Lett.* **47**, 380 (1977).
67. P. S. Skell and R. R. Engel, *J. Am. Chem. Soc.* **88**, 4883 (1966).
68. D. Tremblay and S. Kaliaguine, *Ind. Eng. Chem., Process Des. Dev.*, **11**, 265 (1972).
69. K. J. Klabunde, unpublished results.
70. L. Eng, Ph.D. Thesis, Pennsylvania State University, University Park (1970).
71. P. S. Skell and R. F. Harris, *J. Am. Chem. Soc.* **87**, 5807 (1965).
72. P. S. Skell, J. E. Villaume, J. H. Plonka, and F. A. Fagone, *J. Am. Chem. Soc.* **93**, 2699 (1971).
73. S. R. Prince and R. Schaeffer, *Chem. Commun.* p. 451 (1968).
74. J. E. Dobson, P. M. Tucker, R. Schaeffer, and F. G. A. Stone, *J. Chem. Soc. A* p. 1882 (1969).
75. P. S. Skell and P. W. Owen, *J. Am. Chem. Soc.* **94**, 1578 (1972).
76. P. S. Skell, F. A. Fagone, and K. J. Klabunde, *J. Am. Chem. Soc.* **94**, 7862 (1972).
77. P. S. Skell and L. D. Wescott, Jr., *J. Am. Chem. Soc.* **85**, 1023 (1963).
78. R. H. Parker and P. B. Shevlin, *Tetrahedron Lett.* No. 26, p. 2167 (1975).
79. P. B. Shevlin and A. P. Wolf, *Tetrahedron Lett.* No. 46, p. 3987 (1970).
80. S. Kammula and P. B. Shevlin, *J. Am. Chem. Soc.* **96**, 7830 (1974).
81. S. Kammula and P. B. Shevlin, *J. Am. Chem. Soc.* **95**, 4441 (1973).
82. P. B. Shevlin, *J. Am. Chem. Soc.* **94**, 1379 (1972).
83. P. B. Shevlin and S. Kammula, *J. Am. Chem. Soc.* **99**, 2627 (1977).
84. J. M. Figuera, P. B. Shevlin, and S. D. Worley, *J. Am. Chem. Soc.* **98**, 3820 (1976).
85. P. S. Skell and P. W. Owen, *J. Am. Chem. Soc.* **94**, 5434 (1972).
86. P. W. Owen and P. S. Skell, *Tetrahedron Lett.* No. 18, p. (1972).
87. M. J. McGlinchey and T. S. Tan, *Inorg. Chem.* **14**, 1209 (1975).
88. P. L. Timms, private communications.
89. A. Bos and J. S. Ogden, *J. Phys. Chem.* **77**, 1513 (1973).
90. A. Bos, J. S. Ogden, and L. Orgee, *J. Phys. Chem.* **78**, 1763 (1974).
91. J. Ogden, *in* "Cryochemistry" (M. Moskovits and G. A. Ozin, eds.), p. 231. Wiley (Interscience), New York, 1976.
92. T. O. Murdock, unpublished results from this laboratory.
93. D. F. Dickinson and C. A. Gottlieb, *Astrophys. Lett.* **7**, 205 (1971).
94. A. L. Parson, *Mon. Not. R. Astron. Soc.* **105**, 244 (1945).
95. P. Harteck and R. Reeves, *Bull. Soc. Chim. Belg.* **71**, (11-12), 682 (1962).
96. D. Buhl, *Sky & Telescope* p. 156 (1973).
97. W. P. Wood and J. Heicklen, *J. Phys. Chem.* **75**, 854 and 861 (1971).
98. W. Doran and A. E. Gillam, *J. Chem. Soc., Chem. Ind.* **47**, 259 (1928).
99. G. Porter, *Proc. R. Soc. London, Ser. A* **200**, 284 (1950).
100. R. G. Norrish, *Z. Elektrochem.* **56**, 705 (1952).
101. M. DeSorgo, A. J. Yarwood, O. P. Strausz, and H. E. Gunning, *Can. J. Chem.* **43**, 1886 (1965).
102. K. J. Klabunde, C. M. White, and H. F. Efner, *Inorg. Chem.* **13**, 1778 (1974).
103. C. M. White, Master's Thesis, University of North Dakota, Grand Forks (1974).
104. P. L. Timms, R. A. Kent, T. C. Ehlert, and J. L. Margrave, *J. Am. Chem. Soc.* **87**, 2824 (1965).
105. P. L. Timms, *Acc. Chem. Res.* **6**, 118 (1973).
106. S. A. Kriche, L. Herman, and R. Herman, *J. Quant. Spectrosc. & Radiat. Transfer.* **4**, 863 (1964).
107. H. A. Wiebe and J. Heicklen, *Can. J. Chem.* **47**, 2965 (1969).
108. D. Solan, *U.S. C. F. S. T. I.*, *PB Rep.* **PB-187819** (1969).
109. J. Dewar and H. O. Jones, *Proc. R. Soc. London, Ser. A* **83**, 527 (1910).
110. J. Dewar and H. O. Jones, *Proc. R. Soc. London, Ser. A* **85**, 574 (1911).
111. L. C. Martin, *Proc. R. Soc. London, Ser. A* **89**, 127 (1913).

112. M. A. P. Hogg and J. E. Spice, *J. Chem. Soc.* **158**, 5196 (1958).
113. A. B. Callear, J. A. Green, and G. J. Williams, *Trans. Faraday Soc.* **61**, 1831 (1965).
114. R. W. Field and T. H. Bergeman, *J. Chem. Phys.* **54**, 2936 (1971).
115. A. B. Callear, *Proc. R. Soc. London, Ser. A* **276**, 401 (1963).
116. A. B. Callear and R. G. Norrish, *Nature (London)* **188**, 53 (1960).
117. I. Norman and G. Porter, *Proc. R. Soc. London, Ser. A* **230**, 399 (1955).
118. L. P. Blanchard and P. LeGoff, *Can. J. Chem.* **35**, 89 (1957).
119. A. G. Gaydon, G. H. Kimbell, and H. B. Palmer, *Proc. R. Soc. London, Ser. A* **279**, 313 (1964).
120. A. Tewarson and H. B. Palmer, *J. Mol. Spectrosc.* **27**, 246 (1968).
121. G. Hancock and I. W. M. Smith, *Trans. Faraday. Soc.* **67**, 2586 (1971).
122. I. W. M. Smith, *Trans. Faraday Soc.* **64**, 3183 (1968).
123. H. Wiedemeier and H. Schafer, *Z. Anorg. Allg. Chem.* **326**, 235 (1963).
124. D. E. Milligan and M. E. Jacox, *J. Chem. Phys.* **49**, 4269 (1968).
125. M. Schmeisser and K. P. Ehlers, *Angew. Chem.* **76**, 781 (1964).
126. D. L. Perry, P. F. Meier, R. H. Hauge and J. L. Margrave, *Inorg. Chem.* **17**, 1364 (1978).
126a. D. L. Perry and J. L. Margrave, *J. Chem. Educ.* **53**, 696 (1976).
127. P. F. Antipin and V. V. Sergeev, *Zh. Prikl. Khim. (Leningrad)* **27**, 784 (1954).
128. G. Maass, R. H. Hauge, and J. L. Margrave, *Z. Anorg. Allg. Chem.* **392**, 295 (1972).
129. A. I. Belyaev, L. A. Firsamova, L. P. Egorov, Y. Pinchuk, V. N. Chechentsev, A. N. Kochubeev, S. A. Vysotskii, and Z. M. Aledseeva, *Metody. Poluch. Anal. Veshchestv. Osoboi. Chist., Tr. Vses. Konf., 1968* p. 87 (1970).
129a. T. O. Sedgwick, *J. Electrochem. Soc.* **112**, 496 (1965).
130. V. I. Zubkov, M. V. Tikhomirov, K. A. Andrianov, and S. A. Golubtsov, *Dokl. Akad. Nauk SSSR* **188**, 594 (1969).
131. U. Pick, *J. Sci. Instrum.* **44**, 70 (1967).
132. C. E. Drumheller, *Natl. Symp. Technol. Trans.* p. 306 (1960).
133. H. Schafer and R. Hornle, *Z. Anorg. Allg. Chem.* **263**, 261 (1950).
134. I. V. Ryabchifov, M. S. Krushchev, and Y. S. Shchedrovitskii, *Dokl. Akad. Nauk SSSR* **167**, 155 (1966).
135. M. D. Carithers, *Rev. Sci. Instrum.* **39**, 920 (1968).
136. D. L. Hildenbrand, *High. Temp. Sci.* **4**, 244 (1972).
137. G. Haas and J. B. Ramsey, *Appl. Opt.* **8**, 1115 (1969).
138. J. P. Dauvergne, *Rev. Int. Hautes Temp. Refract.* **4**, 155 (1967).
139. B. F. Yudin and A. K. Karklit, *Zh. Prikl. Khim. (Leningrad)* **39**, 537 (1966).
140. G. P. Adams, J. L. Margrave, R. P. Steiger, and P. W. Wilson, *J. Chem. Thermodyn.* **3**, 297 (1971).
141. N. Bartlett and K. C. Yu, *Can. J. Chem.* **39**, 80 (1961).
142. H. Huber, E. P. Kuendig, G. A. Ozin, and A. Vander Voet, *Can. J. Chem.* **52**, 95 (1974).
143. C. W. Moulton and J. G. Miller, *J. Am. Chem. Soc.* **78**, 2702 (1956).
144. K. Matsumoto, N. Kiba, and T. Takeuchi, *Talanta* **22**, 321 (1975).
145. R. J. Isabel, G. R. Smith, R. K. McGraw, and W. A. Guillory, *J. Chem. Phys.* **58**, 818 (1973).
146. J. Drowart, F. Degreve, G. Verhaegen, and R. Colin, *Trans. Faraday Soc.* **61**, 1072 (1965).
147. G. A. Bergman, *Zh. Neorg. Khim.* **3**, 2422 (1958).
148. V. I. Davydov, *Zh. Neorg. Khim.* **2**, 1460 (1957).
149. G. N. Sosnovskii and M. A. Abdeev, *Izv. Akad. Nauk Kaz. SSR, Ser. Metall., Obogashch. Ogneuporov* No. 2, p. 2 (1961).
150. J. Drowart, A. Patloret, and S. Smoes, *Proc. Br. Ceram. Soc.* **8**, 67 (1967).
151. S. G. Karbanov, V. I. Belousov, V. P. Zlomanov, and A. V. Novoselova, *Vestn. Mosk. Univ., Khim.* **23**, 93 (1968).

152. S. G. Karbanov, M. I. Karakhanova, A. S. Pashinkin, V. P. Zlomanov, and A. V. Noveselova, *Izv. Akad. Nauk SSSR, Neorg. Mater.* **7**, 1914 (1971).
153. E. Shimazaki and T. Wada, *Bull. Chem. Soc. Jpn.* **29**, 294 (1956).
154. V. I. Davydov and N. P. Diev, *Zh. Neorg. Khim.* **2**, 2003 (1957).
155. V. A. Ivanchenko, O. P. Pchelyakov, and S. I. Stenin, *Izv. Akad. Nauk SSSR, Neorg. Mater.* **12**, 12 (1976).
156. C. P. Marino, J. D. Guérin, and E. R. Nixon, *J. Mol. Spectrosc.* **51**, 160 (1974).
157. E. Shimazaki and N. Matsumoto, *Nippon Kagaku Zasshi* **77**, 1089 (1956).
158. K. F. Zmbov, J. W. Hastie, and J. L. Margrave, *Trans. Faraday Soc.* **64**, 861 (1968).
159. A. S. Buchanan, D. J. Knowles, and D. L. Swingler, *J. Phys. Chem.* **73**, 4394 (1969).
160. C. G. Maier, *U.S., Bur. Mines, Tech. Pap.* **360**, 1 (1925).
160a. D. J. Knowles, A. J. C. Nicholson, and D. L. Swingler, *J. Phys. Chem.* **74**, 3642 (1970).
161. C. L. Hoenig and A. W. Searey, *J. Am. Ceram. Soc.* **49**, 128 (1966).
162. R. F. Porter, *U.S. A. E. C.* **UCRL-2416** (1953).
163. M. Lemarchands and M. Jacob, *Bull. Soc. Chim. Fr.* **2**, 479 (1935).
164. T. Suntola and J. Antson, German Patent 2,553,048 (1974).
165. D. N. Klushin and V. Y. Chernykh, *Zh. Neorg. Khim.* **5**, 1409 (1960).
166. A. S. Tumanev and L. N. Filina, *Izv. Vyssh. Uchebn. Zaved., Tsvetn. Metall.* **8**, 82 (1965).
167. Y. B. Fuks, S. M. Kozhakhmetov, M. T. Chokaev, and N. O. Ospanov, *Tr. Inst. Metall. Obogashch., Akad. Nauk. Kaz. SSR* No. 45, p. 62 (1972).
168. I. G. Murgulescu and E. Ivana, *Rev Roum. Chim.* **18**, 1667 (1973).
169. M. N. Spiridonova, Y. A. Aleksandrov, and B. V. Emelyanov, *Tr. Khim. Tekhnol.* No. 1, p. 212 (1971).
170. H. Schaefer and M. Binnewies, *Z. Anorg. Allg. Chem.* **410**, 251 (1974).
171. H. Bloom, J. Bockris, N. E. Richards, and R. G. Taylor, *J. Am. Chem. Soc.* **80**, 2044 (1958).
172. K. Matsumoto, N. Kiba, and T. Takeuchi, *Talanta* **22**, 695 (1975).
173. A. Buechler, J. L. Stauffer, and W. Klemperer, *J. Am. Chem. Soc.* **86**, 4544 (1964).
174. D. M. Chizhikov, E. K. Kazanas, and Y. V. Tsvetkov, *Izv. Akad. Nauk SSSR, Met.* No. 5, p. 57 (1969).
175. A. N. Nesmeyanov, L. P. Firsova, and E. P. Isakova, *Zh. Fiz. Khim.* **34**, 1200 (1960).
176. E. K. Kazenas, D. Chizhikov, and Y. V. Tsvetkov, *Termodin. Kinet. Protsessov Vosstanov. Met., Mater. Konf., 1969* p. 14 (1972).
177. R. A. Isakova, *Vestn. Akad. Nauk Kaz. SSR* No. 6, p. 30 (1975).
178. P. J. Dyne and D. A. Ramsay, *J. Chem. Phys.* **20**, 1055 (1952).
179. P. J. Dyne, *Can. J. Phys.* **31**, 453 (1953).
180. R. Steudel, *Z. Naturforsch., Teil B* **21**, 1106 (1966).
181. S. Silvers, T. Bergeman, and W. Klemperer, *J. Chem. Phys.* **52**, 4385 (1970).
182. M. A. P. Hogg and J. E. Spice, *J. Chem. Soc.* p. 4196 (1958).
183. J. Dewar and H. O. Jones, *Proc. R. Soc. London, Ser. A* **83**, 526 (1910).
184. A. Klemenc, *Z. Elektrochem.* **36**, 722 (1930).
185. J. J. Thomson, *Philos. Mag.* [6] **24**, 209 (1912); *Chem. Abstr.* **6**, 3223 (1912).
186. L. C. Martin, *Proc. R. Soc. London, Ser. A* **89**, 127 (1913); *Chem. Abstr.* **8**, 8 (1914).
187. R. Steudel, *Angew. Chem., Int. Ed. Engl.* **6**, 635 (1967).
188. R. Steudel, *Z. Anorg. Allg. Chem.* **361**, 180 (1968).
189. V. M. Khanna, G. Besenbruch, and J. L. Margrave, *J. Chem. Phys.* **46**, 2310 (1967).
190. J. L. Margrave, J. W. Hastie, and R. H. Hauge, *Prepr., Div. Petrol. Chem., Am. Chem. Soc.* **14**, E11 (1969).
191. J. W. Hastie, R. H. Hauge, and J. L. Margrave, *J. Am. Chem. Soc.* **91**, 2536 (1969).
192. J. W. Hastie, R. H. Hauge, and J. L. Margrave, *J. Mol. Spectrosc.* **29**, 152 (1969).
193. H. Takeo, R. F. Curl, Jr., and P. W. Wilson, *J. Mol. Spectrosc.* **38**, 464 (1971).

194. J. Drowart, R. Colin, and G. Exsteen, *V. S. C. F. S. T. I.*, *AD Rep.* **612780** (1964); *U.S. Gov. Res. & Dev. Rep.* **40**, 20 (1965).
195. R. Steudel, *Z. Naturforsch.*, *Teil B* **21**, 1106 (1966).
196. R. Steudel, *Angew. Chem.*, *Int. Ed. Engl.* **6**, 635 (1967).
197. R. Steudel, *Z. Anorg. Allg. Chem.* **361**, 180 (1968).
198. M. de Sorgo, A. J. Yarwood, O. P. Strausz, and H. E. Gunning, *Can. J. Chem.* **43**, 1886 (1965).
199. G. Pannetier, P. Goudmand, O. Dessaux, and I. Rebejkow, *Bull. Soc. Chim. Fr.* **12**, 2808 (1963).
200. G. Hancock and I. W. M. Smith, *Chem. Phys. Lett.* **3**, 573 (1969).
201. I. Norman and G. Porter, *Proc. R. Soc. London, Ser. A* **230**, 399 (1955).
202. J. L. Margrave and P. W. Wilson, *Acc. Chem. Res.* **4**, 145 (1971).
203. J. C. Thompson and J. L. Margrave, *Science* **155**, 669 (1967).
204. J. C. Thompson and J. L. Margrave, *Inorg. Chem.* **11**, 913 (1972).
205. A. Orlando, C. S. Liu, and J. C. Thompson, *J. Fluorine Chem.* **2**, 103 (1972).
206. P. L. Timms, D. D. Stump, R. A. Kent, and J. L. Margrave, *J. Am. Chem. Soc.* **88**, 940 (1956).
207. J. L. Margrave and P. L. Timms, U.S. Patent 3, 485, 862 (1969).
208. J. C. Thompson and J. L. Margrave, *Chem. Commun.* p. 566 (1966).
209. C. S. Liu, J. L. Margrave, J. C. Thompson, and P. L. Timms, *Can. J. Chem.* **50**, 459 (1972).
210. C. Liu, J. L. Margrave, and J. C. Thompson, *Can. J. Chem.* **50**, 465 (1972).
211. J. C. Thompson and C. S. Liu, *Inorg. Chem.* **10**, 1100 (1971).
212. P. L. Timms, *Acc. Chem. Res.* **6**, 118 (1973).
213. C. S. Liu and J. C. Thompson, *J. Organomet. Chem.* **38**, 249 (1972).
214. F. D. Catrett and J. L. Margrave, *J. Inorg. Nucl. Chem.* **35**, 1087 (1973).
214a. F. D. Catrett and J. L. Margrave, *Synth. Inorg. Met.-Org. Chem.* **2**, 329 (1972).
215. P. L. Timms, T. C. Ehlert, J. L. Margrave, F. E. Brinckman, T. C. Farrar, and T. D. Coyle, *J. Am. Chem. Soc.* **87**, 3819 (1965).
216. J. L. Margrave, P. L. Timms, and T. C. Ehlert, U.S. Patent 3, 379, 512 (1968).
217. J. M. Bassler, P. L. Timms, and J. L. Margrave, *Inorg. Chem.* **5**, 729 (1966).
218. J. W. Hastie, R. H. Hauge, and J. L. Margrave, *J. Am. Chem. Soc.* **91**, 2536 (1969).
219. D. E. Milligan and M. E. Jacox, *J. Chem. Phys.* **49**, 4269 (1968).
220. H. P. Hopkins, J. C. Thompson, and J. L. Margrave, *J. Am. Chem. Soc.* **90**, 901 (1968).
221. P. L. Timms, R. A. Kent, T. C. Ehlert, and J. L. Margrave, *Nature (London)* **207**, 187 (1965).
222. J. L. Margrave and D. L. Perry, *Inorg. Chem.* **16**, 1820 (1977).
223. J. L. Margrave, K. G. Sharp, and P. W. Wilson, *J. Am. Chem. Soc.* **92**, 1530 (1970).
224. J. L. Margrave, K. G. Sharp, and P. W. Wilson, *J. Inorg. Nucl. Chem.* **32**, 1817 (1970).
225. J. L. Margrave, K. G. Sharp, and P. W. Wilson, *J. Inorg. Nucl. Chem.* **32**, 1813 (1970).
226. K. G. Sharp and J. L. Margrave, *Inorg. Chem.* **8**, 2655 (1969).
227. D. Solan and P. L. Timms, *Inorg. Chem.* **7**, 2157 (1968).
228. D. Solan, *U.S. C. F. S. T. I.*, *PB Rep.* **PB-187819**, 1 (1969) *U.S. Gov. Res. & Dev. Rep.* **70**, 61 (1970).
229. J. L. Margrave, D. L. Williams, and P. W. Wilson, *Inorg. Nucl. Chem. Lett.* **7**, 103 (1971).
230. R. W. Kirk and P. L. Timms, *J. Am. Chem. Soc.* **91**, 6315 (1969).
231. D. L. Smith, R. Kirk, and P. L. Timms, *J. Chem. Soc., Chem. Commun.* p. 295 (1972).
232. J. L. Margrave, K. G. Sharp, and P. W. Wilson, *Inorg. Nucl. Chem. Lett.* **5**, 995 (1969).
233. K. G. Sharp and J. L. Margrave, *J. Inorg. Nucl. Chem.* **33**, 2813 (1971).
234. C. Lau and J. C. Thompson, *Inorg. Nucl. Chem. Lett.* **13**, 433 (1977).
235. K. G. Sharp and J. F. Bald, *Inorg. Chem.* **14**, 2553 (1975).
236. D. Solan and A. B. Burg, *Inorg. Chem.* **11**, 1253 (1972).

237. G. R. Langford, D. C. Moody, and J. D. Odom, *Inorg. Chem.* **14**, 134 (1975).
238. J. L. Margrave, and D. L. Perry, *Inorg. Chem.* **16**, 1820 (1977).
239. O. F. Zeck, Y. Su, and Y. Tang, *J. Chem. Soc., Chem. Commun* No. 5, 156 (1975).
240. R. A. Ferrieri, E. E. Siefert, M. J. Griffin, O. F. Zeck, and Y. Tang, *J. Chem. Soc., Chem. Commun.* No. 1, p. 6 (1977).
241. P. L. Timms, *Inorg. Chem.* **7**, 387 (1968).
242. E. T. Schaschel, D. N. Gray, and P. L. Timms, *J. Organomet. Chem.* **35**, 69 (1972).
243. P. Riviere, J. Satge, and A. Boy, *J. Organomet. Chem.* **96**, 25 (1975).
244. R. Riviere,. Satge, and A. Castel, *C. R. Hebd. Seances Acad. Sci., Ser. C* **284** (10), 395 (1977).
245. L. W. Yarbrough, G. V. Calder, and J. G. Verkade, *J. Chem. Soc., Chem. Commun.* p. 705 (1973).
246. G. A. Ozin, private communications.

CHAPTER 9

Arsenic, Antimony, Bismuth, Selenium, and Tellurium (Metals of Groups VA and VIA)

I. Arsenic, Antimony, Bismuth, Selenium, and Tellurium Vapors (As, Sb, Bi, Se, Te)

A. Occurrence, Properties, and Techniques

These metals do not vaporize monatomically, and because the vapor compositions are complex, a wide variety of studies of their vaporizations have appeared in the literature. Carrying out the vaporizations is not difficult as their heats of vaporization are quite low.

Table 9-1 summarizes some of the properties of these elements and groups the vaporization studies together.[1–29]

Vaporization of arsenic from Knudsen cells or open surfaces leads almost totally to As_4 molecules.[1] Even heating of Cd_2As_2 leads mainly to As_4 in the vapor above the bulk.[2] Small amounts of As_6 and As_8 have also been detected.[3] Also, it has been shown that the rate of thermal decomposition of GaAs films is dependent on the rate with which As (or Ga) vaporizes.[4]

Antimony has been vaporized by a variety of methods including laser and e-beam methods. In each case, Sb telomers are observed, with Sb_3 most abundant by laser evaporation.[5] However, thermal vaporization yields appreciable amounts of Sb_2 at $1000°C$,[7] vaporizations of Sn–Sb mixtures yield only Sb_4 in the vapor at $445°–545°C$.[8]

Bismuth is also very easily vaporized but does not yield atoms. Lasers[12] and normal resistive heating methods have been employed. Although there is some disagreement in the literature,[13,14,16] the energy involved for Bi vaporization to give Bi_2 is only about 25 kcal/mole. However, if the $Bi_2(g)$ is considered, also about 25 kcal/mole, the energy required to generate atoms by Bi–metal vaporization is about 50 kcal.[13,16] During normal thermal vaporization, 65% Bi_2 and 35% Bi are formed,[15] in the $580°–680°C$ range.

TABLE 9-1

Vaporization Data for As, Sb, Bi, Se, and Te

Element	mp (°C)	bp (°C)	ΔH vap (kcal/g atom)	Techniques	Vapor composition	References
As	817	613 subl	38.5	Res. heating by Knudsen cell or crucibles, from GaAs	As mainly As_6, As_8	1–4
Sb	631	1750	28.1 (38)	Laser e-beam Res. heating of crucibles	Sb_3	5 6 7–11
Bi	271	1560	24.7[a]	Laser	Bi_2 (mainly) Bi, Bi_4 (small)	12, 13 14–17
Se	217	685	23 (11)	Crucible from CdSe, Laser Res. heating of crucible	Se_2 (all) Se_5 (mainly) Se_2–Se_9	2, 18–26
Te	452	1390 (992)	40 (27)	Laser Arc-plasma Res heating	Te_5	24, 26–29 5

[a] Bi–Bi dissociation energy in Bi_2 = 24.9 kcal/mole.

And, as expected, as the temperature of the Knudsen cell is increased the relative amount of atomic Bi is increased.[15]

Bismuth atoms are believed to exist in small amounts in the upper atmosphere.[30] Although the method of detection is not a direct one, it is believed that the light emitted (and detected) in the reaction Bi + O → BiO + hv is very characteristic of that reaction. Lead atoms have also been detected in a similar way in the upper atmosphere.[30]

Selenium vaporizations have been studied extensively.[18,23] Selenium atoms have never been detected, and the vapor consists of Se_2–Se_{10} species with Se_5–Se_{10} molecules predominating. In laser evaporations, Se_5 was found to be in highest concentration[19,22] whereas direct sublimation near 175°C yielded Se_6 ($\sim$60%), Se_5($\sim$30%), and Se_7($\sim$10%).[20] The average molecular formula for the vapor species formed by field ion vaporization was $Se_{5.2}$ from a free surface and $Se_{6.2}$ in a vapor equilibrium state.[21] The vapor species are believed to be mainly ring compounds, and ring formation in the heated condensed phase may be the rate-determining step in the vaporization.[21]

Tellurium has been vaporized by normal resistive heating means as well as by arc plasma methods.[27,28] The predominant species are Te_2 and Te_4.

B. Chemistry

Only a single series of experiments has been carried out where Te vapor was codeposited with solvents by the normal codeposition methods.[31] Upon warming, Te–solvent slurries were formed. Fine, black slurries in pentane and THF were prepared, and their activities in reactions with alkyl iodides measured. It was determined that Te–pentane slurries reacted with CH_3I and CH_3CH_2I at pentane reflux temperatures.[31]

$$Te_n/\text{pentane} + CH_3I \longrightarrow (CH_3)_2TeI_2$$
$$80\%$$

$$Te_n/\text{pentane} + CH_3CH_2I \longrightarrow (CH_3CH_2)_2TeI_2$$
$$30\%$$

Alkyltellurium iodides were formed in yields higher than those obtained by conventional higher temperature sealed-tube reactions with normal Te powder and RI. Thus, the storable Te–pentane slurries were reasonably reactive and allowed milder, less hazardous conditions for the synthesis of these compounds. A variety of other experiments using more temperature-sensitive halides would be worthwhile.

II. Vapors of Arsenic, Antimony, Bismuth, Selenium, and Tellurium Subhalides, Oxides, and Sulfides

A. Occurrence, Properties and Techniques

Very little has been reported in the literature regarding the detection of subhalide, oxide, and sulfide vapors of these elements under natural conditions. Of course, in stars and similar high-temperature bodies, a variety of these materials might exist as vapors, especially in the cooler regions.

In the laboratory a few studies of vaporizations have been reported and Table 9-2 summarizes these reports.[32–55] It should be noted that direct vaporization of the normal state of the oxides and sulfides is readily carried out, but extreme decomposition is typical. The subhalides have not been studied except for a gas-phase ESR investigation of SeF[44,45] which was prepared by passing the effluent of a microwave discharge on CF_4 into a stream of carbonyl selenide. The ESR spectrum of SeF was obtained with SeF in the gas phase. A similar study of S–F was also carried out.[44]

$$CF_4 \xrightarrow{\text{discharge}} [F\cdot] \xrightarrow{O=C=Se} [SeF]$$

Essentially nothing has been reported dealing with other possible subhalides and oxides.

TABLE 9-2

Properties Vaporization, and/or Synthetic Methods for Groups VA and VIA Metallic Subhalides, Oxides, and Sulfides

Species	mp (°C)	bp (°C)	Method of formation	Vapor composition	ΔH_{vap} (kcal/mole)	References
As_2O_3	315		As_2O_3 evap		76	32–34
As_2S_3	300	707	Light-enhanced evap or simple evap of As_2S_3		84	33–35
Sb_2O_3	656	1550 subl	Sb_2O_3 evap. (290–425°C)	SbO, Sb_2O_2, Sb_3O_3, Sb_4O_4, Sb_2O_4, Sb_2O_5, Sb_2O_6		34, 36
Sb_2S_3	550	1150	Sb_2S_3 evap	SbS, Sb_2S_3, Sb_3S_2 S_2, Sb_3S_3 Sb_3S_4 (450°–640°C)	35	34, 37, 39
Bi_2O_3	820	1855	Bi_2O_3 evap by heating or laser	Bi_4O_4, Bi_4O_2, Bi_3O_3, Bi_2O_2, BiO, Bi, O_2		34, 40–42
Bi_2S_3	d 685		Bi_2S_3 evap	Bi, S_2, BiS, (550°–650°C)	30	34, 37, 43
SeF				ESR studies		44, 45
$SeCl_2$				Stabilized by coordination (wet chemistry)		46
SeO_2		350 subl	SeO_2 evap	SeO_2 $(SeO_2)_2$ (small)	25.5	34, 47–49
SeO_3	118	d 180	SeO_3 evap	SeO_3	11	34, 48
SeS	d 118		Overheating Se–S mixtures (1000°C)	SeS + others		50
SeS_2	>100	d	SeS_2 evap			34, 51
TeO	d 370	d	TeO evap	Te, O_2, TeO		34, 52
TeO_2	733	1245	TeO_2 evap, (930°–1130°C)	TeO_2 (10) TeO (1) Te_2O_4 (1)	55 (59)	34, 52–55
TeS			Overheating Te–S mixture (1000°C)	TeS + others		50

B. Chemistry

No chemistry has been reported.

References

1. C. C. Herrick and R. C. Feber, *J. Phys. Chem.* **72**, 1102 (1968).
2. J. B. Westmore, H. Fujisaki, and A. W. Tickner, *Adv. Chem. Ser.* **72**, 231 (1968).
3. J. S. Kane and J. H. Reynolds, *J. Chem. Phys.* **25**, 342 (1956).
4. B. Goldstein, D. J. Szostak, and V. S. Ban, *Surf. Sci.* **57**, 733 (1976).
5. V. S. Ban and B. E. Knox, *J. Chem. Phys.* **51**, 524 (1969).
6. M. Burden and P. A. Walley, *Vacuum* **19**, 397 (1969).
7. G. M. Rosenblatt, *J. Phys. Chem.* **66**, 2259 (1962).
8. G. M. Rosenblatt and C. E. Birchenall, *Trans. AIME* **224**, 481 (1962).
9. F. Myzenkov and D. N. Klushin, *Zh. Prikl. Khim.* **38**, 1709 (1965).
10. V. Muradov, *Dokl. Akad. Nauk SSSR* **221**, 379 (1975).
11. B. Caband, A. Hoareau, P. Nounou, and R. Uzan, *Int. J. Mass Spectrom. Ion Phys.* **11**, 157 (1973).
12. A. M. Bonch-Bruevich and Y. A. Imas, *Exp. Tech. Phys.* **15**, 323 (1967).
13. P. A. Rice and D. V. Ragone, *J. Chem. Phys.* **45**, 4141 (1966).
14. G. F. Voronin, *Zh. Fiz. Khim.* **40**, 1381 (1966).
15. J. H. Kim and A. Cosgarea, Jr., *J. Chem. Phys.* **44**, 806 (1966).
16. A. K. Fischer, *J. Chem. Phys.* **45**, 375 (1966).
17. I. D. Apostol, C. Grigoriu, I. Morjan, I. Mihailescu, and V. Batonov, *Rev. Roum. Phys.* **21**, 371 (1976).
18. J. Berkowitz and W. A. Chupka, *J. Chem. Phys.* **48**, 5743 (1968).
19. B. E. Knox, *Adv. Mass Spectrom.* **4**, 491 (1968).
20. H. Fujisaki, J. B. Westmore, and A. W. Tickner, *Can. J. Chem.* **44**, 3063 (1966).
21. H. Saure and J. Block, *Int. J. Mass Spectrom. Ion. Phys.* **7**, 145 (1971).
22. B. E. Knox, *Mater. Res. Bull.* **3**, 329 (1968).
23. R. Yamdagni and R. Porter, *J. Electrochem. Soc.* **115**, 601 (1968).
24. Y. S. Chernozubov, B. P. Kuznetsov, A. A. Klimenko, and E. V. Podmogilnyi, *Zh. Fiz. Khim.* **46**, 275 (1972).
25. A. G. Sigai, *J. Vac. Sci. Technol.* **12**, 958 (1975).
26. K. H. Grupe, K. Hellwig, and L. Kolditz, *Z. Phys. Chem.* **255**, 1015 (1974).
27. G. P. Ustyugov and E. N. Vigdorovich, *Izv. Akad. Nauk. SSSR, Neorg. Master.* **4**, 2022 (1968).
28. M. A. Luzhnova and Y. D. Raikhbaum, *Teplofiz. Vys. Temp.* **7**, 313 (1969).
29. L. Malaspina, R. Gigli, and G. Bardi, *Rev. Int. Hautes Temp. Refract.* **9**, 131 (1972).
30. R. R. Reeves, E. W. Albers, and P. Harteck, NASA Accession No. N65-17694, Rep. No. NASA-CR-60442; *Sci. Tech. Aerosp. Rep.* **3**, 1279 (1965).
31. C. King, K. J. Klabunde, and K. Irgolic, unpublished results; also *cf.* K. J. Klabunde, *Ann. N. Y. Acad. Sci.* **295**, 83 (1977).
32. V. S. Ban and B. E. Knox, *J. Chem. Phys.* **52**, 248 (1970).
33. M. Richnow, *Met. Erz.* **38**, 32 (1941).
34. "Handbook of Chemistry and Physics," 5th ed. CRC Press, Cleveland, Ohio, 1975–1976.
35. M. Janai and P. S. Rudman, *Phys. Status Solidi, A* **42**, 729 (1977).
36. E. K. Kazenas, D. M. Chizhikov, and Y. V. Tsvetkov, *Zh. Fiz. Khim.* **47**, 1547 (1973).
37. G. G. Gospodinov, B. A. Popovkin, A. S. Pashinkin, and A. V. Novoselova, *Vest. Mosk. Univ., Ser., II* **22**, 54 (1967).

38. L. C. Sullivan, J. E. Prusaczyk, and K. D. Carlson, *J. Chem. Phys.* **53**, 1289 (1970).
39. A. S. Shendyapin, V. N. Nesterov, and E. T. Ibragimov, Inst. Metall. Obogashch, Alma Ata., USSR, *Deposited Doc., VINITI* p. 1037 (1975).
40. V. S. Ban and B. E. Knox, *J. Chem. Phys.* **52**, 243 (1970).
41. V. Ilin, *Zh. Neorg. Khim.* **21**, 1645 (1976).
42. E. K. Kazenas, D. M. Chizhikov, Y. U. Ysvetkov, and M. V. Olshevskii, *Dokl. Akad. Nauk SSSR* **207**, 354 (1972).
43. G. A. Komlev and M. V. Olshevskii, *Issled. Protsessov Soversh. Tekhnol Proizvod. Polim. Mater. Stekla* p. 107 (1974).
44. A. Carrington, G. N. Currie, T. A. Miller, and D. H. Levy, *J. Chem. Phys.* **50**, 2726 (1969).
45. J. M. Brown, C. R. Byfleet, B. J. Howard, and D. K. Russell, *Mol. Phys.* **23**, 457 (1972).
46. K. J. Wynne and P. S. Pearson, *J. Chem. Soc. D* p. 293 (1971).
47. N. N. Dyachkova, E. N. Vigdorovich, G. P. Ustyugov, and A. A. Kudryavtsev, *Izv. Akad. Nauk SSSR, Neorg. Mater.* **5**, 2219 (1969).
48. P. J. Ficalora, J. C. Thompson, and J. L. Margrave, *J. Inorg. Nucl. Chem.* **31**, 3771 (1969).
49. M. Spoliti, V. Grosso, and C. S. Nunziante, *J. Mol. Struct.* **21**, 7 (1974).
50. T. L. Sheredina and A. A. Maltsev, *Ural. Konf. Spektrosk., Tezisy Dokl., 7th, 1971* vol. 3, 178 (1971).
51. V. A. Umilin, I. L. Agaforov, L. N. Pornev, and G. G. Devyatykh, *Zh. Neorg. Khim.* **9**, 2492 (1964).
52. G. H. Staley, *J. Chem. Phys.* **52**, 4311 (1970).
53. V. Piacente, L. Malaspina, G. Bardi, and R. Gigli, *Rev. Int. Hautes Temp. Refract.* **6**, 91 (1969).
54. J. R. Soulen, P. Sthapitanonda, and J. L. Margrave, *J. Phys. Chem.* **59**, 132 (1955).
55. D. Kunev, K. Vasilev, and T. Nikolov, *God. Vissh. Khimikotekhnol. Inst. Sofia* **14**, 83 (1971).

CHAPTER 10

Lanthanides and Actinides

I. Lanthanide and Actinide Metal Atoms

A. Occurrence, Properties and Techniques

The term "rare earths" is not really accurate, for the rarest of the lanthanides, thulium, has a larger earth crustal abundance than mercury.[1] This alone is sufficient to provide for the further development of the chemistry of these elements, and as seen later in this chapter, the use of metal atoms has provided some inroads that seemed improbable only a few years ago.

Atoms, ions, and compounds of the lanthanides and actinides exist in stars,[2,3] in the sun and solar atmosphere,[4,5] and in sun spots.[6] Detection of these elements as atoms in the earth's atmosphere apparently has not been reported.

The lanthanides are not particularly difficult to vaporize, and resistive heating of $W–Al_2O_3$ crucibles is usually satisfactory.[7-15] A variety of other vaporization techniques, such as lasers and e-beams could be employed, although very little has been reported on these methods.[7-15] The vapors are assumed to be monatomic.[16]

The actinides, Th, Pa, U, Np, Pu, and Am, are much more difficult to vaporize. Resistive heating of W wires directly supporting U pieces or wire has been employed with limited success.[17] However, some W was also evaporated at the same time, due to the extreme high temperatures necessary. Again, laser or e-beam methods are promising techniques for these metals, but little has been published.[18-20]

Table 10-1 summarizes the vaporization data and references for the lanthanides and actinides.[7-15,18-21] It would appear from these data that a great deal remains to be done with vaporization studies. Also, since the lanthanides are conveniently vaporized, these elements constitute a rich area for metal vapor chemistry investigations on synthetic scale.

B. Chemistry

1. OXIDATIVE ADDITION PROCESSES

It would be of great interest to study lanthanide metal atom–alkyl (aryl) halide reactions, and apparently some studies along these lines have been initiated.[22]

TABLE 10-1

Vaporization Data for the Lanthanides and Antinides

Element	mp (°C)	bp (°C)	ΔH_{vap} (kcal/g atom)	Techniques and vapor composition[a]	References
La	921	3457	97 (101)		9–11, 13, 21
Ce	799	3426	96 (105)		9–11, 13, 21
Pr	931	3512	78 (83)		9–11, 13, 21
Nd	1021	3068	76 (75)		9–11, 13, 21
Pm					
Sm	1077	1791	49 (40)	Resistive heating, 650°–900°C	7–13, 21
Eu	822	1597	42 (41)		9–11, 13, 21
Gd	1313	3266	84 (90)		9–11, 13, 21
Tb	1360	3123	87 (88)		9–11, 13, 21
Dy	1412	2562	70 (69)	Resistive heating	9–11, 13, 14, 21
Ho	1474	2695	70 (69)		9–11, 13, 21
Er	1529	2863	61 (79)	Resistive heating	8–11, 13, 14, 21
Tm	1545	1947	58 (57)		10, 11, 13, 15, 21
Yb	819	1194	35 (40)	Resistive heating, 500°–650°C	8, 10–13, 15, 21
Lu	1663	3395	94 (99)		9–11, 13, 21
Th					13, 20, 21
Pa	<1600				13, 20, 21
U	1132	3818	131	Resistive heating, e-beam	13, 18–21
Np	640	3902			13, 20, 21
Pu	641	3232			13, 20, 21
Am	994	2607			13, 20, 21

[a] Vapors are probably monatomic.[16]

Studies of lanthanide and actinide metal atom reactions with alkenes, dienes, and polyenes will be discussed in the next section, Simple Orbital Mixing, even though in some instances the formation of a M–C σ-bond may be a better description of the reaction mode than π-complexation (orbital mixing). This is perhaps evident from the initial experiments carried

$$M + \begin{array}{c} C \\ \| \\ C \end{array} \longrightarrow M \begin{array}{c} \diagup C \\ | \\ \diagdown C \end{array} \text{ vs } M - \begin{array}{c} C \\ \| \\ C \end{array}$$

out by Skell and co-workers[23,24] where Dy and Er atoms were deposited with propene, followed by treatment of the matrices with D_2O. It was anticipated that σ-bound propene would take up deuterium whereas π-bound propene would not. In the Dy and Er work, deuterium was taken up in substantial amounts, implying M–C σ-bonding.

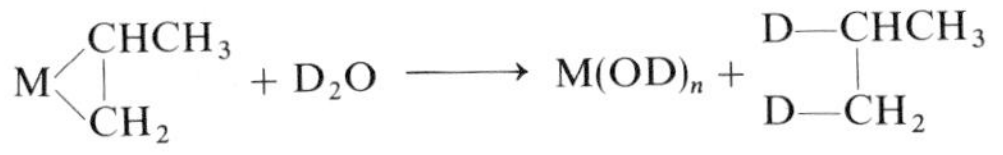

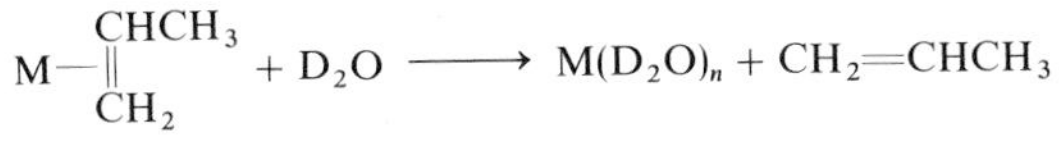

$$M \underset{CH_2}{\overset{CHCH_3}{<}} + D_2O \longrightarrow M(OD)_n + \underset{D-CH_2}{\overset{D-CHCH_3}{|}}$$

$$M - \underset{CH_2}{\overset{CHCH_3}{||}} + D_2O \longrightarrow M(D_2O)_n + CH_2 {=} CHCH_3$$

A very brief report concerning the reaction of Er, Dy, and Ho with 2,4-pentanedione appeared in 1976.[25] In these reactions, oxidative addition–elimination apparently took place to generate H_2 and M (acac)$_3$ derivatives.

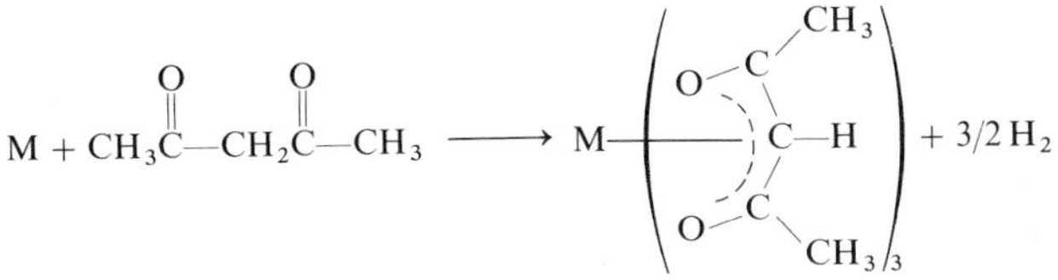

$$M + CH_3\overset{O}{\overset{||}{C}} - CH_2\overset{O}{\overset{||}{C}} - CH_3 \longrightarrow M \left(\begin{matrix} O^{-C} \diagdown \\ {)}C-H \\ O^{-C} \diagup \end{matrix} \overset{CH_3}{} \overset{}{\underset{CH_3}{}} \right)_3 + 3/2\,H_2$$

2. SIMPLE ORBITAL MIXING PROCESSES

On a microscale, a variety of lanthanide metal atoms and actinide metal atoms have been codeposited with CO in varying concentrations. In this way $v_{C=O}$ values for $M(CO)_{1-6}$ have been determined. Table X-2 summarizes[26-30] these results, due mainly to the work of Sheline, Slater, and co-workers,[26-28] and Moskovits and Ozin.[29]

From the $v_{C=O}$ values listed in Table 10-2, it is evident that there is a strong interaction between lanthanide metal atoms and CO, and as expected, the strength of the interaction with each CO is lessened as more CO molecules are coordinated. Thus, $v_{C=O}$ increases with increasing coordination number, indicating that π-backbonding is strongest in the M–(CO)$_1$ complexes and weakest in the $M(CO)_6$ complexes. So even with the lanthanides, which possess low-energy 4f orbitals, the normal Dewar-Chatt $d\pi$–$p\pi$ bonding scheme may be adequate.

Each lanthanide metal coordinates to CO with very similar strengths of interaction. However, a slight weakening of the M–CO interaction is evident in passing from Pr through Gd, but then this trend is reversed on progressing to Ho. Perhaps there is a very slight minimum in the strength of interaction near the center of the lanthanide group.

On moving to the actinide U, it should be noted that a significantly greater M–(CO)$_{3-6}$ interaction, as compared with the lanthanide series, is evident. Thus, $v_{C=O}$ is much lower in the $U(CO)_{3-6}$ complexes when compared with the lanthanides, while the $U(CO)_{1-3}$ complexes exhibit similar values when

TABLE 10-2

Lanthanide and Actinide–CO Complexes (Microscale)

Complex	$v_{C=O}(cm^{-1})$	Reference
Pr(CO)	1835	26–29
Pr(CO)$_2$	1858	26–29
Pr(CO)$_3$	1885	26–29
Pr(CO)$_4$	1940	26–29
Pr(CO)$_5$	1965	26–29
Pr(CO)$_6$	1989	26–29
Nd(CO)	1840	26–29
Nd(CO)$_2$	1861	26–29
Nd(CO)$_3$	1891	26–29
Nd(CO)$_4$	1940	26–29
Nd(CO)$_5$	1965	26–29
Nd(CO)$_6$	1990	26–29
Eu(CO)		26–29
Eu(CO)$_2$	1873	26–29
Eu(CO)$_3$		26–29
Eu(CO)$_4$	1968	26–29
Eu(CO)$_5$	1974	26–29
Eu(CO)$_6$	2000	26–29
Gd(CO)	1841	26–29
Gd(CO)$_2$	1864	26–29
Gd(CO)$_3$	1901	26–29
Gd(CO)$_4$	1945	30
Gd(CO)$_5$	1967	30
Gd(CO)$_6$	1986	26–29
Ho(CO)	1830	26–29
Ho(CO)$_2$	1859	26–29
Ho(CO)$_3$	1902	26–29
Ho(CO)$_4$	1929	26–29
Ho(CO)$_5$	1961	26–29
Ho(CO)$_6$	1982	26–29
U(CO)	1832/1817	26–28
U(CO)$_2$	1855/1846	26–28
U(CO)$_3$	1893	26–28
U(CO)$_4$	1919	26–28
U(CO)$_5$	1938	26–28
U(CO)$_6$	1961	26–28

compared with the lanthanides. Perhaps the greater size of U allows the higher coordination-number complexes to have a stronger interaction.

In comparing a typical transition metal M–(CO)$_n$ complex (cf. Chapters 4 and 5) it can be seen that U apparently perturbs and weakens the CO bond even more than V or Ni, which reemphasizes the apparently remarkable strength of the U–(CO)$_n$ interaction. It would be worthwhile to investigate further actinide atom–CO chemistry with regard to actual thermal stabilities of the $M(CO)_n$ species formed.

Two research groups, those of DeKock and Evans, have recently shown that lanthanide metal atom chemistry has great potential for the macroscale synthesis of low-valent M–ligand complexes.

Ely, Hopkins, and DeKock[31] have codeposited Nd atoms with cyclooctatetraene (COT) initially yielding $Nd_2(COT)_3$, which is similar to the well-known $U(COT)_2$ and $KLn(COT)_2$ compounds prepared by more conventional means by Streitwieser and co-workers.[32,33] Upon treatment with THF, and Soxlet extraction into THF, bright green crystals of $[Nd(COT)\cdot(THF)_2][Nd(COT)_2]$ were obtained.[31] The molecular and crystal structure

of this novel compound was determined, demonstrating the π-electron aromatic nature of the COT^{2-} rings, as the bond distances and angles agreed well with those found in $[Ce(COT)Cl\cdot 2THF]_2$.[34] The structure of this Nd complex is quite unique, particularly with regard to the asymmetrically bound COT ring to both Nd atoms. The COT rings are not equidistant or parallel from the Nd atom. This sharing of COT allows a donation of 14π-electrons to Nd(I) and 10π- and 4σ-electrons to Nd(II).

A series of these lanthanide–COT complexes has been prepared, with the general structures of the La, Ce, Nd, and Er analogs being very similar.[35] In each case, the COT exists as COT^{2-}. Also in each case, THF solvation was taken on when THF extraction was carried out. However, these THF molecules were readily removed. For example, on standing at room temperature, the Ce complex gave up its THF molecules.[35]

Evans, Engerer, and Neville[36] have attempted to extend the low-valency chemistry of the lanthanides by codeposition of 1,3-butadiene and 2,3-dimethyl-1,3-butadiene, with several lanthanide metals. Extraction of the resultant products yielded compounds that analyzed for $Ln(C_4H_6)_3$ or $Ln[(CH_3)_2C_4H_4]_2$. Hydrolysis of these materials yielded mainly the starting diene plus some oligomers of the diene. According to the analyses, magnetic susceptibilities, and hydrolysis data, the bonding in these materials was best described as π-type, and they are obviously not monomeric complexes.

The DeKock[31,35] and Evans[36] work suggests that a variety of low-valency $M–(L)_n$ ligands may be available via metal vapor techniques.

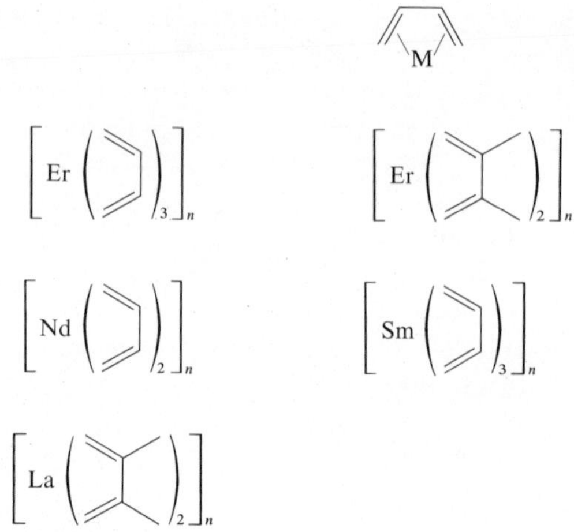

3. CLUSTER FORMATION PROCESSES

Although no work has been reported dealing with the use of metal vapor methods for generating clusters and/or active metal slurries of the lanthanide or actinide metals, such would be quite worthwhile. For example, Streitwieser and co-workers have shown that uranocene $[U(COT)_2]$ can be prepared by the interaction of COT with active U powder prepared by the decomposition of uranium hydride.[37]

In another example, Evans and co-workers[38] have shown that Pr powder, prepared by potassium metal reduction of $PrCl_3$, is useful for the conversion of 1,5-cyclooctadiene to cyclooctatetraene (COT). An intermediate species is possibly $K[Pr(COT)_2]$, which upon oxidative decomposition yields C_8H_8.

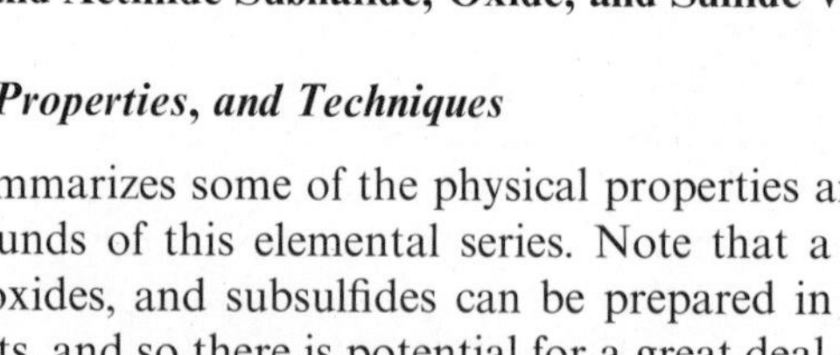

$$PrCl_3 + K \longrightarrow Pr_n^* \cdot KCl$$

$$\downarrow 1,5\text{-}C_8H_{12}$$

$$C_8H_8 \xleftarrow{\text{ox}} K[Pr(COT)_2] \xrightarrow{UCl_4} U(COT)_2$$

II.   Lanthanide and Actinide Subhalide, Oxide, and Sulfide Vapors

A.   Occurrence, Properties, and Techniques

Table 10-3 summarizes some of the physical properties and vaporization data for compounds of this elemental series. Note that a wide variety of subhalides, suboxides, and subsulfides can be prepared in the vapor state for these elements, and so there is potential for a great deal of new chemical investigations. However, it should also be noted that for the oxides, a great

TABLE 10-3

Vaporization Data for the Lanthanide and Actinide Subhalides, Oxides, and Sulfides

Compound[a]	mp (°C)	bp (°C)	ΔH vap (kcal/mole)	Techniques[b]	Vapor composition	References
$LaCl_3$	860	>1000				21, 39, 40
LaF_3	772			Isolated in matrix[c]	LaF_3 some La_2F_6	40–43
LaI_3						21, 40
La_2O_3	2315	4200		Resistive heating, dc-arc	La, LaO, O	21, 44–51
La_2S_3	2100			$1320°–1920°C$		21, 47
LaS					La, S	52, 53
$CeCl_3$	848	1727				21, 39, 40
CeF_3	1460	2300				21, 40, 41, 43
Ce_2O_3	1692			Resistive heating, dc arc	Ce, CeO, CeO_2, O	21, 44, 46, 48, 51
CeO_2	d 2100				M, MO, MO_2	48
Ce_2S_3				$1320°–1920°C$		21, 47
CeS			~70	~1100°C	Ce, S	52, 53
$PrBr_3$	691	1547				21, 40, 54
$PrCl_3$	786	1700				21, 39, 40
PrF_3				Isolated in matrix[c]		40, 41, 43
PrI_3	737					21, 40
PrO_2	d 350					21
Pr_2O_3	d			Resistive heating, dc arc	Pr, PrO, PrO_2, O	21, 44, 46, 48, 51
Pr_2S_3	d			$1320°–1920°C$		21
PrS			~70	~1100°C	Pr, S	52, 53
$NdBr_3$	684	1540				21, 40, 54

(continued)

TABLE 10-3 (continued)

Compound[a]	mp (°C)	bp (°C)	ΔH vap (kcal/mole)	Techniques[b]	Vapor composition	References
$NdCl_3$	784	1600				21, 39, 40
NdF_3	1410	2300		Isolated in matrix[c]		21, 40, 41, 43
NdI_3	775	1370				21, 40
Nd_2O_3	1900			Resistive heating, dc arc	Nd, NdO, O	21, 44–46, 51
Nd_2S_3	d					21
NdS				1320°–1920°C	Nd, S	52, 53
$SmBr_2$	508	1880				21
$SmCl_2$	740 (859)	1950	57	Reduce $SmCl_3$ subl, isolated in matrix[d]	$SmCl_2$	21, 39, 40, 58
$SmCl_3$	678	d				29,
SmF_2	1306	>2400	44	Isolated in matrix[d]		21, 57
SmF_3	1306	2323		Isolated in matrix[c]		21, 40, 41, 43
SmI_2	527	1580				21
SmI_3	850					21, 40
Sm_2O_3				Resistive heating, dc arc	Sm, SmO, SmO_2, O	21, 44, 46, 48, 51
Sm_2S_3	1900					21
SmS				1320°–1920°C	Sm, S	52, 53
$EuBr_2$	677	1880				21
$EuBr_3$	702	d	~70	~1100°C		21, 40, 54
$EuCl_2$	727 (738)	2190	55	Reduce $EuCl_3$ subl, isolated in matrix,[d]	$EuCl_2$	21, 55–57
$EuCl_3$	850					21, 39, 40
EuF_2	1380	>2400		Isolated in matrix[d]		21, 57
EuF_3	1390	2280		Isolated in matrix[c]		21, 40–42
EuI_2	527	1580				21

Compound				Conditions	Products	References
EuI₃	877					21, 40
Eu₂O₃				Resistive heating, dc arc	Eu, EuO, O	21, 44, 46, 51
EuS				1320°–1920°C	Eu, S	22, 53
GdCl₃	609					21, 39, 40
GdI₃	926		1340			21, 40
GdF₃						21, 40, 43
Gd₂O₃				Resistive heating, dc arc	Gd, GdO, O	21, 44, 46, 48, 51
Gd₂S₃						21
GdS				1320°–1920°C	Gd, S	52, 53
TbBr₃	827	~70	1490	~1100°C		21, 40, 54
TbF₃	1172		2280(?)			21, 40, 43
TbI₃	946		>1300			21
Tb₂O₃				Resistive heating, dc arc	Tb, TbO, TbO₂, O	21, 44, 46, 48
DyBr₃	881		1480			21, 40, 54
DyCl₃	718	~70	1500	~1100°C		21, 39, 40
DyF₃	1360		>2200			21, 40, 43
DyI₃	955		1320			21, 40
Dy₂O₃	2340			Resistive heating, dc arc	Dy, DyO, O	21, 44, 46, 48, 51
HoBr₃	914		1470			21, 40, 54
HoCl₃	718	~70	1500	~1100°C		21, 39, 40
HoI₃	989		1300			21, 40
HoF₃	1143		>2200			21, 40, 43
Ho₂O₃				Resistive heating, dc arc	Ho, HoO, O	21, 44, 46, 51
ErF₃	1350		2200			21, 40, 43
ErI₃	1020		1280			21, 40
Er₂O₃				Resistive heating, dc arc	Er, ErO, O	21, 44, 46, 48, 51

(continued)

TABLE 10-3 (*continued*)

Compound[a]	mp (°C)	bp (°C)	ΔH vap (kcal/mole)	Techniques[b]	Vapor composition	References
$TmBr_3$	952	1440	~70	~1100°C		21, 40, 54
TmF_3	1158	>2200				21, 40, 43
TmI_3	1015	1260				21, 40
Tm_2O_3				Resistive heating, dc arc	Tm, TmO, O	21, 44, 46, 51
$YbBr_2$	677	1800				21
$YbBr_3$	956	d	~70	~1100°C		21, 40, 54
$YbCl_2$	702	1900	56	Reduce $YbCl_3$ subl, isolated in matrix[d]	$YbCl_2$	21, 55–57
	(708)	(2105)		isolated in matrix[d]		
YbF_2	1052	2380				21, 57
YbF_3	1157	2200				21, 40, 43
YbI_2	780					21
YbI_3	d 700	d 1300				21, 40
Yb_2O_3				Resistive heating, dc arc	Yb, YbO, O	21, 44, 46, 51
YbS				1320°–1920°C	Yb, S	53
$LuBr_3$	1025	1400	~70	~1100°C		21, 54
$LuCl_3$	905	750 subl				21, 39, 40
LuF_3	1182	2200				21, 40, 43
LuI_3	1050	1200				21, 40
Lu_2O_3				Resistive heating, dc arc	Lu, LuO, O	21, 44, 46, 51
$ThBr_4$	610 subl	725				21
$ThCl_4$	770	d 928				21
ThF_4	>900					21
ThI_4	566	839				21
ThO_2	3050	4400		1500°–2000°C subl		21, 59
ThS_2	1925					21
$PaCl_4$	400 subl					21

Species						Ref.
UBr₄	516					21
UBr₃	730	792				21
UCl₅	d 300	volatile				21
UCl₄	590	792				21
UCl₃	842					21
UF₄	960	1456	51	1018°–1302°C, isolated in matrix^e		21, 44, 60, 60a
UF₃	d 1000					21
UO₂	2500	1400–2300 subl	149	Resistive heating, laser, matrix isolated,	UO₂ mainly, U, UO	21, 47, 59, 61–65
U₃O₈	d 130					21
US₂	d > 1100					21
US	2735		150	1550°–2130°C, subl	U mainly, US, US₂	21, 66, 67
NpCl₄	800 subl					21
NpCl₄	538					21
NpCl₃	800					21
Np₂O₃	d 500					21
PuBr₃	681					21
PuCl₃	760					21
PuF₄	1037					21
PuI₃	777					21
AmBr₃	subl					21
AmCl₃	850 subl					21

^{a} See discussion in text for a series of M₂O species also available.

^{b} Unless otherwise noted, resistive heating was employed.

^{c} In Ar, Kr, and N₂ matrices. LaF₃, CeF₃, SmF₃, and EuF₃ are planar molecules while PrF₃ is pyramidal.[41]

^{d} In Ar, Kr, and N₂ matrices at 21 K. The dihalides were deposited by simple vaporization, while the difluorides were prepared by the M + MF₃ reaction.[57] SmCl₂ and EuCl₂ have 130° bond angles while YbCl₂ is about 140°.

^{e} In Ne, Ar, Kr, Xe, O₂, N₂, NO, CO matrices. UF₄ tetrahedral.[60]

10   Lanthanides and Actinides

deal of decomposition is prevalent upon evaporation. This is not generally true of the halides, however.

In addition to the species shown in Table X-3,[21,39-67] a series of lanthanide suboxides have been prepared in the gas phase by high-temperature Knudsen cell disproportionation processes.[50,68] Thus, M_2O[68] and MO[49] have been prepared and MO (M = La, Ce, Pr, Nd, Pm, Sm, Gd, Tb, Dy, Ho, Er, Tm, Lu) trapped in low-temperature matrices.[49] In this work, CeO_2, PrO_2, and TbO_2 were also trapped.

B. Chemistry

1. SIMPLE ORBITAL MIXING PROCESSES

The only chemistry yet studied for these species, in particular the low-valency fluorides, has dealt with microscale MF_3- or MF_4–CO interactions. Upon complexation of CO to UF_4, a shift in $\nu_{C=O}$ to higher frequency ($\rightarrow 2182\ cm^{-1}$) was observed.[60] An interaction of CO consisting primarily of σ-electron donation can be expected to show this type of behavior. Thus, removal of electrons from the slightly antibonding 5σ orbital has the effect of increasing the bond strength and hence the frequency.[69] It is analogous to CO complexing to an M^{4+} site, and π-backbonding or any type of π-interaction is not important.

Similar results have been obtained by DeKock, Wesley, and Radtke,[57] wherein $LnCl_3$ vapors were codeposited with CO (and inert gas) on microscale, and upon formation of Cl_3Ln–CO a significant $\nu_{C=O}$ shift to higher frequency was observed, indicating a strictly ionic type of interaction without the usual d_π–p_π backbonding.

Further extentions of this work to F_3Ln–CO microscale studies have shown that the increase in $\nu_{C=O}$ may be predictable based on the molecular ionic character of the MF_3 molecule in question, if the MF_3 molecules being compared are of the same geometry. Thus, Hauge, Gransden, and Margrave[70] have codeposited a wide variety of MF_2 and MF_3 species with CO, among them LaF_3, NdF_3, GdF_3, HoF_3, and LuF_3. An excellent correlation between ionic radius and $\Delta\nu_{C=O}$ was observed for these F_3Ln–CO complexes. Therefore, the CO shift can serve as a qualitative probe of molecular ionic character, and will perhaps serve as a quantitative measure once the behavior of CO in high nonlinear fields is well documented.

References

1. J. E. Huheey, "Inorganic Chemistry," 2nd ed., p. 778. Harper, New York, 1978.
2. T. Ohnishi, *Nature (London)* **249**, 532 (1974).
3. S. J. Adelman, *Astrophys. J.* **183**, 95 (1973).

4. N. Grevesse and G. Blanquet, *Sol. Phys.* **8**, 5 (1969).
5. O. Hauge and O. Engvold, *U.S. A. E. C.* **NP-18857** (1970); from *Nucl. Sci. Abstr.* **25**, 3809 (1971).
6. H. Molnar, *Astron. Astrophys.* **20**, 69 (1972).
7. C. C. Herrick, *J. Less-Common Met.* **7**, 330 (1964).
8. A. A. Kruglikh, G. P. Kovtun, and V. S. Pavlov, *Ukr. Fiz. Zh.* **10**, 432 (1965).
9. D. White, P. N. Walsh, H. W. Goldstein, and D. F. Dever, *J. Phys. Chem.* **65**, 1401 (1961).
10. C. E. Habermann and A. H. Daane, *J. Chem. Phys.* **41**, 2818 (1964).
11. A. A. Kruglykh and V. S. Pavlov, *Izv. Akad. Nauk SSSR, Met.* No. 1, p. 178 (1966).
12. A. Desideri, V. Piacente, and N. Vincenzo, *J. Chem. Eng. Data* **18**, 140 (1973).
13. L. J. Nugent, J. L. Burnett, and L. R. Morss, *J. Chem. Thermodyn.* **5**, 665 (1973).
14. J. M. McCormack, P. R. Platt, and R. K. Saxer, *J. Chem. Eng. Data* **16**, 167 (1971).
15. W. R. Savage, D. E. Hudson, and F. H. Spedding, *J. Chem. Phys.* **30**, 221 (1959).
16. E. B. Owens and A. M. Sherman, *U.S., Dep. Commer., Off. Tech. Serv., AD* **275**, **468** (1962).
17. W. Kennelly and K. J. Klabunde, unpublished results from this laboratory.
18. K. A. Gingerich, *J. Chem. Phys.* **51**, 4433 (1969).
19. A. Pattoret, J. Drowart, and S. Smoes, *Bull. Soc. Fr. Ceram.* **77**, 75 (1967).
20. I. Krivy, AEC Accession No. 38876, Rep. No. UJV-1598; from *Nucl. Sci. Abstr.* **20**, 4721 (1966).
21. "Handbook of Chemistry and Physcis," 56th ed., CRC Press, p. B-67. Cleveland, Ohio, 1975–1976.
22. M. L. H. Green, private communications.
23. P. S. Skell, *Proc. Int. Congr. Pure Appl. Chem.* **23**, 215 (1971).
24. P. S. Skell, D. L. Williams-Smith, and M. J. McGlinchey, *J. Am. Chem. Soc.* **95**, 3337 (1973).
25. J. R. Blackborow, C. R. Eady, E. A. Koerner von Gustorf, A. Scrivanti, and O. Wolfbeis, *J. Organomet. Chem.* **108**, C32 (1976).
26. R. K. Sheline and J. L. Slater, *Angew. Chem., Int. Ed. Engl.* **14**, 309 (1975).
27. W. Weltner, Jr., J. L. Slater, R. K. Sheline, and K. C. Lin, *J. Chem. Phys.* **55**, 5129 (1971).
28. J. L. Slater, T. C. DeVore, and V. Calder, *Inorg. Chem.* **13**, 1808 (1974); **12**, 1918 (1973).
29. M. Moskovits and G. A. Ozin, *in* "Cryochemistry" (M. Moskovits and G. A. Ozin, eds.), p. 261. Wiley (Interscience), New York, 1976.
30. A. Bos, *J. Chem. Soc., Chem. Commun.* p. 26 (1972).
31. S. R. Ely, T. E. Hopkins, and C. W. DeKock, *J. Am. Chem. Soc.* **98**, 1624 (1976).
32. K. O. Hodgson, F. Mares, P. F. Starks, and A. Streitwieser, *J. Am. Chem. Soc.* **95**, 8650 (1973).
33. A. Streitwieser, U. Müller-Westerhoff, G. Sonnichsen, F. Mares, D. G. Morrell, K. O. Hodgson, and C. A. Harmon, *J. Am. Chem. Soc.* **95**, 8644 (1973).
34. K. O. Hodgson and K. N. Raymond, *Inorg. Chem.* **11**, 171 (1972).
35. C. W. DeKock, S. R. Ely, T. E. Hopkins, and M. A. Brault, *Inorg. Chem.* **17**, 625 (1978).
36. W. J. Evans, S. C. Engerer, and A. C. Neville, *J. Am. Chem. Soc.* **100**, 331 (1978).
37. D. F. Starks and A. Streitwieser, Jr., *J. Am. Chem. Soc.* **95**, 3423 (1973).
38. W. J. Evans, A. L. Wayda, C. W. Chang, and W. W. Cwirla, *J. Am. Chem. Soc.* **100**, 333 (1978).
39. J. L. Moriarty, *J. Chem. Eng. Data* **8**, 422 (1963).
40. C. E. Myers and D. T. Graves, *J. Chem. Eng. Data* **22**, 440 (1977).
41. R. D. Wesley and C. W. DeKock, *J. Chem. Phys.* **55**, 3866 (1971).
42. H. B. Skinner and A. W. Searcy, *J. Phys. Chem.* **75**, 108 (1971).
43. K. F. Zmbov and J. L. Margrave, *Adv. Chem. Ser.* **72**, 267 (1968).
44. D. White, P. N. Walsh, L. L. Ames, and H. W. Goldstein, *Thermodyn. Nucl. Mater., Proc. Symp., 1962* p. 417 (1963).

45. H. W. Goldstein, P. N. Walsh, and D. White, *J. Phys. Chem.* **65**, 1400 (1961).
46. A. V. Karyakin, N. V. Laktionova, an L. I. Pavlenko, *Zh. Prikl. Khim.* (*Leningrad*) 42, 751 (1969).
47. J. Drowart, A. Pattoret, and S. Smoes, *Proc. Br. Ceram. Soc.* **8**, 67 (1967).
48. S. A. Shchukarev and G. A. Semenov, *Dokl. Akad. Nauk SSSR* **141**, 652 (1961).
49. W. Weltner, Jr. and R. L. DeKock, *J. Phys. Chem.* **75**, 514 (1971).
50. L. L. Ames, P. N. Walsh, and D. White, *J. Phys. Chem.* **71**, 2707 (1967).
51. G. Benezech and M. Foex, *C. R. Hebd. Seances Acad. Sci., Ser. C* **268**, 2315 (1969).
52. R. L. Wu and P. W. Gilles, *J. Chem. Phys.* **59**, 6136 (1973).
53. S. P. Gordienko and B. V. Fenochka, *Zh. Fiz. Khim.* **48**, 493 (1974).
54. A. Makhmadmurodov, G. P. Dudchik, and P. G. Polyachenok, *Zh. Fiz. Khim.* **49**, 2714 (1975).
55. O. G. Polyachenok and G. I. Novikov, *Zh. Neorg. Khim.* **8**, 2631 (1963).
56. V. K. Ilin, A. D. Chervonnyi, A. V. Baluev, V. A. Krenev, and V. I. Evdokimov, *Deposited Publ., VINITI* p. 5688 (1973).
57. C. W. DeKock, R. D. Wesley, and D. D. Radtke, *High Temp. Sci.* **4**, 41 (1972).
58. A. S. Pashinkin, D. V. Drobot, Z. N. Shevtsova, and B. G. Korhsunov, *Zh. Neorg. Khim.* **7**, 2811 (1962).
59. N. M. Voronov, A. S. Danilin, and I. T. Kovalev, *Thermodyn. Nucl. Mater., Proc. Symp., 1962* p. 789 (1963).
60. K. R. Kunze, R. H. Hauge, D. Hamill, and J. L. Margrave, *J. Phys. Chem.* **81**, 1664 (1977).
60a. S. Langer and F. F. Blankenship, *J. Inorg. Nucl. Chem.* **14**, 26 (1960).
61. R. J. Ackermann, P. W. Gilles, and R. J. Thorn, *J. Chem. Phys.* **25**, 1089 (1956); **29**, 237 (1958).
62. J. F. Babelot, G. D. Brumme, P. R. Kinsman, and R. W. Ohse, *Atomwirtsch., Atomtech.* **22**, 387 (1977).
63. P. E. Blackburn and P. M. Danielson, *J. Chem. Phys.* **56**, 6156 (1972).
64. S. Abramowitz, N. Acquista, and K. R. Thompson, *J. Phys. Chem.* **75**, 2288 (1971).
65. R. W. Ohse, *J. Chem. Phys.* **44**, 1375 (1966).
66. E. D. Cater, E. G. Rauh, and R. J. Thorn, *J. Chem. Phys.* **44**, 3106 (1966).
67. E. D. Cater, P. W. Gilles, and R. J. Thorn, *J. Chem. Phys.* **35**, 608 and 619 (1961).
68. J. Kordis and K. A. Gingerich, *J. Chem. Phys.* **66**, 483 (1977).
69. D. Tevault and K. Nakamoto, *Inorg. Chem.* **15**, 1282 (1976).
70. R. H. Hauge, S. E. Gransden, and J. L. Margrave, *J. Chem. Soc.*, Dalton 745 (1979).

INDEX

A